MICHAEL MANLEY

The Making of a Leader

By the same author

The Prados of Sao Paulo Brazil: An Elite Family and Social Change

DARRELL E. LEVI

MICHAEL MANLEY

The Making of a Leader

The University of Georgia Press
Athens

Published in the United States
of America in 1990 by the
University of Georgia Press
Athens, Georgia 30602

First published in Great Britain 1989
by André Deutsch Limited
105-106 Great Russell Street London WC1

ISBN 0-8203-1221-5

Printed in Great Britain by
Ebenezer Baylis & Son, Worcester

For Bobbi, with love

TABLE OF CONTENTS

ABBREVIATIONS

ACP	African-Caribbean-Pacific group of nations
AID	Agency for International Development
ALCAN	Aluminum Company of Canada
ALCOA	Aluminum Company of America
API	Agency for Public Information
BITU	Bustamante Industrial Trade Union
CAIB	Covert Action Information Bulletin
CARICOM	Caribbean Common Market
CAST	College of Applied Sciences and Technology
CBI	Caribbean Basin Initiative of the Reagan Administration
CEO	Community Enterprise Organization
CIA	Central Intelligence Agency of the United States
CLC	Caribbean Labour Congress
CWP	Committee of Women for Progress
DG	*Daily Gleaner*
EPP	Emergency Production Plan
FBI	Federal Bureau of Investigation (United States)
GROW	Growing and Reaping Our Wealth
IBA	International Bauxite Association
ICFTU	International Confederation of Free Trade Unions
IAPA	Inter-American Press Association
IDB	Inter-American Development Bank
IMF	International Monetary Fund
JALGO	Jamaica Association of Local Government Officers
JAMAL	Jamaica Adult Literacy program
JBC	Jamaica Broadcasting Company
JBPA	Jamaica Banana Producers' Association
JCF	Jamaica Constabulary Force (police)
JDF	Jamaica Defense Force (army)
JDP	Jamaica Democratic Party
JFW	Jamaica Federation of Women
JLP	Jamaica Labour Party
JMA	Jamaica Manufacturers' Association
JWG	*Jamaican Weekly Gleaner*
JTA	Jamaica Teachers' Association
KSAC	Kingston and St Andrew Corporation (city government)
LSE	London School of Economics
MP	Member of Parliament
NAM	Non-Aligned Movement
NEC	National Executive Council (People's National Party)
NIEO	New International Economic Order
NJM	New Jewel Movement (Grenada)

NWU	National Worker's Union
OAS	Organization of American States
OPEC	Organization of Petroleum Exporting Companies
OPIC	Overseas Private Investment Corporation (USA)
ORIT	Inter-American Regional Labor Organization
PAJ	Press Association of Jamaica
PNP	People's National Party
PNPYO	PNP Youth Organization
PNPWM	PNP Women's Movement
PPP	People's Political Party (Jamaica)
PPP	People's Progressive Party (Guyana)
PSOJ	Private Sector Organization of Jamaica
RCAF	Royal Canadian Air Force
RJR	Radio Jamaica Rediffusion
SG	*Sunday Gleaner*
SI	Socialist International
SMA	Sugar Manufacturers' Association
STC	State Trading Corporation
TUAC	Trade Union Advisory Council
TUC	Trade Union Council
UNIA	United Negro Improvement Association
UWI	University of the West Indies
WCC	World Council of Churches
WLL	Workers' Liberation League
WPJ	Workers' Party of Jamaica
YDA	Youth Development Agency
YSL	Young Socialist League

ACKNOWLEDGEMENTS

The list of those to whom I owe thanks is a long one. In addition to Manley himself, Beverley Anderson-Manley, Matthew Beaubrun, Tony Bogues, David Coore, D. K. Duncan, Carlyle Dunkley, Barclay Ewart, Glynne Ewart, Douglas Manley, Edna Manley, Joseph Manley, Thelma Manley, John Maxwell, Corina Meeks, O. K. Melhado, Rex Nettleford, Pamela O'Gorman, Paul Robertson, Hugh Lawson Shearer and Carl Stone kindly shared their experiences and knowledge with me. Marva Roberts and Howard Aris of Manley's staff were very helpful in making local arrangements. Olga Hammond and Kenneth Allen of the PNP's library facilitated my research. Shirley Lloyd and Miss Una Wright helped me to have comfortable stays on two research trips. The staff of the Institute of Jamaica and the National Archive were efficient and helpful. Vicky Gee, Melissa Sundstrom, and Fred Zawalicki provided research help at Florida State University. Tony Bogues and Jon Urbach read portions of the manuscript and made valuable suggestions. Among my Florida State University colleagues, Rodney Anderson, Peter Garretson, Jim Jones, and Peter Ripley gave special encouragement. Some financial support was provided by the History Department of Florida State University. Naturally I accept full responsibility for any unintentional errors this book may contain.

Introduction

Michael Norman Manley was born in Kingston, Jamaica in 1924. His parents Edna Manley (1900–1987) and Norman Washington Manley (1893–1969) belonged to Jamaica's small middle class, most of whose members were of brown skin and intermediate social standing between the even smaller Euro-Jamaican minority at the top of Jamaica's power structure and the black masses at its base. The second of two sons (Douglas is two years older), Michael Manley's early childhood was apparently uneventful, but the young boy demonstrated the same impetuosity, aggressiveness, and temper that continued into adult life. In the mid-'30s Manley[1] enrolled at Jamaica's most exclusive secondary school, Jamaica College. At first he suffered bullying by older boys, but then rose to school leadership due to his athletic prowess and organizing skills. At the end of his school career, however, he came into conflict with his teachers and the new headmaster. These experiences significantly affected his personality and outlook.

After resigning from Jamaica College in 1943, Manley joined the Royal Canadian Air Force, spending the last period of World War II training as a wireless operator/gunner. He then enrolled at the London School of Economics which was one of the centres of engagement in the global decolonization struggle against Britain and the other western imperialist powers.

Following graduation, Manley spent a year in Britain following the fortunes of the West Indian cricket team (of all the sports, he had a particular passion for cricket), whilst training as a journalist.

Manley was back in Jamaica in 1952, a fateful year for the People's National Party (PNP), the nationalist and nominally socialist party headed by his father. That year, under pressure from both domestic politics and the Cold War, the PNP expelled several prominent leaders of its left wing, who were among the party's best labour organizers. Manley, launched into the breach as a trade unionist for the PNP-affiliated National Workers' Union (NWU), was an active trade unionist for twenty years, which profoundly affected his life and political career. In 1962, the year Jamaica gained its independence from Britain, Manley was appointed senator in Jamaica's parliament by his father. He was elected to the more powerful House of Representatives in 1967 and assumed leadership of the PNP in 1969, shortly before the death of Norman Manley.

As leader of the PNP, Manley inherited a bitter political struggle with the opposition Jamaica Labour Party (JLP), founded by Norman Manley's cousin, Alexander Bustamante (1884–1977). The colourful, conservative Bustamante had dominated Jamaican politics from 1944, the year of Jamaica's first elections based on universal suffrage, until 1955 when Norman Manley assumed Jamaica's leadership till independence in 1962 thus culminating the nationalist programme of Norman Manley and the PNP. Ironically, Bustamante and the JLP—which had at first opposed independence—led the first post-independence government. The JLP ruled until 1972, while Bustamante gradually withdrew from direct leadership and the ruling party struggled to replace him. The old rivalry between Norman Manley and Bustamante was gradually replaced by a new, more bitter feud between Michael Manley and the JLP's Edward Seaga, who served as Minister of Culture and Development and later Minister of Finance in the 1962–72 JLP administration.

In 1972 Michael Manley was overwhelmingly elected Prime Minister of Jamaica, following a campaign which capitalized on widespread popular disaffection with the JLP and on Manley's personal appeal and organizational abilities. Significantly, his campaign attracted voters from across Jamaica's highly unequal social strata. After two years of reforms variously described by scholars as 'liberal', 'populist', or 'social democratic', Manley's PNP recommitted itself to 'democratic socialism', a term the party did not define until 1979. Manley's government attempted to enact broad social and economic changes. Many forms of opposition developed, including normal political resistance, emigration, disinvestment by Jamaican and foreign capitalists, organized violence, and pressure and destabilization by the United States and the International Monetary Fund (IMF). An election in 1976, explicitly run by the parties on the choice between 'socialist' and 'capitalist' paths to 'development', resulted in another solid victory for Manley. But Jamaica had become more polarized, the poor majority supporting Manley more strongly, while the rich, the middle classes, and some workers, peasants and unemployed backed the JLP, headed by Seaga. After 1976, because of a declining economy and controversial dealings with the IMF, Jamaica became even more polarized.

In the '70s, Manley emerged as an important Third World spokesman, leader of the Non-Aligned Movement, and advocate of the New International Economic Order (NIEO), a proposal to change the devastatingly unfair trade relations between the rich, developed nations and the poor nations of the Third World. His government's bauxite levy of 1974 temporarily improved Jamaica's return from multinational aluminium companies. Manley's advocacy of African liberation struggles, uncompromising opposition to apartheid, support for Third World solidarity, and friendship with Cuba, Eastern Europe and the Soviet Union (while attempting to maintain cordial ties with the United States, the United Kingdom, and Canada), earned Manley the praise of some on the left and the damnation of some on the right in Jamaica and abroad. He was awarded a United Nations gold medal for his resistance to apartheid.

The second Manley administration (1977–80) faced a prolonged, agonized descent from the triumph of 1976. The troubled economy forced the government to turn to the IMF, nominally for 'financial aid'. As Jamaica's economy steadily declined, the IMF became the major arbiter of Jamaican politics, forcing the government to adopt unpopular policies. Internal government and party squabbles compounded this situation. The JLP, led by Seaga, orchestrated a campaign of opposition on many fronts. Political violence took more than eight hundred lives in the election year of 1980 which resulted in a landslide victory for Seaga over Manley. Jamaica's first experiment in 'democratic socialism' had ended.

For Manley, the '80s have been a political wilderness. A snap election called by Seaga in 1983, boycotted as 'bogus' by Manley's PNP, gave the JLP embarrassing total control of the House of Representatives.[2] In parliamentary terms, at least, Jamaica became a one-party democracy. However, even with the Reagan Administration's strong backing, Seaga's government was unable to prevent the further deterioration of the economy and society, although the former improved somewhat after 1985. Seaga's implementation of IMF requirements extracted a heavy cost from the Jamaican people, especially the poor majority, stimulating support for Manley and the PNP. In July 1986 the PNP won a crashing victory in local elections, significant mainly as a referendum on Seaga's unpopular policies and performance. Then in February 1989, Manley and the PNP returned to power with fifty-seven percent of the vote and a commanding number of seats in the Parliament. The Manley story is far from complete, but now is an appropriate time for an interim historical assessment of the man, his life, and his politics.

The drama and significance of the Manley administrations' attempt to free Jamaica of neocolonialism, dependency, underdevelopment, and poverty via 'democratic socialism' have already resulted in several serious studies.[3] Manley's own books also illuminate aspects of his thought and career.[4] This book seeks to fill a gap in the literature by systematically and comprehensively tracing the development of Manley's life and career, devoting roughly equal treatment to the periods before and after he became Prime Minister in 1972.

Historians should say where they stand. This book proceeds from a point of view. Like everyone's perspective, mine began to form at an early age, and has largely resulted from the fact that my father was a granulator operator in a sugar refinery in San Francisco. An active member of Harry Bridges' International Longshoreman's and Warehouseman's Union, he lost his job—the centre of his life and friendships—because of a corporate merger when I was still a young boy. Trade unionism and sympathy for workers were convictions I absorbed at the family dinner table. San Francisco was a liberal influence. The

University of California campuses at Berkeley and Davis, San Francisco State University and Yale University took me far from my roots without breaking them and made me into a professional historian. Over twenty years of studying and teaching Latin American and Caribbean history have moved me leftward from liberalism to a position broadly sympathetic to 'democratic socialism'.[5] This stance is based on my study and personal observations of the human misery wrought by the dependent form of capitalism[6] in Latin America and the Caribbean and the failure of liberal democracy to significantly reduce the suffering. At the same time, I am wary of the authoritarian politics and cultures of Marxist-Leninist regimes, although I acknowledge their achievements in science, education, health, and nutrition. As a professional historian I recognize my obligation to seek 'the truth', but I also believe that it is beyond the ability of limited mortals—burdened by prejudices of nationality, race, gender, religion, education, family, and culture—to escape the essential subjectivity of life. This book will be my truth; others will find their own.

It is somewhat ironic that the first biography of Manley—who has challenged the economic, political, and cultural relationships between the First and Third Worlds as unjust—should be written by a First World historian. Third World people are justifiably suspicious of the intentions and biases of First World historians who write Third World history. When I arrived in Brazil in 1971 to begin my doctoral research on São Paulo's Prado family, I soon encountered a popular magazine which warned of 'The History of Brazil, written in the USA', and featured on its cover heroes of Brazilian history mouthing their most famous lines in English.[7] Similarly, West Indians, including leading politicians such as Trinidad and Tobago's Eric Williams, have warned against foreign, imperial, and neocolonial biases in the writing of history.[8] There are obvious hazards in a foreign scholar attempting to understand a culture which can appear at times inscrutable even to its own intellectuals. The foreigner can never see or understand some things, but he can see and understand others. Perhaps this first biography of Manley will help stimulate others by Jamaican and West Indian scholars. In any case, history should be written from many perspectives.

I encountered several difficulties in writing this biography. Michael Manley is a complex private person and public figure. The volume of his writings and speeches is very large and cannot be analyzed in detail here; there is a need for an intellectual biography, but in this first attempt I thought it more important to use a comprehensive approach, sacrificing a more profound treatment of Manley's ideas. Jamaican political culture also presents problems. The society is highly polarized by PNP-JLP 'tribalism', and it is difficult for the outsider to evaluate sharply contrasting accounts of the same events. The newspaper of record, the *Gleaner*, is a generally conservative newspaper with an anti-Manley bias, which became extremely reactionary when Manley was in power; yet the *Gleaner* cannot be simply ignored. Sorting through the sometimes wildly different PNP, JLP, and *Gleaner* accounts (and others), I found the academic and journalistic writings of Carl Stone extremely helpful.

Honesty requires an explanation of my relationship with Michael Manley. This book began in April 1983, when Michael Manley lectured at Florida State University, thanks to the efforts of Graciela Cuervo and the university's Centre for Participant Education. Like countless others who have heard Manley, I was engaged by his speech. As a scholar and teacher, I was fascinated by his technique and by his explanations of such things as the unfairness in international trade. I was impressed by his handling of some hostile Jamaican students in the audience. Afterwards, in a small gathering of students with Manley at a local bar, I felt that this public man and Third World leader had a warm, reflective private side. Despite his late-night fatigue and constant badgering by a local communist (since turned Christian fundamentalist!), Manley was patient, informative, and forthright.

Nearly a year later I proposed the idea of a biography to Manley. We met in New York in March 1984 to discuss the proposal. From the start Manley reckoned that we were bound to see some things differently, but he was obviously pleased and flattered to be the subject of this book. We both understood from the beginning that neither of us wanted to do an 'authorized' or 'official' biography. At the same time, I have healthy suspicion of the dangers of biography based only on the public record of politicians, which as Montaigne observed in the sixteenth century, often has the quality of theatre rather than life. My approach is to deepen understanding of a public man by examining both his public and his private sides, even though Manley rarely shares the latter with more than a few intimates. Nonetheless, Manley has helped me enormously by facilitating interviews with himself and others, furnishing his private correspondence for the crucial '70s and '80s, and giving me unrestricted access to the PNP's files. Because I was studying an active politician in a heated political culture where important issues and even matters of life and death were at stake, I did agree not to quote from private sources without Manley's approval. In practice this meant that on a few occasions Manley requested me not to quote a candid remark from his correspondence or our recorded interviews, a restriction which has not materially altered the book. In no case was permission to quote from papers in the PNP library denied. Overall, Manley and his associates have been extremely open and generous to me. Although I have immersed myself in the views of Manley, members of his family, and his associates, the resulting book is mine alone.

Michael Manley is a powerfully charismatic man. I like him. I sympathize with him and wish him well. I believe that his politics, while sometimes impetuous and inadequately executed and supported by his party, are, in my view nonetheless correct in their direction and courageous in their advocacy. Democratic socialism is essential to the solution of Jamaica's forbidding human problems. At the same time, because I am a scholar, and because I take Manley and his politics seriously, I have tried to blend my admitted sympathy for the subject with as much critical detachment as possible. The reader can judge the extent to which I have succeeded.

Darrell E. Levi
Department of History
Florida State University
Tallahassee, Florida.

CHAPTER 1

Confrontation and Compassion

Michael Manley's life has been filled with confrontations: those of the nationalist facing the British and United States empires, those of a trade unionist in conflict with corporate power, those of a politician committed to the politics of change. These struggles have often been bitter, and the 'tribalism' of Manley's PNP and the rival JLP has poisoned the Jamaican politics and society which form the backdrop to this biography. To Manley's great credit, in a life of confrontations he has maintained a reputation amongst the Jamaican people as a caring person, which is his greatest political asset. The two episodes that follow provide introductory glimpses of the confrontations and the compassion that have marked Manley's life.

A Confrontation at Jamaica College (1942–3)

At the age of eighteen, Manley bitterly confronted the authorities of the most prestigious secondary school in colonial Jamaica, Jamaica College.[1] Manley had lacked the discipline and motivation required to be a really outstanding student, but prided himself on being a successful sports captain who had temporarily ended the rampant bullying at Jamaica College, of which he himself had been a victim. As he recalled more than forty years later, at the age of eighteen he was becoming very conscious of himself intellectually and very nationalistic. His mother Edna had contributed to his outlook through the nationalistic themes of her art, and in 1938 his father Norman Washington Manley had helped establish the People's National Party which pressured Britain for self-government. Michael Manley's confrontation at Jamaica College had the deep overtones of a young colonial in conflict with the educational establishment of the British Empire.

Jamaica College had long been under the benign headmastership of Reggie Murray, 'a brilliant mathematician and amazing fellow', 'one of those—very much in the English tradition—figures of awe in a school for boys'. In late 1942 it was announced that a new headmaster, J. W. S. Hardie, a man in his thirties, son of the Anglican Bishop of Jamaica, was to replace the beloved Reggie Murray. Michael was very conscious of the fact that his Aunt Vera Moody, 'a big figure in education', endorsed Hardie's selection.

Once ensconced in the headmaster's office, Hardie offered Michael the position of headboy, upon the unanimous recommendation of the school staff, even though it meant advancing him above David Coore, who had better academic credentials.[2] Michael declined the post ostensibly on the grounds that he did not attend chapel.

> So when I put all this to Hardie, Hardie got a little sarcastic about the whole thing and said, oh, was I some sort of rebellious atheist? And I said no, not as extreme as that, but I was going through an agnostic phase, and I therefore think it not appropriate for me to accept the headboyship because this is an Anglican school. And I think your headboy must be very much part of your tradition. And go to Church and have lots of things to do and so on and so forth, that are related to the traditions of the mainstream of the school.

Behind this stance was the fear of Michael—the spellbinding crowd-pleaser of the '70s—that as headboy in the '40s he might have to make a speech, a terrifying prospect. Hardie was very upset by Michael's decision, but David Coore became headboy.

In January 1943, headmaster Hardie made his first address to the assembled school. He reckoned that Jamaica College was very badly run down, that it had almost gone to the dogs. He said that together: staff, students and himself would shape it up, turn it once more into what it should be.

As Manley recalled much later,

> *I was fit to be tied.* I was *bitter* about this *facety*[3] little Englishman coming here to suggest that our school and our headmaster, that we'd all loved. . . . and I led a protest movement and demanded audience, and told him that we all bitterly resented this remark about the school.

Tensions mounted. Reggie Murray had treated young Manley 'almost as a semi-adult', allowing him great freedom to do such things as go to the gym if he didn't have class. In contrast, Hardie treated Manley imperiously, shaming him in one incident before younger boys whom Manley, as a monitor, was supposed to keep in order. Manley responded with a protest letter, which angered Hardie who could not imagine, in Manley's words, 'the impertinence of a schoolboy thinking he could pen a letter to his headmaster'. By this time, Manley was thinking of Hardie, 'You're really just a jerk.'

Manley shared his troubles with other senior teachers who 'were just spies and went right back to [Hardie] with it.' His sixth-form English teacher, Englishman Wingfield Digby (who later headed the British Museum, another symbol of English hegemony), assigned him an extra essay that other sixth-formers did not have to do. When Manley inquired why, he was told: 'Because it is an order.' Manley refused to do the extra work. Digby threatened to report him to the headmaster. Manley lost his temper, threw his books at the blackboard, said, 'Do your damnedest!', and stormed out.

In due course Hardie summoned Manley to his office. The headmaster gave him the option of taking a caning—'imagine a boy of eighteen who was probably capable of breaking his neck with one finger, taking a caning from this little Englishman'—or of being expelled. The athletic Manley, looking very strong, stood up and said, 'With respect to the caning, would you care to try it?' Hardie said, 'I suppose you wouldn't take it. Then you leave me no choice but to expel you.' Manley responded, 'You know something, Mr Hardie, you'll never live to expel me. I resign!'

Manley returned to his dormitory, burst into tears, and cried for an hour. It was a 'terrible moment' in his life. He packed his bags and walked out the door to catch a bus home. When his schoolmates saw him they burst out: 'We-want-Manley. Down-with-Hardy.'[4] He was borne shoulder high out to the bus, and the school stayed on strike for two weeks in protest.

Manley explained what had happened to his father, Jamaica's leading lawyer and the principal leader of its nationalist movement. Norman Manley told Hardie that if Michael were expelled he would take the school to court. Hardie backed down, and it was left that Michael had resigned.

With the passage of time, we all embellish, embroider and selectively remember important events in our lives. Still, to hear Manley tell that story, to watch him as he tells it, is to be aware that his latter-day assessment is not far off target:

> That story, to this day, evokes very deep, disturbed emotion in me, just as a confession. And that story is a lot of what I am. Now you make whatever you like of that story. But that story is me, raw, me as a man. That's how I am.[5]

'Thanks, Michael Manley' (1970s)

To demonstrate Manley's quality of compassion is the following story recorded by a former JLP candidate who expressed his happiness that Manley 'went across the political barrier to prove that out of many we are really one people.'[6]

Shortly after the 1972 election which led to Manley's selection as Prime Minister of Jamaica, Mrs Esme Grant, a member of the rival JLP who had lost her parliamentary seat in the election, tried to phone the new Prime Minister. He was in a meeting, but returned her call within five minutes. Grant told Manley, 'Look I have lost my seat and I have no job.' He told her to name the school of her choice to teach in, but she was not interested in returning to the classroom. She mentioned the literacy programme, JAMAL, and a job there was agreed within minutes.

While at JAMAL Mrs Grant had her legs amputated as a result of diabetes. Some at JAMAL wondered whether she was capable of continuing her work. Word of this reached Manley 'who enquired if because she had lost her legs she had also lost her mind. Upon being told that she had not lost her mind, [he] said to them that the work must be placed in a car and sent home, and that as long as he remained Prime Minister she would remain in her job.' This story came to light shortly after Esme Grant died in 1987.

CHAPTER 2

What Sort of Man?

Odd this business of knowing and yet not knowing people. You each of you turn a section of yourself to your friends, rather like the tip of an iceberg, but the real bulk of the person is hidden. . . .[1]

This is how history is written. Not the truth about people—but just the records of man's prejudices and distortions.[2]

Myth and History

Any attempt to understand the personality of Michael Manley must reckon with several obstacles. Some of his supporters have created a heroic myth about him, an image sustained partly by political public relations machinery. Critics and influential enemies have designed an anti-myth to discredit Manley as politician and as a person. Manley is an extremely complicated person. He appears to be many things to many people. Even friends and family members disagree about such fundamentals as his attitude toward politics and power. Clues as to Manley's character are implicit throughout this book, but it is useful to address the matter explicitly at the outset.

The positive, heroic myth of Michael Manley is rooted in Jamaican political culture, his family background, and Manley's dramatic personality and struggles as a trade unionist in the '50s and '60s and politician in the '70s and '80s. Manley's father, Norman Washington Manley, bore the political nickname 'Moses', and in the tumultuous '60s Manley became 'Joshua'. In the campaign of 1972 which made him Prime Minister, Manley threatened to punish the corrupt and wicked enemies of 'the people' with 'Joshua's rod', a walking stick allegedly presented personally to him by Emperor Haile Selassie of Ethiopia, the living god of Jamaica's Rastafarian religion. Some of Manley's followers, perhaps many, attached almost god-like stature to him, and he voiced deep-seated yearnings in a long-suffering, impoverished, mostly black people. In 1980, one constituent wrote to Manley's wife Beverley that 'the destiny of a people lies in the hands of those whom God appoints to rule at that particular era. Your husband, the Hon Michael Manley was appointed in 1972 when there was a definite plan for the destruction of the Race.'[3] Such sentiments, expressed after the eight years of dramatic, bitter and often violent political struggle during the Manley administrations, testify to the strength of the myth in some minds.

In the '70s, Manley's political enemies created an anti-myth of him.

Jamaica's Michael Manley: Messiah, Muddler, or Marionette?, a polemic by a Jamaican businessman angered by Manley's policies, is the fullest expression of this view. It argued that Manley was a false messiah, a real muddler and a puppet controlled by dark, conspiratorial, communist forces, and asked the question whether Manley's 'disastrous experiment with Democratic Socialism' had 'wrecked Jamaica's economy beyond repair?'[4] The muddler/marionette anti-myth was propagated for years in the reactionary, monopolistic, vastly influential monolith of Jamaican journalism, *The Daily Gleaner*. According to the *Gleaner* Manley was a dangerous, powerful messianic leader recklessly pursuing a hidden agenda, or a crypto-communist, a tool of Fidel Castro, a puppet of the left wing of the PNP, Manley's party. The international press, largely ignorant of Jamaican realities and deeply suspicious of attempts to rock the global status quo, generally echoed the anti-myth. Manley was pictured as an incompetent who single-handedly destroyed the Jamaican economy, with no help from international economic conditions such as the twin oil shocks of the '70s and stagflation, or from middle-class Jamaicans who withdrew their capital and skills from the economy, or workers who resisted Manley's calls for discipline and self-sacrifice, or from the International Monetary Fund whose stringent policies not only failed to right Jamaica's economy, but undermined Manley's political support.

Myths and anti-myths gain credibility only if they capture some people's perceptions (however limited and erroneous) of reality. Both the myth and the anti-myth are based on elements of Manley's background and character. Manley's parents, Norman Washington Manley and Edna Manley are themselves icons of twentieth-century Jamaican history, Norman an official national hero, Edna a central figure of twentieth-century Jamaican art, sometimes dubbed 'the mother of the nation'. Like their son, their public stature concealed more fragile, shy, human qualities. Manley strenuously denies purposefully creating 'a cult of the leader', but in the late '60s and early '70s he responded to messianic urges deep in the Jamaican people for a hero to solve their problems. Myth-making involves desires from below and guidance from above. Manley himself long had a deep attraction for heroes, the romantic, the dramatic. Politicians must have big egos. Manley's public behaviour—which along with rumours about his private life fed the myths and anti-myths about him—was often enough impetuous, theatrical, emotional, colourful, visionary, genuinely heroic. His language drew from the resonant rhetoric of national liberation, socialism, and Biblical prophecy.

In their simplicity, neither the 'Joshua' myth nor the 'Muddler/Marionette' anti-myth do Manley, the man justice. Nonetheless, it is important to realize that the difference between the myths of daily life and of politics and the myths woven by historians is more a matter of complexity and motivation than anything else. According to William H. McNeill,

> Myth and history are close kin inasmuch as both explain how things got to be the way they are by telling some sort of story. But our common parlance reckons myth to be false while history is, or aspires to be, true. Accordingly, a historian who rejects someone else's conclusions calls them mythical, while claiming that his own views are true. But what seems true to one historian will seem false to another, so one historian's truth becomes another's myth, even at the moment of utterance.[5]

The Joshua myth encapsulated the hopes and repressed aggression of Jamaica's 'sufferers', just as the 'Muddler/Marionette' anti-myth encased the fears of Jamaica's upper and middle classes. While this book attempts a more complex interpretation of Michael Manley, it does so with the recognition that even carefully crafted history has the quality of myth about it.

Family, Interests, Friends

Michael Manley is, most importantly, the product of a remarkable family. His father, Norman Washington Manley, was a champion schoolboy athlete, Rhodes Scholar, Oxford-trained lawyer, one of the West Indies' leading lawyers, an intellectual with deep knowledge of literature, music and art, a founder of the nationalist, sometimes 'socialist' People's Nationalist Party and its leader from 1938–69, the island's political leader during the last years of British colonialism (1955–62). Many people remember Norman Manley as an austere, awesome figure, but that public visage concealed a shyer, more humane side. Michael Manley reveres and deifies his father's memory. Norman had a profound influence on Michael's personal and political development. His mother Edna, a tremendously talented visual artist driven by a sometimes semi-mystical desire to portray the essence of things (the owliness of the owl, for example), had an even more profound effect on Manley. Difficult to specify, this maternal influence lay in the realms of artistic and human sensitivity, personal and public integrity, openness to persons of all social classes. With a suspicion of political 'isms', it is likely that she moderated the political tendencies of her son. Edna Manley, as wife of a Jamaican chief minister and mother of a prime minister, was a living repository of Jamaica's political struggles and experience in the half-century after 1938, and her relationship to Michael Manley was exceedingly close. She died in 1987, when Manley was sixty-four years old.

To be the child of truly exceptional parents is a mixed blessing. Though politics cost his father most of the money he earned as a successful lawyer, Manley never had to worry about those things that have framed the existence of most Jamaicans: daily food, clothing, shelter, education. His parents provided stimulation and exposure to art and politics far beyond that of even most privileged Jamaicans. Manley admitted in 1987 that he was, by virtue of his family's spiritual, professional, and material qualities, 'the most privileged man in

Jamaica'.[6] He confessed that such advantage, in the midst of Jamaica's widespread suffering, inspired guilt that made it difficult, in later years, to control undisciplined youth and PNP leftists when he was Prime Minister. Having an illustrious father had obvious benefits, but it was also an intimidating element, as Manley believed he could never live up to his father's record, be it in sports or politics. All of us must struggle to overcome parental authority and establish an independent personality; Manley's struggle with his father's record and aura was like a guerilla war of liberation against a superpower: winnable, but protracted and costly. Although Manley's early youth was apparently 'normal' in most respects (given the abnormality of extremely intense, intellectual, driven parents), and although he is remembered as a cheerful and happy child, his early years were also marked by outbursts of his famous temper. Anger can form about Manley's brow like a thunderstorm building in Jamaica's Blue Mountains; lightning flashes in reply to what he considers to be unfair criticism, in response to the frustration of his racehorse instincts, or in reaction to the simple incompetence of subordinates.

Descendant of a special family, struggler and activist, Manley is also a 'man-of-words', a trait with deep echoes in African, Afro-American, and Afro-Caribbean culture.[7] As a young man he wrote long reflective letters to his parents, commenting on art and politics. One of his earliest goals was to be a journalist, and in the early '50s he worked for the Jamaican progressive weekly *Public Opinion*, writing a column, 'Root of the Matter', a title suggesting a desire, like that of his mother, to get at the essence of things (see Chapter 8). Trade unionism and politics interrupted Manley's writing until the '70s, but thereafter he authored a series of books: *The Politics of Change* (1974), a preliminary survey of his views on Jamaica and his strategy for change; *A Voice at the Workplace: Reflections on Colonialism and the Jamaican Worker* (1975), a more substantial, largely autobiographical account of his trade union activities and reflections on the limitations of Jamaican trade unionism; *Jamaica: Struggle in the Periphery* (1982), his account of his years in power, 1972 to 1980; *Up the Down Escalator, Development and the International Economy: A Jamaica Case Study; and A History of West Indies Cricket* (1988). Manley has written dozens of articles for journals such as *Foreign Affairs, The Caribbean Review, The Nation* of New York, *The Nation* of Barbados, *South*, and *Express* of Trinidad and Tobago. Manley says he writes largely to organize his thoughts, as reflection on the preparation for political action.[8] Manley writes interestingly and at times with poetic eloquence and drama.[9] His thought and writing distinguish him from other politicians in Jamaica and elsewhere. But writing is only one aspect of this man-of-words. His skills at oral communication are unsurpassed, and in Jamaica and perhaps abroad more people may 'know' Manley via his sometimes electrifying speeches than through his penned words.

Manley is more than just an academic intellectual. He is also a man of action, as expressed in his twenty years of vigorous trade unionism, eight years as Prime Minister and two decades as party leader. As a trade unionist, Manley developed remarkable skills as a negotiator between labour and capital. As Prime Minister, Manley tried, with mixed results, to move the rather moribund liberal democracy of a poor Third world country with massive economic and social problems in the direction of social reform, greater equity, self-reliance, and Third World solidarity. As a party leader, Manley has struggled to balance competing factions of the PNP, to produce the compromises necessary to unite the shaky alliance of upper, middle and lower class elements upon which the party is based. The union, political, and party stories are told in detail below.

Manley is also a man of cultural refinement, inherited from his parents. In music, he is most knowledgeable about jazz and art music. Of the latter, he favours the romantics, Beethoven and Brahms, for example, probably because of the emotion, drama, and theatricality of their music. Although he finds reggae on the whole a bit repetitive and monotonous, he regards Bob Marley (1945–80) as a great revolutionary artist. Manley is not above enjoying a sentimental Lionel Richie ballad. He is also passionately devoted to theatre and dance; his former wife Thelma, a dancer, and Professor Rex Nettleford, a dancer and director of Jamaica's renowned National Dance Theatre Company, regard him as a very knowledgeable observer of the art.[10]

Pamela O'Gorman, head of the Jamaica School of Music, finds Manley to be one of the few people she can seriously listen to music with, since he has 'a tremendous capacity to listen, to absorb, and to assimilate very quickly,' both in music and everyday life. She finds him to be diffident in expressing opinions about music, but his observations tend to be well conceived, intelligent, sincerely felt, and serious. However, despite his passion for music, Manley gave up listening to it when he was Prime Minister because, as he told O'Gorman in the '70s, it just sensitized him too much, and he couldn't afford to be that vulnerable. Sacrificing his beloved music to politics in that way, O'Gorman surmised, provided a great clue to Manley's character. O'Gorman believes that personal integrity is one of Manley's most important characteristics, despite the fact that people tend to overlook it, perhaps because of Manley's flamboyant, impulsive 'Joshua' image; in the Manley family, she observes, 'personal integrity and public integrity were accepted as a way of life.' Among Manley's personal characteristics O'Gorman lists a tremendous capacity and love of life, intensity, profound sensitivity and compassion, a willingness to cry when listening to music (although this could be interpreted as weakness in a politician), a personal generosity of spirit. She observes that Manley normally has great self-control, despite occasional shows of fiery temper. Contrasting with Manley's enthusiasm is—deep, deep down—a very real sadness that may come from what he has seen politics do to his family.[11]

Another side of Manley is his enormous enthusiasm for sport. In his childhood, boxing and horse racing were important to his family. Becoming captain of the swimming team in secondary school was a major event in Manley's adolescence. Manley became a fan of virtually all sports, but most especially cricket. Throughout his days as a part-time journalist, in the '50s and again in the '80s, a portion of Manley's columns were reserved for sport. Like other fans, Manley was attracted by the sheer drama and vicarious thrills of sport, but unlike most fans he would also be aware of its subtle interweaving with social history, politics, and nationalism. Sport is also another arena for Manley's passion for the heroic.

Throughout his life Manley has been at odds with Jamaica's conservative political and economic establishment. His closest friends have come from the progressive middle-class realms of the professions, the arts, and academe. David Coore, a distinguished lawyer, was Manley's best friend in secondary school, and has remained closely associated with him for over fifty years. O'Gorman, an Australian by birth, came to Jamaica in 1958 and became a member of the Manley family's inner circle through their shared passion for art music. Professor Nettleford, head of the Joint Trade Union Education Institute at the University of the West Indies (UWI), director of the National Dance Theatre Company, author of books on Jamaican culture and editor of a collection of the writings of Norman Manley, also shares artistic interests with Manley. Corina Meeks is a public relations expert. She had served Manley as Assistant to the Prime Minister during his first administration and headed the Agency for Public Information during the second. O.K. Melhado, who worked for Manley in a similar position and then headed the State Trading Corporation, was an executive of the Desnoes and Geddes brewing company in the '80s. Barclay and Glynn Ewart are self-made capitalists (Barclay's father was a barber) who, like Melhado, share Manley's interests in sports and the arts. Another close friend is Matthew Beaubrun, a physician who attended Manley during his serious illnesses in the '80s, and who closely observed Manley's political behaviour, with the keen eye of a scientist and amateur historian aware of the value of source materials. Hugh Shearer, a remote cousin who was JLP Prime Minister from 1967 to 1972, is a long time trade union and political rival of Manley's. Nonetheless, Shearer and Manley have shared a friendship which has survived sometimes bitter union and political struggles.[12]

Public and Private Sides

Dimensions of Manley's personality are revealed in his consistently expressed public attitudes. He is a champion of social justice, social change, self-reliance, the necessity for sacrifice, non-violence (except against the most tyrannical of regimes, such as South Africa or Pinochet's Chile), the necessity for struggle, the union of thought and action, principled relationships among people and nations, Caribbean regional and Third World solidarity. He believes in the 'free press' despite the generally unfavourable coverage he has received from the *Gleaner* and the international media. Manley accepts liberal democracy, although he has reservations about its utility in effecting social change and social justice, and believes it is necessary to democratize democracy by encouraging mass participation in decisions affecting economic, social and political life. Manley is more reformer than radical, and is ambivalent about the radical left, regarding it as a necessary stimulus to social change, but—in Jamaica—perhaps immature and undisciplined. Manley strongly opposes imperialism and its Jamaican ally, the neocolonial right. He had fought apartheid, and, in general, all forms of domination. New directions in his publicly expressed attitudes—the result of reflection on the '70s—include a suspicion of statism, and an acceptance of 'social democracy' rather than 'democratic socialism' as the most viable path of social change in Jamaica.

Politically, Manley is motivated by a desire to serve Jamaica and to help people, particularly the poor. His personal attitude to power may be ambivalent; he claims to regard it as a burden, but some people close to him maintain that he enjoys the exercise of power. Manley tends to personalize politics: the early struggle of his father versus Bustamante; the more recent struggle of himself versus Edward Seaga. As a political leader he drives his assistants hard, but not harder than he pushes himself. He seeks diverse views, consensus, and compromise, yet his own intellect, preparation and persuasiveness may overwhelm even strong personalities. His socialization and instincts are deeply 'democratic', but the society in which he operates still bears a markedly authoritarian stamp; his political behaviour is driven thus by an ongoing dialectic between democratic idealism and the social reality of authoritarianism.

Manley belongs to a dissident wing of the Jamaican middle class. Some members of the middle class claimed in the '70s that he 'betrayed' their class. They ignored the fact that the Manleys generally, and Michael in particular, had long been critical of smug, conventional middle-class people who, from the comfort of the veranda, have been quite willing to accept as inevitable the suffering, degradation, and poverty of most Jamaicans. For half a century Manley has exhibited a rebellious streak which resists confinement, limitations and some bourgeois conventions. At the same time, Manley's closest friends come from the progressive wing of the middle class. Despite 'charismatic' or 'messianic' appeal, he is not a man of the people, though he is a man for the people. 'Power for the People', a PNP slogan of the '70s, captures the essence of Manley's tutorial leadership style, rather than the more radical 'Power to the People'.

An important aspect of Manley's appeal has been physical. Very light skinned, he is nonetheless acceptably 'brown' in Jamaica's race-conscious, but not now racist, society. He has the 'touch of the tar-brush' necessary in Jamaican politics, as Professor Nettleford has noted.[13] Furthermore, Manley is tall and handsome—even in his seventh decade of life. Although he prefers informal dress, Manley takes pains with his personal appearance, weight and diet; he jogs and plays tennis; he has given up smoking, since he became Prime Minister in 1972, and he has moderated his drinking to small amounts of white wine. Manley has a history of ulcers, flare-ups coinciding with personal traumas such as divorces and the deaths of his third wife and his father. Diverticulitis almost cost him his life in 1987; his colon was removed, but he has since made an excellent recovery.[14]

Manley maintains two homes, one in Kingston and the other—Nyumbani (Swahili for 'Peace in Your Home')—his retreat and 'real home' high in the Jamaican mountains near Mavis Bank, east of Kingston. The homes are neither ostentatious nor palatial, but are, in the style of Jamaica's upper middle class, far superior to those occupied by Jamaica's masses. The Kingston home—called 'modest' by a *New York Times* reporter—is on land formerly occupied by his parents' much larger estate 'Drumblair' (subdivided and sold off during hard times in the '60s). In her final years Manley's mother lived in a neighbouring house, 'Regardless', a refuge for various members of the extended family. Both of Manley's houses are surrounded by well-tended gardens which reflect one of their owner's prime interests. At Nyumbani, Manley grows flowers and coffee for profit and harvests pine. The homes are tastefully furnished and decorated with the works of Jamaican artists, most notably those of his mother. Cricket annals line the shelf of Manley's study in Kingston. A satellite dish brings in sporting events and news from around the world. Both houses boast elaborate stereo equipment testifying to Manley's love of music. At Nyumbani, next to his parent's old retreat, 'Nomdmi', Manley has established a family reserve for his children which expresses his sense of family solidarity.

Many books on contemporary politics, economics and international issues occupy Manley's bookshelves, but he admits that his serious, concentrated reading ended with his university days in the '40s. A busy life of trade unionism and politics has simply not allowed Manley time for sustained reading of academic studies—an admitted but unavoidable problem. Manley keeps abreast of current affairs through a variety of publications. For pleasure, Manley reads thrillers and sports books.

Manley is a reflective, intellectually creative person. He is capable of detachment about himself and his politics. He seeks a comprehensive rationale for his politics in the 'logic of history'. He is more of an idealist than a materialist; he believes in the power of ideas and consciousness. Manley respects Marx as a great intellectual, but is not deeply read in Marxist literature; his father and the British socialist Harold Laski were more important intellectual influences. Manley prefers the universal and sweeping to the particular and mundane. Nonetheless, when motivated (as in union negotiations or important speeches), his almost photographic memory can master a tremendous amount of detail. His writings bespeak a logical mind, with a tendency toward listing: characteristics, causes, and results. Manley is open-minded and flexible, capable of altering sincere beliefs in the light of new information or new circumstances. 'I am grateful,' he wrote once, 'that God has given me the kind of mind that does not assume that what I thought I knew yesterday is an eternal guarantee of truth.'[15]

Manley's interpersonal relations naturally have exhibited a range of behaviour, according to the character of the other party. Some find him egotistical (at least in part), while others find him open and generous, attentive to the needs of others.[16] Although his temper can flare up in the face of incompetence or what he considers unfair criticism, Manley is quick to cool down and (with a few exceptions) doesn't hold grudges. He tends to trust people, perhaps to the point of being taken advantage of by some. At the same time he has the competitive urges of the athlete, the politician, the public figure.

That figure has been the source of considerable public commentary, much of which has been negative. A foreign journalist recorded in 1976 that 'opponents say Michael has a messianic complex, that he searches for ways to outdo his father.'[17] Ramesh Deosaran, a lecturer in psychology and sociology at the UWI Trinidad campus, accused Manley in 1978 of creating 'enemy after enemy as a technique to excuse his growing incapacity to govern.'[18] John Hearne, a longtime friend who turned savagely against Manley in the mid-'70s and wrote at least one libellous column against him, wrote of Manley's government that 'there were too few people ready to say "No!" to a man who felt that his private vision was public reality.'[19]

Others have seen Manley more positively. Hans Massaquoi referred to him in 1973 as a 'pragmatic politician who combines the polish of a Brahmin with a deft common touch,' adding that Manley had charisma 'which women especially find difficult to resist.'[20] George Eaton, a biographer of Bustamante, reckoned that democracy, equality, and social justice constituted the touchstones of Manley's political morality.[21] To Patrick Keatley, writing in 1974, Manley was an 'idealist turned pragmatist' with a 'first-class brain' and 'no touch of the *poseur*'.[22] Max Lerner of *The New York Post* wrote in 1975 that he respected Manley for his earnestness and integrity.[23] An anonymous writer to Kingston's *Star*, 'Seneca', declared that it was a 'sad irony' that Manley had to call for a State of Emergency in 1976 (see Chapter 16), since 'there is no more compassionate heart.' 'Seneca', a conservative, noted that 'Manley's embrace of the exploited made him a trade unionist . . . in his own life and that of Jamaica when his connections could easily have secured him ease and influence without hazard.'[24] Before he turned against him, John Hearne referred to the 'consistency of Michael Manley's philosophy' and said he was 'the first truly national leader to emerge from Jamaica's process of self-discovery since the two giants [Norman Manley and Bustamante] assumed their proper authority in

1938.[25] Andrew Kopkind noted in 1980 that 'to friend and foe alike, Manley is a charismatic personality of heroic stature.'[26] Finally, when he interviewed Manley in 1983 Michael Massing found him 'smiling, outgoing, emotive, his talk at once grand, discursive, philosophical.'[27]

In private, Manley exhibits traits which seem to contrast with his public image. He is at heart a shy person, as he himself and many close friends observe. He has long lived in his father's shadow and striven to establish a separate identity. Manley is self-critical, sometimes self-effacing, and sometimes affects a show of humility which others find unnecessary or unconvincing, although it may reflect real guilt, real shyness, and genuine insecurities. Mild self-effacement and humour is evident in Manley's comment to a friend that he is a 'moron in mechanical matters.'[28] To another correspondent he confided that 'I am slightly claustrophobic and do not cope well in a cramped environment.'[29] As a youth of nineteen, Manley wrote to his father that he was overcoming his 'horrid, weak dependency on the multitude & its good opinions', a habit that had worried him often,[30] but he admitted forty years later that he still had a terrible need to please people. In the '70s he informed a close friend that he was gaining control of his 'fears, guilts, sins, angers, hurt feelings and resentments'.[31] Even as a speech-maker, the eloquent, spellbinding Manley had his doubts, as he confessed to another correspondent:

> Notwithstanding how I may look [on television], I am not all that confident about myself as a performer. I have confidence about what I believe in and in myself in relation to those beliefs. But this is a different thing from feeling that one is good enough to do a particular kind of broadcast. . . .[32]

One suspects that Manley suffers from feeling misunderstood by the society at large, and that to a certain extent he verifies the saying that it is lonely at the top. Like all of us, Manley needs emotional support, approval, recognition and love.

Manley is generally optimistic, despite having been seared by personal trauma and political fire. His optimism and enthusiasm can cause difficulty, most notably when it led him to underestimate local and foreign resistance to the changes he attempted in Jamaica in the '70s. 'One tends,' he said, characteristically using the third person to discuss himself, 'to develop enthusiasm about one's emotions, the things that you feel and would like to see happen, to the point where they slightly outstrip the judgment that comes from the intellect.'[33]

A shy, vulnerable, sensitive soul resides in Manley's inner sanctum, to which only a privileged few are granted access. Childhood bullying by older children, the deaths of his third wife Barbara and his father Norman, and the political scalding of 1980 have left deep scars. In late 1987 Manley was repressing his reaction to the death of his mother the previous February, not allowing himself, yet, to explore that troubled terrain.

Wives and Children

There are also wounds carved by conflict of career and family. Manley has married four times and has five children. While a student in London in 1946, Manley married Jacqueline Kamellard, who already had a daughter, Anita, by a previous marriage to Robert Verity, a Jamaican. Manley's first child Rachel, who became a poet, was born in 1947. The first marriage broke up in 1951. In 1955 Manley, married Thelma Verity, a Jamaican of Afro-Chinese ancestry and dancer who, as a child had been adopted by Robert Verity's family. Manley and Thelma's son Joseph, who would be educated in economics and national planning in Cuba in the late '70s and early '80s, was born in 1958. The second marriage coincided with the tough early years of Manley's trade union activity, when he worked like a man possessed. The energy which he invested in career and politics made it difficult to devote a normal amount of attention to family matters. The second marriage also ended in divorce, in 1960. Manley married Barbara Lewars in 1966. Their daughter Sarah was born in 1967. Tragically, Barbara died of cancer in 1968, after urging Manley to succeed his father as head of the PNP. Her death was a great trauma that almost ended Manley's political career. In 1972, Manley married his fourth wife, Beverley Anderson, daughter of a Jamaica Railway clerk. Whereas Manley's other children were given Old Testament names, the daughter of this marriage, born in 1974, is named Natasha after the character in *War and Peace*. Their son David was born in 1980, in the midst of the awful political violence which contributed to Manley's defeat. Beverley Anderson-Manley, while pursuing a deferred university education, became politically active in the '70s. Leading the PNP Women's Movement, she was prominent and controversial in the left wing of the PNP. Politics probably contributed to the personal estrangement in this fourth marriage, and Manley and Beverley, whilst remaining officially married, have maintained separate residences from 1984 on.

Excluding the tragic third one, Manley's marriages are apparent evidence of conflict, restlessness, and instability in his family life. One source close to Manley suggests that he fell into and out of love quickly.[34] Moreover, Manley was attracted to the kind of bright, talented, strong women who were bound to find it difficult to live and grow in an intimate environment occupied by Manley's own expansive ego. Career and political pressures also undermined 'normal' family life, particularly in the '50s and '60s. In the '70s and '80s, a more mature Manley devoted increased attention to family matters, including the needs of his adult and younger children, and even learned to participate in routine chores such as preparing food.[35]

It is unclear how personal problems may have affected Manley's public life. The tragic death of Barbara Lewars-Manley would have made it impossible for Manley to continue in politics, had his close friends not helped him to put the tragedy behind him. It seems that

Manley is driven mostly by 'large' causes—the struggle for justice for workers, the campaign for a more equitable and democratic Jamaica, the crusade for Third World solidarity—and this may have allowed him to place family problems in perspective. His older children, scarred by the break-up of their parents' marriages, nonetheless came to understand, to approve, and to appreciate the public struggle waged by their father, sometimes at the expense of family life. The fact that Manley married four times also led some members of the middle class to regard Manley as 'undisciplined', a charge that unfairly infiltrated into other areas of public life where Manley actually exhibited great self control. In the end, one is left with the rather unsatisfactory view that, in a complex man, there may be no simple ratio between private and public behaviour, though the two are undoubtedly connected in subtle and mysterious ways.

In his biography of Edna Manley, Wayne Brown commented on the difficulty of trying to capture that exquisite spirit between the covers of a book.[36] Such a difficulty confronts us here, compounded by the simplistic myths and anti-myths that surround the man. The elusive truth may be that Michael Manley is, indeed, many things to many people. The search for understanding such a man must take us back to the historical roots of the Jamaican society from which the Manley family emerged.

CHAPTER 3

Roots

Early Jamaica

Michael Manley was born in 1924, in the 270th year of British imperialism in Jamaica which was to last until 1962. Imperialism—in its classic British form and as United States neocolonialism after 1962—was the most important formative influence on the Manley family. Their struggle for Jamaica's decolonization became the most important theme of the family's history. Although the Manleys fought colonialism and neocolonialism, they themselves were partly products of the colonial system and their lives were marked by its contradictions, ambiguities, and ambivalences.

In the '80s Manley wrote that Jamaican history was marked by six important dates: 1494, 1655, 1838, 1865, 1938 and 1962[1]. Columbus landed in 1494 which event was the beginning of Spanish imperialism. Spanish economic demands, diseases and violence eliminated the peaceful Arawak Indians and increased the need for African slaves. But after the conquest of Mexico and Peru and the eventual creation of a vast empire stretching from California to Chile, Spain's interest in the Caribbean was slight. Jamaica languished until 1655 when it was conquered by Britain. In the late 1600s and 1700s the British molded Jamaica into a classic sugar island to suit their own purposes. The results of this 'sugar revolution'—which occurred at various times in the sweeping tropical arc from Louisiana through the Caribbean to Brazil—are well known. Economically, sugar grown for export to Europe caused an unevenly structured economy featuring 'monocrop' agriculture; the enrichment of the few at the expense of the many; the mass horror and degradation of slavery; dependency on foreign markets, foreign imports and foreign decisions; a weak and timid local bourgeoisie which exchanged goods instead of producing them. Politically, since Britain found it necessary in the mid–1600s to offer some liberty to its white colonists to induce them to settle in the Caribbean, the sugar revolution resulted in a local legislature of considerable power; it was restricted to the dominant plantocracy and its allies, which ruled in their own class interest. Socially, the Jamaica sugar boom formed a mixture of classes and racial castes. Wealth and power were distributed in inverse proportion to the size of the racial castes: most went to the few whites; eventually enough went to intermediate browns (the result of often-informal race mixture) for them to become an incipient 'middle class'; and the huge black population—seventy-five to eighty-five percent of the population in modern times—

was massively poor. Society was marked by material inequality, psychological confusion, and alienation. Although creolization in time created a distinct Jamaican society neither European nor African, 'English' and 'white' values tended to elevate and 'African' and 'black' values tended to degrade, although the overwhelming majority were black and of African origin.[2] Blacks actively helped shape creole society, adapting to white power where necessary, retaining and reshaping African traditions where possible, notably in 'African' and black Christian religions. The ambiguities, ambivalences, and contradictions of the society were most manifest in the intermediate brown ('coloured') caste/class—which like mulattoes and mestizoes throughout the Caribbean and Latin America—was psychologically orphaned by both parent races, degraded by whites and resented by the black masses.

1838 was the year that the slavery regime ended, although its effects are still felt in economic, political, and social behaviour and attitudes. It was a confluence of forces that ended slavery. First, the slaves themselves bitterly resented their situation and rebelled against it often (most notably in the 1831 uprising led by Sam Sharpe which finally resulted in the end of slavery). In 1799 slaves in Kingston sang

> One two tree,
> All de same
> Black, white, brown
> All de same;
> All de same.[3]

And Charles Campbell, in his memoirs of 1828, recorded one slave's protest against the evil system:

> 'Who planted the cane?' said a slave to me one day, when I checked him for [stealing] a lump of sugar. 'Who nourished its growth? Was it not the poor Negro? Negro man work all day in the hot sun: he toil through mud and rain. He have hunger and wet all day, cold all night, yet he plant the cane, he watch over him, he cut him down, carry him to the mill, he make sugar. Shall Buckra [white] man, who do nothing, eat all; and poor Negro man, who do all dem tings, starve?'[4]

By countless acts of daily sabotage—a lump of sugar 'stolen', a hoe broken—the slave helped weaken Jamaican sugar's competitiveness with that of Cuba and Brazil. Jamaica's British market could only be maintained by a tax on other sugar, creating a drain on the exchequer of a Britain committed to 'free trade' (read British commercial supremacy) and penalizing the British consumer. The economic bases of the sugar regime deteriorated badly from the late 1700s onward. English humanitarians, religious dissenters, and—most importantly—Jamaica's black Baptist missionaries increased the pressure on the system. Finally, the intransigence of the Jamaican plantocracy alienated even the British crown. The plantocracy's brutality in suppressing black rebellions disgusted the enlightened public; and its blind lack of creative ability to reform meant that Britain would impose abolition from abroad.

Abolition itself was a great event, but as history throughout the Western Hemisphere has shown, a necessary but not sufficient condition to fully enfranchise the former slaves, much less to provide them with decent, fulfilling lives. Between 1838 and 1865 Jamaica's legislature remained firmly in the hands of the white minority, supported (and sometimes criticized by) members of the ascending brown middle group. The economy changed little. Many social customs and traditions continued as before, although the relative decline of the sugar estates intensified the creation of two Jamaicas, one white and one black, with the browns in ambiguous limbo. The biggest change of these years was the break-up of large estates and the creation of numerous black 'peasants', the owners of small farms, destined to be a significant group in Jamaican life.

The distress of this group led to a local protest—the Morant Bay Rebellion of 1865—which resulted in the deaths of a few white officials and landholders and some brown policemen. Then, in retaliation, hundreds of innocent blacks were slaughtered. Two Jamaican national heroes were eventually martyred: the black Paul Bogle and the prosperous brown George William Gordon. The rebellion so alarmed the local élite that in a moment of temporary panic it asked Britain to dispense with the power of the local legislature and rule Jamaica directly as a crown colony.

Norman and Edna Manley

The other dates of Manley's list—1938 and 1962—are intimately connected with the activities of the Manley family, as will appear below. The Jamaica into which Michael Manley was born in 1924, was the product of one of the longest and most profound experiences of imperialism in modern world history. Counting both the Spanish and British eras, it had lasted 430 years at the time of Manley's birth. It was a society, in the 1920s and for some time thereafter, where class and colour prejudices were strong. Sugar, the principal export and employer, was chronically troubled. Bananas created some prosperity for small growers, but subject to the market manipulations of the Boston-based United Fruit Company—the 'octopus' of Central America—and the caprices of hurricanes. The powerful English governor created a tradition of centralized one-man rule and a compliant legislature that has survived to this day.

Occasionally religious revivals shook the people and messiahs arose from among the godly to redress the wrongs done to them. Whites and browns, in schools and newspapers (most notably the semi-monopolistic, conservative *Daily Gleaner*), and in some churches, sanctified the 'glories' of the British Empire, of the English language, and of whiteness. Art was largely imitative and there were few worthy bookstores.

In any society, Michael Manley's parents, Norman and Edna, would have been exceptional. In Jamaica—small, insular, tropical, conservative, sleepy, colonial; poisoned by racism, class biases, and mass poverty—their extraordinary talents were magnified. In time they became icons, exemplars of the exaggeration of style that some claim is a Jamaican trait.[5]

Norman Washington Manley was born on 4 July 1893, the son of Thomas Albert Samuel Manley and Margaret Ann Manley (born Shearer).[6] Norman's father, Michael's paternal grandfather Thomas Samuel Manley was an illegitimate son of 'a peasant woman, fathered by an Englishman who left at least 2 other sons by different women.'[7] Thomas Samuel—an energetic, risk-taking, litigation-loving entrepreneur who traded produce with the United States—married Margaret Ann Shearer, the 'nearly white' daughter of Elsie Hunter Clarke, 'a coloured woman', and Alexander Shearer, 'a white man of Irish extraction'.[8] Elsie Hunter Clarke was the common ancestor of Norman Manley, his eventual wife Edna, and of Norman's ultimate political enemy William Alexander Clarke, better known as Alexander Bustamante. Margaret Ann Manley, as 'nearly white', was ostracized by white neighbours for marrying the nearly black Thomas, the beginning of a persistent Manley family trait of being at odds with dominant, conservative community values and institutions. Margaret, a small, quiet woman, had a fierce temper which she struggled to restrain—another persistent trait in the Manley family. 'She appeared to be completely without fear and was undaunted by adversity.'[9] She insisted on hard work, which she herself performed after Thomas died in 1900, having wasted his fortune in litigation. After an unsuccessful attempt at work in the United States, Margaret Ann Manley settled Norman and his siblings—Muriel, Vera, and Roy—on a small estate previously purchased by her husband and her father. There she worked hard on the farm with her children and paid her hired hands higher wages than customary, a practice which further irritated the neighbours.

Aged nine Norman went to elementary school. Later he attended secondary school in Spanish Town. The fortunate few who, in those days, received secondary education became separated from the Jamaican masses. Aged thirteen, in 1906, Norman enrolled at the island's most prestigious secondary school, Jamaica College, a 'tough school' of about one hundred boarders and fifty day-school boys, rampant with bullying and 'a good deal of homosexuality, for the most part not carried to extremes.'[10] At first Norman was the ringleader of rowdies who challenged weak teachers. But his mother's death in 1909 caused him to repent and, to the shock of his headmaster, to fix his eyes on a Rhodes Scholarship. His subsequent academic exploits, skill as a

A. The Clarke, Shearer, Manley, and Swithenbank Families

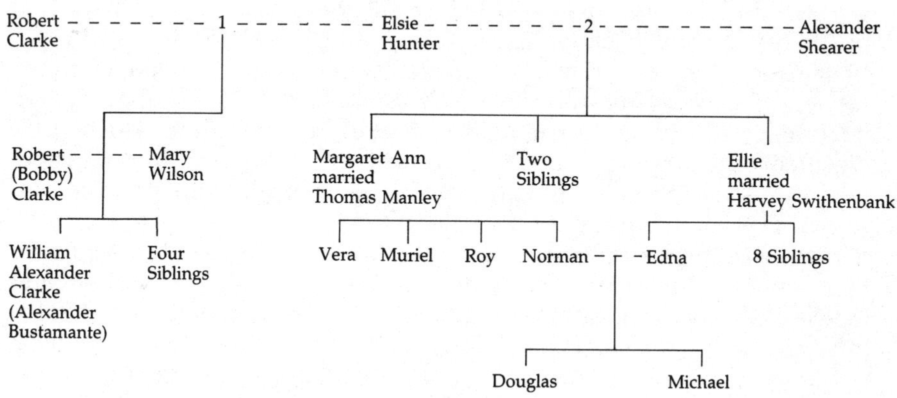

B. Marriages and Children of Michael Manley

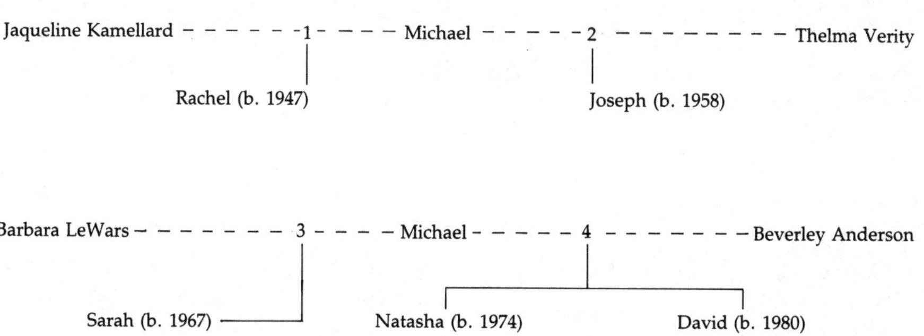

debater (taking the winning affirmative side in 'War is Inconsistent with Christianity') and feats in track and field, shooting, football (soccer) and cricket got him the prize.[11]

Arriving in England in the autumn of 1914, Norman joined his siblings, Vera (a music student on a Commonwealth scholarship), Muriel (a medical student), and Roy (a student in a private secondary school). Norman also met his aunt Ellie Shearer Swithenbank and her English-born daughter of fourteen, Edna, 'a long-legged tempestuous girl' whom he would marry in 1921.[12] After Norman completed a year at Oxford he and Roy volunteered for the British army. As brown Jamaicans, they experienced considerable racism, but also won the respect of their fellow soldiers. Roy was killed in 1916; Norman helped bury him in blankets in the field of battle. Following the war in 1919 Norman had a nervous breakdown which 'would take him years to face. . . . perhaps never to extirpate.'[13] His love relationship with first cousin Edna helped him survive the crisis.

Edna was the fifth of nine children of the Jamaican-born Ellie Shearer Swithenbank, the sister of Norman's mother, and Harvey Swithenbank (1865–1909), an English Methodist minister who met Ellie whilst a missionary in Jamaica. Two of Ellie's and Harvey's children were born in Jamaica; the others, including Edna, were born in England.[14] 'Ellie' (Martha Matilda Elliot Shearer Swithenbank) was born into the Jamaican Shearer family in 1869, four years after the traumatic Morant Bay Rebellion. Her family was white or near-white, 'part of a tiny privileged minority' which 'existed like islands in a rising sea of blacks, free now, mutinous, vengeful.'[15] The young Ellie lived in physical and social isolation, strong family ties offsetting a near-suspicion of the wider community. At twenty-three, she fell in love with the young missionary, Swithenbank. They were married and had two children in Jamaica before returning to England in 1897.

Harvey Swithenbank was the son of a self-made industrialist from Leeds, Oliver Swithenbank, who was a 'rigid opponent of the Church of England'. Harvey declined the family business and chose the ministry, setting himself apart by his 'uncompromising spirituality' from his materialistic age. An 'exciting preacher', he was said to be loved by the 'ordinary people' of his district in England. His was a demanding faith:

> We have no latitude of choice. We must fulfill the purpose of our being or pass away. . . . The popular idea is that our Saviour came into the world primarily to make men happy. The popular idea is a popular illusion. His *primary* object was to make men holy.[16]

In September 1909, about two months before his death on 18 November, Harvey Swithenbank sermonized about 'aimlessness', saying that though people were 'not masters of their environment yet they were masters of their purpose in life', and that 'life without aim was like a ship without a rudder. . . .' Intriguing premonitions of the later Michael Manley reside in the exciting preacher loved by the ordinary folk of Cornwall, though Michael's spirituality would be of a rather more secular type, involving such high causes as anti-imperialism, Third World solidarity and a new international economic order.

Manley's mother Edna was born 1 March 1900.[17] The young Edna was said to be 'gentle, sensitive, obstinate, combative, needing to be understood, to be attended to.' She found the attention she craved difficult to get in a large family, and the understanding hard to come by after her father died. Edna strived for self-control but was wretchedly 'driven to disruption'. Her anarchic tendency competed with her need for approval. She suffered rages that exceeded normal childhood tantrums. She was a 'weird child'. Two worlds struggled within her, 'the one Christian, rational, the other animist, instinctual. . . .' When they first met in September 1914 Norman Manley's impression was of 'a strange, shy, and highly individualistic person'. A headmistress, Miss Hanna, observed of Edna at sixteen that 'She wears glasses; but that is really to blind outsiders. She can see as far as most or farther. . . . Conventions are light & airy things to her, capable of destruction at a moment's notice.'[18] Many of these traits were inherited by her son Michael.

At sixteen Edna began to direct her schooling toward art and literature. In 1917 she began a series of war-related jobs: with the Pensions Office in the overwhelming 'first city of the world', London; harvesting flax in the countryside; breaking half-wild Canadian horses destined for combat. Edna's oldest brother Leslie was killed in the war, as were Edna's sister Lena's husband of six months, another sister's fiancé, and cousin Roy Manley. In response to such tragedies, Edna's behaviour became more erratic and reckless. By eighteen, she was regarded as the family's problem child. Her mother encouraged her to become a secretary. She herself considered the relatively conventional career of art teacher. However, with Norman Manley's encouragement, she became a professional artist who was to make a supreme contribution to Jamaican national identity. As a young artist, Edna was a child of her age, influenced by Freud, by the camera (which made representational art somewhat obsolescent), and by the contending influences of Post-Impressionism, Cubism, and Futurism.[19]

Edna's and Norman's love was forged during the disillusionment of the post-war era. World War I had destroyed the myth of a superior Europe in charge of its progressive destiny and stimulated anti-imperialism in Asia, Africa, and (less intensely) the Caribbean. Norman wrote of himself in 1919, the year of his breakdown, as 'a speck floating about or swimming in, a liquid mass—in it but not of it. . . .'[20] Norman and Edna were temperamental opposites, and their love, spanning half a century, was intense, complex, creative, at times stormy. Norman was rational, analytical, capable of withdrawing into storklike aloofness, often perceived as arrogant, a stance which concealed a deep shyness. Edna in 1919 was 'bubbling, anarchic, seven years younger' and would 'swarm all over' Norman.[21]

The letters of William McCulloch, a Jamaican friend, provide glimpses of Norman's struggles in 1921, when he completed law studies and married Edna. In March McCulloch asked if Norman was reading any biology or psychology: 'Keep off the abnormal side of the thing; you haven't any time to go really deeply into it, and it merely gives one a very low idea of one's own mind.' A few weeks later McCulloch noted that Norman had written of 'scientific politics and of being weary of idealism'. McCulloch suspected that, despite Norman's disclaimer, 'you are as full of idealism as any red hot miner (so they say) but you have accepted a pose of the hard man of the world, full of self and shrewd commonsense which seems to me a d— selfish chap.' McCulloch had caught one of Norman Manley's enduring traits, the hard, aloof exterior camouflaging an exquisitely sensitive spirit. McCulloch went on to call Manley 'the greatest poseur of J.C.' (Jesus College, Oxford) and reckoned Norman's impending marriage to 'the cyclone' would do him good: 'You are bound to lose a lot of the cynicism of your *expressions* (not beliefs observe).'[22]

In May 1921 Norman won the prestigious Lee Prize at Oxford for his essay on Samuel Butler, the author of *Erewhon*. In June he passed his written law exams and he and Edna were married on 25 June 1921. These happy events, however, did not erase the deep malaise caused by the war and its aftermath and by racism in Europe. In May McCulloch wrote of Norman's 'tirade against the vagueness of idealism'. In August Manley's friend and sometime confidant observed:

> It is inconceivable that a man who makes his love of a woman dominate his life will be able to remain apparently self-centered and 'harder than ever'. That is a little pose my lad to blind yourself and others to the ignoble fear of a surrender of what is bad for something that is good.

This scolding apparently provoked Norman to reveal a bit of himself, for McCulloch wrote later that he was glad that 'for once, you have been human and thawed sufficiently to give me a very good glimpse at the Manley one suspects is there, but is never sure of.' McCulloch, however, worried about Norman Manley's reading of psychology—'I mistrust your keenness for the subject'—and recommended biology and metaphysics as better for part-time reflection.[23]

Norman Manley was deeply alienated in England, except in the company of fellow West Indians. Even his love and marriage with Edna strained bonds within an extended family which crossed cultural and racial barriers. Although Edna's grandmother was said to be a Jamaican 'coloured' woman,[24] Edna's mother and Edna herself could and did 'pass' for white, and Edna's father was English. Edna's Jamaican-born mother Ellie Swithenbank worried about the 'undesirability of children in a mixed marriage'.[25] In response Norman asked Edna's permission 'to point out [to Edna's family] one or two moral trifles they don't seem to grasp.'[26] Edna's and Norman's small wedding occurred in a registry office, witnessed only by Vera Manley and Lena Swithenbank. The lack of a church service caused Edna's former headmistress, Miss Hanna, to scold: 'Your father & your forebears will, I think[,] be sorry that one of their children thinks so little of the church they loved. Why?'[27] Perhaps the answer was that the church which Edna's father had served as minister had done too little to dispel the injustice and irrationality of racism in its followers.

Edna and Norman spent their last year in England finishing their apprenticeships. Edna completed her artistic training while Norman did the compulsory observation of a barrister and of the courts to qualify for the bar. Their first child—Douglas Ralph Manley—named after Norman's war-martyred brother Roy Douglas and Edna's baby brother Ralph—was born on 30 May 1922. On 18 August they sailed on an Elders and Fyffes banana boat, *Camito*, for Jamaica. Norman warned the apprehensive Edna that the island was a place without art.

Jamaica in the 1920's

Jamaica was a provincial colony, inhospitable to such high-strung, intense, serious, artistic, unconventional, and intellectual souls as Norman and Edna. They were trying to create demanding careers and raise first one and soon, two, sons. Norman Manley summarized the situation:

> I had a very narrow view of Jamaica. School and a very derelict isolated property in very deep rural Jamaica and much reading and thinking about life in general than in any particular place. Then the war and in all eight years when I made no contact at all with my homeland! Then back home—expecting what?
>
> To be frank, I was surprised and shocked. Few, if any, in Jamaica, were interested in anything except the daily round of events in Jamaica, and even as to that range of interest was extremely narrow. You talked, or tried to talk[,] about important events in other countries, but hardly anyone was interested. Books and drama were barren gambits—music was worse. Politics locally was of little importance. . . . The people seemed content and quiet. Though sugar was down, bananas was rising rapidly. The economy was turgid and there was a general air of emptiness.
>
> I did not study it. I merely met it. It rolled over me and left me feeling almost in despair.[28]

Norman at least had deep, if neglected, roots in the homeland he faced with dismay. Edna's ties were mere rootlets—nostalgic tales of Jamaica told by her mother—and in October 1923 she fled back to her 'home' in England with baby Douglas. Norman wooed her back by writing 'carefully casual, practical, newsy letters'.[29] He also purchased Drumblair, a somewhat rundown two-story house with extensive grounds, where Edna might exercise her need for privacy and personal freedom. Edna sailed back from England in February 1924. Shortly after her return a second child was conceived. It was intended to be a girl named June Patricia.

CHAPTER 4

Youth 1924-43

One never changes in life—the things that stimulated or wrecked you as a child stay the same—one builds all sorts of defenses to the bad ones, but it's just the same, and the challenges that call to one are as much music as ever.

Edna Manley[1]

The Manleys and Colonial Jamaica

Contrary and exuberant from the start, the future Prime Minister of Jamaica, Michael Manley, was born on 10 December 1924. The '20s and '30s were years of achievement and conflict for his parents. Along with the influences of a stormy era in Jamaica, young Michael absorbed the tensions, triumphs, and defeats of his parents. For Edna, this meant deciding where 'home' was and wrestling with identity. 'Home' as late as 1929 was England,[2] though in the '30s Edna began trying 'to think like a peasant'[3] in her art and to identify more and more completely with Jamaica. Her personal inner identity was never fully resolved, as she struggled with profound personal and artistic concerns into old age. During the Depression she produced great sculpture, the prophetic 'Negro Aroused' and the more militant 'Prophet' (both 1935) and the intriguingly contemplative 'Strike' (1938). She was the artist as prophet of history. She had, however, long felt the conflict of art and the duties of wife and mother, conflicts similar to those which Michael would also experience years later.[4]

For Norman Manley, these were momentous and troubled years. He rapidly rose in the small circle of law, winning the prestigious title of King's Counsel in 1932 and gaining the reputation of the West Indies' best lawyer, the man to save you from hanging if you had committed murder. Norman achieved high distinction by what he called 'an absolutely egotistic determination to win every case I was engaged in, which I regard as the most important attribute of an advocate . . . a total will to win, so that you stop at nothing but deal with maximum concentration, maximum observation, maximum study, maximum everything, and the reward was winning.'[5] Yet, the successful and driven lawyer was a man in conflict. He developed a stomach ulcer shortly before being appointed King's Counsel. Three months later, addressing the Jamaica College Old Boys' Association, he stated that lawyers 'like himself, were parasites,' members of 'a profession which made them eke out an existence on murder and other crimes.'[6] In 1933, Norman's old friend William McCulloch chided him:

> What is the legal quibble, by the way, which permits you people to defend a confessed criminal? . . . Do you ever feel ashamed, or do you put all ethics behind you, and just play a fine game of a battle of wits? Is it a social performance or anti-social?[7]

Other conflicts stirred Jamaican society and the Manley family. Marcus Garvey's messianic call for black power in the world and in his native Jamaica was one. Garvey (1887–1940), a black man born in humble circumstances in St Ann's Bay, arguably the national hero with the most appeal to Jamaica's black majority, enjoyed a phenomenal rise to worldwide prominence as a radical activist. The year 1917 found him addressing thousands of blacks in Harlem, and the following year he founded the weekly *Negro World*, which at its peak had a circulation of over a hundred thousand. His United Negro Improvement Association (UNIA) was the largest black organization up to its time in the United States, and UNIA branches existed throughout the Caribbean. United States authorities opposed Garvey's activities, especially in the climate of the Red Scare of 1919. In that year the FBI under J. Edgar Hoover, began a campaign to deport Garvey. In 1922 he was convicted of a trumped-up mail fraud charge, and was deported from the United States in 1927. He returned to Jamaica in 1928 and launched his People's Political Party (PPP).

Norman Manley recorded that 'it is one of the legends that the Jamaica Labour party have sought to embarrass me with over the years when they say I tried to injure Garvey and indeed claimed that I partly succeeded.'[8] On three occasions the elder Manley and Garvey were adversaries in court, 'the successful brown man of the élite against the black anti-establishment man.'[9] The first, in 1929, stemmed from Garvey's three-month imprisonment because of a 1929 PPP call for the impeachment and jailing of unfair judges. Garvey had been elected to the Kingston and Saint Andrew Parish Council, but missed three meetings because of his imprisonment. A majority of the council's members held that Garvey's seat should be declared vacant. Unsure of themselves, they asked their counsel, Norman Manley, to render an opinion, and he upheld the council's decision. Norman Manley recalled that he had a 'furious row' with Garvey over the issue.[10] The second incident, also in 1929, involved a libel case a poor woman brought against Garvey for an article in his Jamaican newspaper *Blackman*. Garvey represented himself and the plaintiff was represented by Norman Manley, who reported on his adversary's courtroom manner.

> He defended himself and did much to enhance his own reputation. He made a brilliant closing address to a Special Jury hearing the case. He was witty and amusing, took every advantage open to the litigant who defends himself and though I won the case, which was really quite indefensible,

> he got that jury to award the lady a bare half the damages we had expected to win. It was a fine performance.[11]

Norman Manley referred to his third confrontation with Garvey as a 'run-of-the-mill' manner involving an American UNIA employee, George Mark, who sued the Jamaican UNIA for back wages. Manley, representing Mark, won the case, but the local UNIA and Garvey won on appeal.[12] Norman Manley had acted strictly in his capacity as a lawyer in all three cases, but the JLP bedevilled him for years with the charge that he had helped frustrate Garvey. Edna Manley's biographer, Wayne Brown, observed that the legal system Norman Manley served

> was placing him on the wrong side of history, as of conscience, threatening him with the patronage of a class whose real hostility to Garvey he can hardly have shared, and it is unlikely that he could have failed to sense this. (It is perhaps significant that it was immediately after the second Garvey case that the clamour began that Norman should be appointed a K.C. [King's Council].[13]

Brown speculates that Norman's political awakening began in the aftermath of the Garvey court clashes.

Norman also came into conflict with a foreign economic giant, the United Fruit Company. As counsel for the Jamaica Banana Producers' Association (JBPA)—an experience which he said drew him into the 'real life of the country' among the small, rural banana producers[14]—Norman Manley worked out a compromise with the Boston-based banana giant. United Fruit insisted that the JBPA be not organized as a cooperative; in exchange the company agreed to set up a welfare fund to benefit mainly Jamaica's rural poor. This 1936 agreement, an aspect of welfare capitalism that left the basic power structure firmly in North American hands, was barely accepted by the JBPA's members, leaving some with the feeling that Manley had sold out the country.[15] A final major conflict was the labour protest of 1938 which led to a movement for colonial reform, the nationalistic People's National Party (PNP) founded by Norman Manley and others, and his long, bitter struggle for political supremacy with his cousin, Alexander Bustamante, matters treated more fully below.

Childhood

Michael Manley's youth was dominated by his remarkable, enigmatic, strongly individualistic parents. Being the child of such parents had its privileges, but also its tensions. Douglas, being the first-born, suffered from the inexperience of his parents, as perhaps all first-borns do. Michael theorized nearly sixty years after the fact that, in the '20s, his parents had adopted 'all kind of crazy, wrong theories about child raising'. As very cerebral people, they were likely to read a book and decide on the basis of their reading how to raise children. 'It was a time when there was a lot of theory about how you mustn't spoil children. You know, if children start to cry, leave them to cry it out.'[16] This approach and the extreme intensity of both Manley parents led to Douglas going to live with his Aunt Vera at the age of four or five. Douglas and Michael spent much of their youth separated, which accounts for their contrasting personalities, Douglas being much more soft-spoken than his more flamboyant brother, and their reticence with one another in adulthood.

Michael remembers his early childhood, before he entered Jamaica College in 1935, as a mostly happy time. One of his earliest memories is of being in his mother's studio, watching her

> with ferocious concentration, hour after hour, chipping away, and the excitement of the chips. I can remember the chips, rising, rising, and this is a very, just early, almost visual, sensual, tactile memory . . . and you'd watch the thing go on for hours. I remember her face utterly intent, this obsessed thing that an artist gets when they're working, where they are totally unconscious of everything else.[17]

Norman Manley also worked legendary long hours in his growing law practice. The Manleys were able to compensate, in part, for their own shortage of time with their sons by employing domestic helpers. By the late '30s the domestic staff included Miss Boyd, the housekeeper/governess; Vivian Dacres, bodyguard, chauffeur, children's confidant and disciplinarian; Charlotte Walters, the 'butleress'; Wright, the gardener; and 'Dreamer', a handyman.[18] Boyd and Dacres were surrogate parents. Miss Boyd was a devout Methodist who worried constantly about the Manleys' spiritual well-being. She stood four foot eight, loved artistic things like arranging dry branches with bronzed leaves in large jars on the floor and was like a second mother to Douglas and Michael.[19] Vivian Dacres, under the fighting name 'St Andrew Pup' had been the feather- and bantamweight boxing champion of Jamaica in the early '30s. Although spankings were not unknown, the more usual forms of discipline in the Manley family were verbal exhortations and the withholding of privileges.

During most of Michael's youth, the family residence in Kingston was Drumblair, a 'rambling, deep verandahed house, set among wide, tree-studded lawns and stands of high grass.'[20] Drumblair was comfortably, but not ostentatiously decorated, furnished as T. A. Marryshow, a pioneer Grenadian labour leader, recorded, with 'just a few bits of furniture in good taste, no costly divans, carpets and such things.' Marryshow reckoned that 'there was what I call character pervading the home.'[21] Drumblair was open to people of all the varied hues and stations of Jamaican society, from the white governor to the poor, black, struggling intuitive artists Edna inspired and nourished. She in return often visited them, unaccompanied, at bars in the depths of Kingston's slums, scandalizing the conservative élite.

George Campbell, Miss Boyd's nephew, later dubbed 'Poet of the Revolution' by Norman Manley, was like a member of the family at Drumblair in those days. Campbell recollected that Michael's and his childhood days were filled with good times, sports and games. He recalled that as a young boy, Michael had quite a temper and was supremely confident. Campbell remembered that when Norman Manley was appointed King's Counsel in 1932, Michael took to a soapbox and announced 'My father is Norman Manley, KC,' a performance that gave father and older brother Doug a laugh.[22]

Boxing appealed very much to the assertive, aggressive Manleys. Douglas defended himself at Munro College with his size and boxing skill. Michael recalled that he 'grew up in local boxing's Golden Age of Juan Herrera, Kid Silver, Lefty Flynn, St Andrew Pup, et al when Puckoo de Souza was a local Tex Ricard and Kid Chocolate went ten rounds with America's Petey Sarron three months before the latter became Featherweight Champion of the world.'[23] Norman Manley took an extraordinary interest in the sport, training Vivian Dacres and judging local events. The elder Manley set up a ring at Drumblair to relax by sparring with Dacres. Michael may have been trying to emulate his father when he sparred with the older George Campbell. When Campbell knocked him flat, Michael responded by throwing rocks and chasing him until Dacres intervened. Campbell recalls that young Michael was 'interested in everything': sports, music, art. They played cricket together, creating elaborate fantasies based on its historical minutia. 'One side of us was the Australians and one side was the English and we were playing for The Ashes.' Michael and George acted out the roles of individual cricketers, based on statistical records, registered these games in scorebooks, and had mock award ceremonies with their own container of ashes. At other times the Napoleonic Wars were re-fought with an army of six hundred lead soldiers arrayed along battle lines that stretched from the doorway of Norman and Edna's second-story bedroom, down the stairs, and out to the lawn, where soldiers in trenches were attacked with rubber-band weapons.

Campbell recalled that Norman Manley at that time was generally considered very elusive, austere, and, above all, arrogant. George sometimes feared Norman, as did Michael. But Norman could also be very warm and caring. Once, when Norman's fast driving made George car-sick, Norman took pains to comfort him, despite being late to court on account of the young boy's illness. After Michael's displays of temper, Norman would pat him on the head and say 'All right, son'.

Campbell developed an intense lifelong friendship with Edna Manley. He recalls her during Michael's childhood as 'very friendly, giving you a sense of independence.' 'From very young Michael could gallop on horses when she was doing that, and if you fell down or anything she didn't make you cry, or if you cried, didn't notice, but was there.' The Manley parental values stressed 'a lot of liberty, but don't get out of control.' There were no forbidden questions, and questions were always answered. In summary, said Campbell, 'the greatest quality with that family was understanding.'[24]

Douglas Manley recalls that Michael's early life before he entered Jamaica College was quite happy, normal for their social class, and that Michael was a very cheerful child. However, 'he was an extremely aggressive child with an appalling temper . . . and a very impulsive sort of person. But there weren't any terrible problems.'[25]

In 1985, Edna Manley recalled some images from Michael's youth. He and Douglas used to like to climb a stinking-toe tree, so high that it scared Edna, but they refused to come down when she told them to. However, they lowered a bucket for food!

When Michael was about eight, there was a terrible family problem of children laughing whenever anybody got hurt. Edna and Norman made rules to prevent this. Nonetheless, when a wasp stung Doug so badly that his face swelled, Michael persisted in laughing at him. Edna warned Michael that something terrible was going to happen to him. Michael later got wasp stings around his eyes, causing them to swell shut. Reminded that he was being punished, Michael refused to admit that he couldn't see. He rushed out of the front door and—unable to see—missed the top of the stairs and fell down them.

Under parental command not to smoke, Douglas sneaked puffs behind Drumblair's horse-shed. His father was under doctor's orders to give up smoking because of his ulcers. One night when Doug went to have a last smoke, he discovered his father behind the horse-shed smoking: at that point Douglas began to realize his father might not be a god![26]

Whether Michael ever experienced a similar awakening is debatable. In the '80s he continued to deify his father, and did not speak easily of the psychology between an enormously gifted father and his talented son in a small island society. He regarded his father more as an Old than a New Testament god. Two stories Michael told, however, suggested the 'god fallen', imperfect and vulnerable, rather than enthroned and omnipotent. The first involved the accidental naming of the Manley's mountaintop retreat. Norman was painting a 'No Admittance' sign there whilst absorbed in discussion with Michael. Lost in conversation, Norman painted N-O-M-D-M-I. When Edna came by, she said 'My God, Norman, what have you done?' On second thought, the half-mystical Edna added, 'But leave it, it's a wonderful name.' Thus the family hideaway, Nomdmi, received its puzzling name by dint of the inattentiveness of the West Indies leading lawyer![27]

A second story Manley told involved an incident with Norman Manley's close friend Leslie Clerk. At Nomdmi, Edna asked Norman to clear a large branch obscuring the view of St Catherine's Peak. Norman mounted the offending branch and proceeded to saw away. Meanwhile, he became fascinated with clerk's description of a clairvoyant who, unable to speak a word of the language, had proceeded to write a novel in French. Long interested in the psychology of the paranormal, Norman became more and more preoccupied with Clerk's account, to the point where, Manley swears, he sawed himself off and fell to the ground amidst the children's howls of laughter![28]

Edna Manley's letters from the '30s provide further glimpses of Michael's early years. From another family refuge at Arthur's Seat where Edna sometimes withdrew to have the freedom to do her art, she wrote to Norman:

> Take [the boys] out in the boat on Sunday and try to see a little of them & remember—be tolerant & humorous—& good natured with all their faults & lacks—they are very young & far from perfect—but they're human & can't live at a high tension of effort. Just don't be critical in your own mind & yet don't treat them obviously as if they were children. All this you know better than I do—only I mention it because when you're busy you may forget and I want us four to be friends.[29]

In 1936, when Norman had gone to the United Stated to negotiate the banana producers' cause with the United Fruit Company's Samuel Zemurray ('Sam the Banana Man'), Edna wrote that 'Michael and I are a calm domestic menage—and he is [quarrelling?] at school like HELL.' She asked Norman to send Doug and Michael 'any little American journalism on the fight. They are both crazy about boxing. Douglas's letter today is nothing but the fight-& there is so little we can give him that he really gets a kick out of.' She gave Douglas a year's subscription to *The Ring* for his birthday. 'Michael sends to say he weeded the pitch yesterday.'[30] In another letter written 'whilst the rain lashes & the gullies roar' and 'with Doug in the bush,' Edna wrote that eleven-year-old Michael was being 'the real man of the house!'[31] When letters arrived from Norman, Michael was 'literally all beside himself.' He and Vivian Dacres sat with eyes bulging while Edna read Norman's letter to him: 'He won't sleep a wink tonight—he feels the world is coming to an end[.]'[32]

When Edna left Jamaica for four months in 1937 to present a one-woman exhibition of twenty sculptures and painting at the French Gallery in London, she wrote to Norman to see to the boys. 'Please do, they're such friends of mine and I do hate them growing up even four months without seeing them.' In another letter, she wrote: 'I don't like the sound of Michael. . . . he wants taking in hand properly'; she added 'do please give up a lot of time to the children on holiday—whatever the sacrifice—you simply must kid & watch them—observantly & write & tell me the truth. And let them live for me please.'[33]

Later, back in Jamaica, Edna wrote that 'Michael is busily painting & quite happy. In the evenings there is complete silence & they read steadily—we all do.' 'I am trying to teach them to play games in a different spirit—without the overwrought excitement & legpulling that borders on gloating. It's easier to win & lose with humour in a quieter atmosphere & I tell them to save the emotional tension for more important things—it may hurt you to lose a race or fail an exam, but it's not necessary to take that spirit into monopoly or rummy. They are old enough to learn a difference & save their nerves a silly strain. Also I won't have this carping nagging at table—so you see—we've set sail for general improvement with a capital G & I!!! They are the sort of yearlings that want calming not stimulating!![34]

Adolescence

The economic, social, and political protest that swept over Jamaica in 1938 was anything but calming. Jamaica's neocolonial economy and the Great Depression intensified suffering which led to protests and death in rural areas and Kingston. Norman Manley and his cousin and rival Bustamante responded by establishing competing movements that would dominate Jamaican politics and initiating (especially in Norman Manley's case) demands for colonial reform and eventual independence for Jamaica. Michael was aware of such events, but he experienced them indirectly, as a self-absorbed adolescent, rather than directly as his parents did.

As her sons approached manhood in the midst of Jamaica's late-colonial convulsions, Edna Manley began to fear how Douglas and Michael might fare. Her apprehensions reflected her own continuing struggle with her 'dark side'. 'Doug must not be lonely or unhappy, it's just death to him. He & Michael will help each other greatly. That generation are of no use in an atmosphere of loneliness & worry. When they are happy they see straight.'[35] In a letter in which Edna referred to herself as a 'debit', she postulated that 'poor Michael is going to be just like me & Doug will be like you.'[36]

One of the most profound parental influences on young Michael was his parents' intense interest in art, literature, and music.

> We were really very lucky. Mother and Dad read a tremendous amount and encouraged, from early on, both my brother Douglas and myself to read. They introduced us first to C.A. Henty and Fennimore Cooper. We got accustomed to reading. . . . Both my parents were voracious readers. It was part of the intellectual tradition of their times. They almost made a fetish of it. . . . Music was listened to with almost religious fervour and discussed after. . . .[37]

Such exposure was exceptional in conservative, colonial Jamaica and contributed to the Manley brothers' awareness of being unusual, at odds with the torpid drift of society. At Munro College, which he attended from 1935 to 1940, Douglas's knowledge of classical music was among the factors that made him an 'odd man out' among his rather aggressively ignorant classmates.[38]

By the early '40s Michael Manley's interests had turned strongly to poetry, the arts and literature. He was much impressed by his childhood playmate George Campbell's poem, 'The Last Negro'.

Green flaming wilderness!
White jagged rocks of faith!

The last Negro moves across the world
In his flesh Time's loins
By his side Time's children.

Way back in 1940
There was murder.

Way back
Dawn bent its rose face
Kissed a black animal
Woman of the negro race
Lovely black animal.

Way back in 1930
There was lynching
He was dangling
And survived his tree.

Spirit in physical
Death is no end of faith!

The last Negro looks into the sun
Into the gold flames
Feeling the heat of stars
And close is God

In creation
In destruction
For Time is God is Man
And peace is chaos.[39]

Fifteen-year-old, precocious Michael wrote to his father that he thought the poem was the finest thing ever written by a Jamaican, that it was an enormous mental stride forward, that it was 'poetry of the first order'. With the self-assurance of youth, Michael added that 'T.S. Elliott [*sic*] dismissed Shelley as immature but I think he was guilty of a bad error of judgment. Shelley was youthfully mature.' He noted further that he was 'literally dying' to hear about a court case Norman Manley had lost. Michael had also decided that his future lay, he thought, in architecture. He closed noting that 'I am not a Pantheist but I can feel an affinity between myself and them.'[40]

Inevitably Michael was moved to try his own poetry. 'Poem' linked nature, dream, and destiny in a shakily romantic 'literary' style:

Oh! What a beautiful day,
Warm yet refreshing.
The music of the sunlight
Playing, stimulating
on my ears.
And I felt the moving beauty of that song
Heard the music of that day.

And as I dreamt these dreams
The wind came stealing on me
And sweeping, swept on through me
Urgent, ever urging till it died
Leaving new within me
Courage for the future
My destiny revealed.[41]

The notion of a 'destiny revealed' and the essential romanticism of the poem reveal enduring aspects of young Michael's maturing character. From his mother he received an appreciation for the free, the wild, the non-rational, the mystic, the unconventional.

In early 1941 Michael wrote to his father that he was reading Thomas Wolfe's *The Web and the Rock,* which he found less intense than *Look Homeward Angel*. 'He is great, Wolfe. Reading Wolfe is an experience.' Foreshadowing Michael's later trade unionism, he was also reading Zola's *Germinal,* 'a clever novel meant as an exposé of mining conditions in France in 1860.' Young Manley also read a great deal of D. H. Lawrence's poetry.[42]

Michael Manley was the child of enormously gifted, enigmatic, and sometimes deeply conflicted parents. The Manleys, although intimately involved in Jamaica's professional, artistic, and sporting circles—and through Jamaica Welfare in the broader society of the poor—were set apart by the quality of their minds, spirits, and values. There was an underlying sense of being at odds with the society of which they were privileged bourgeois members. Michael grew up as a happy, feedom-loving child. But he also could be aggressive, rebellious, impulsive temperamental, stubborn and angry, traits he would strive to control throughout his life, with mixed success. In his teen years he developed an outlet for his passions in the arts. From 1930 on he would have been increasingly aware of the nationalistic political struggle being waged by his father's People's National Party. A further important, but ambivalent influence was Jamaica College.

CHAPTER 5

Jamaica College 1935-43

De lessons and de poems dey write an' send from England
Impress me they were trying to cultivate comedians
Comic books made more sense,
You know it is fictitious with pretense,
C. O. Cutteridge want to keep me in ignorance.

Mighty Sparrow
Trinidadian Calypsonian[1]

. . . We intend to work towards a non-élitist, egalitarian, single-stream educational system with heavy emphasis on self-reliance, heavy emphasis on teaching all people the value and dignity of all kinds of work. Heavy emphasis on making education a preparation for patriotic and co-operative service.

Michael Manley (1976)[2]

Colonial Education

Historically, the Jamaican educational system simultaneously provided social mobility for the few and maintained the colonial status quo. It was both an instrument of change and a political tool of great conservative force.[3] Colonial schooling consolidated British rule by preparing a small middle class élite for intermediate positions in commerce and administration and by providing the masses with basic literacy and ciphering. Like colonial education everywhere, it also prepared a tiny minority, including Michael Manley, to become critics of the system which produced them. Inevitably, their criticism would bear marks of their colonial education.

One of the severest critics of colonial British West Indian education was Eric Williams (1911–1981), a giant of West Indian nationalism and political chief of Trinidad from 1957 until his death. Williams won a colonial scholarship to Oxford as Trinidad's top schoolboy and graduated in 1935 with a first class degree in history. He described how in his youth the slogan of both legislature and school in Trinidad was 'Be British'. He noted that West Indian history was excluded from the Trinidadian secondary curriculum until 1939; then it was included only for the 'weaker boys' who could not handle the supposed rigours of British and European history. 'If anything is needed to underline the inferiority and disparagement of things West Indian.' Williams wrote, 'it is that the West Indian environment should, if studied at all, be studied by weaker boys.' At Oxford Williams concluded that 'my twenty-four years' preparation for an English University had ended by unfitting me totally, in the eyes of those best qualified to judge, for life in Trinidad! The Trinidad people's investment in me was to bring dividends to England!' Thus Williams later became a critic of the educational system similar in Trinidad and Jamaica, that did not prepare students for life in the West Indies.'[4]

The Jamaican colonial educational system likewise reflected the realities of power in the society. 'The hereditary rights of the big land owners made education irrelevant [for their offspring]; the professional, bureaucratic and clerical needs of a colonial economy were supplied by the middle class who were content with a system based on the English public school of the late nineteenth century; and the three R's which were made available to some of the children of the masses exactly reflected the latter's position in the social and economic ladder.'[5] Most Jamaican parents of the lower and middle classes developed strong faith in the power of education to advance their children. The schoolteacher, along with the preacher, became a figure of awe and veneration. But the system, while delivering some upward mobility to the few most talented scholars from the masses, remained deeply hierarchical and inegalitarian, a colonial legacy unfavourable for the building of Jamaica's liberal democracy after 1938.

Schools could be profoundly violent and tyrannical, as revealed in Morris Cargill's memoirs of Munro College. Cargill—conservative descendent of the white planter class (and Manley family foe)—recalled that the college was supposed to be run like an English public (private) school, but 'the trouble was that the English public school system it copied was that of a century earlier, as it was imagined by a Colonial overseer who had never been to one.' Munro's headmaster, 'Wagger' Harrison, was a 'drunken bully who believed in flogging as an essential aid to teaching.' He taught Scripture 'cane in hand, stalk[ing] threateningly between the ranks of his terrified Christian soldiers.' Slight mistakes in answering questions might result in a thrashing, so Wagger's students achieved prodigious feats of memorization, but often without understanding.

Cargill depicts Munro College, a school for the privileged, as a house of horrors. Boys were inducted by initiation ceremonies such as 'candle greasing': dripping candle wax all over a boy's head; 'it took days of combing and cutting to get the stuff out'; and 'crowning': a chamber pot full of urine was placed over a boy's head like a cap and he was forced to get up and declare, '*Rex poniarum sum*!,' ('I am the king of the pony pot'). 'Serious' offenses such as smoking could be punished by a public flogging before the school's 120 boys and 12 staff. The 'sinner' would be pinned across a table by four fifth- and sixth-form boys. Before the flogging Harrison would deliver a short speech, a 'mixture of audience flattery and intimidation'.

By the time he was finished, he had practically everyone on the school on

> his side, screaming for blood. The friends of the boy to be beaten would betray their friendship by agreeing that he deserved it. Wagger made public beatings the most popular entertainment at Munro College. All the herd instincts—the cruelty, the need for hero-worship, the admiration for physical strength—which boys and savages have in common, were turned by Wagger into a cult. Even the unlucky victims of his cane would more often than not go back to worshipping when their wounds healed, in return for a crumb or two of kindliness carelessly thrown in their direction.[6]

Sadists such as 'Wagger' Harrison represented an extreme, to be sure. Nonetheless, the violence and brutality Cargill described persisted at a slightly reduced level for decades throughout élite secondary schools. Douglas Manley attended Munro College from 1935 to 1940. He received six blows with a cane his first night at the school, nominally for accidentally upsetting a jug, but probably also as a 'welcome'. Harrison died two years or so after Douglas began at Munro, but bullying by students continued to be 'pretty rough'. Douglas, however, was shielded from much of it by his boxing ability. At times he got into fights about his mother's art; the students were mostly from the very conservative plantocracy and regarded drawing from a live model as scandalous. Douglas was a bit of an oddball because the Manley boys shared their parents' knowledge of art and music while their classmates were aggressively lowbrow. Douglas had a fair juvenile knowledge of Beethoven and Brahms, but 'a lot of the boys didn't know what classical music was.' To Douglas, his secondary school years were a decidedly negative experience, in contrast with his subsequent happier time at Columbia University.[7] Once again, the image of a Manley rather isolated from the surrounding 'society' comes forth.

Bullying and Coventry

Secondary school at Jamaica College in Kingston had a huge influence on Michael Manley. His college was an élite institution for the favoured few. In 1941, for example, Michael Manley's sixth form consisted of fifteen boys, mostly drawn from the white and brown commercial and professional sectors of Kingston.[8] In the '50s, well after Manley's tenure there, Jamaica College was a place where a student could be expelled for incorrectly placing a knife and fork and where the occasional rural winner of a parish scholarship, like D. K. Duncan, was told he was being trained to be part of the island's leading classes.[9] John Maxwell, who previously attended Calabar College, found the more prestigious Jamaica College a more brutal place. Its staff consisted of mostly expatriate whites, and during Michael's time there the subject matter was almost exclusively that of the English public school.[10]

The bullying Manley received during his first years at Jamaica College affected him deeply. Manley enrolled at the college at the psychologically crucial age of eleven. Earlier he had spent 'very little time really relating to boys; I'd been brought up very often just alone. . . . I'm sure that when I went to boarding school, I just could not understand how to deal with bullying. It was, I remember, at the time as if my entire world had collapsed, and I just couldn't understand what was happening to me.' The bullies were a few boys whose names, Michael later said he would carry to the grave. They would surround him, tease him, partly for being Norman Manley's son, force him to go without lunch, and generally make his life miserable.

> It was my first recollection of agony. I still get very upset when I remember the years as a boarder early, and being very much a part of the tradition of boys that you just don't tell. And I had a terrible period when I was losing weight and my parents were going out of their minds, who didn't know what the hell was wrong with me. And I wouldn't tell anybody, and this went on for years. Oh, Jesus . . . it was awful.

Earlier and in other situations, Manley had been a happy, aggressive child with a flashing temper. Tremendous pressures built up as the bullying denied him his more normal instincts.

The bullying stopped suddenly when Michael was thirteen. Vivian Dacres, Manley family chauffeur and bodyguard, an ex-boxing champ, found out about what was happening and threatened the Murrays and the Grants with 'a bitch of an iron pipe'. Michael did not find out about Dacres' intervention until later.

The harassment of Michael then took a subtler form in 1938 or thereabouts, when Jamaica was exploding in protest and Norman Manley and associates were founding the People's National Party. Denied the possibility of physically abusing Michael, the school bullies 'placed him in Coventry': no one would speak to him for a whole term. This was a tremendous punishment for a person who was beginning to attach great importance to words. The exception to 'Coventry' was David Coore, Manley's best friend (later Deputy Prime Minister in the Manley government). Seeing Michael's great upset, his obvious depression, Coore would invite him to walk up and down the driveway of the college, and they would talk about philosophy, the Romantic poets, and 'this, that, and the other.'

When Norman Manley found out about Michael's problems he railed about 'this butchery of babies' and called Jamaica College, his *alma mater*, 'a filthy school'. He felt that Michael had 'guts and far more sensitive feelings than I can pretend to' and wanted to send him to a boarding school in England. Edna opposed this since, to do so, while keeping Douglas in Jamaica, would place a 'drawn sword' between them. Michael stayed on at Jamaica College, while Edna reproached Norman: 'Once a child knows its parent is dissatisfied they cease to make any effort at adjustment. . . . [Michael] needs a good deal of licking into shape.'[11] The difference in parental attitudes is striking.

Body and Mind

Manley later believed that the bullying gave him a taste of oppression and even hunger which served him well later on as trade unionist and political leader. A more immediate result was that aged fourteen, he began to take athletics much more seriously. Sports, a major Manley family interest, became a lifelong passion with Michael. With the bullying experience fresh in his mind, he began to frequent the college's gym 'first of all to be able to defend myself.' He advanced to doing a hundred push-ups and a hundred sit-ups and skipping with a rope for an hour. Because of a damaged knee which kept him from following his father and brother on the track, his best sport was swimming, and he eventually became team captain. Setting himself up 'like a Robin Hood,' he prevented bullying by having team members report bad cases to him during his last years at the college.

Although athletics helped restore Michael's confidence and chanelled his aggressiveness, it also had a negative side. Manley was 'terribly nervous' as a swimmer, an athlete who did less well under pressure, and who had 'a terrible fear of failure'. He performed well as a relay swimmer, but not so well individually. 'Every time it was me on the line I became so obsessed . . . that [I] might fail and they would laugh at [me] and say your father was never a failure.'

> Dad's thing was so legendary, you just couldn't move anywhere at Jamaica College—bowling average, three years; scored the most goals in soccer, three years; champion athlete, champion athlete; vice-captain rifle-shooting team. Everywhere you turned you just saw his picture. It was just too much for Doug and myself.

In 1941 Douglas equalled his father's Jamaican record in the hundred yard dash and never ran again, an event that, like Michael's reflections, reveals a great deal about the psychological relationship between Norman Manley and his sons.

Where Michael felt he shone in his own right was as swimming team captain. In 1941 the college magazine attributed the team's success—it won the Simpson Shield that year and the next—to 'the able coaching of Mr. A. A. Chaplain' and noted that 'credit . . . should go also to M. N. Manley, the vice-captain, who put in some very hard work training the team when the captain fell ill.'[12] With no false modesty, Manley later recalled that he was a 'great' sports captain, one that knew how to inspire and lead his team. Although his bad knee had kept him from the track, on the basis of his performance as swim captain he was unanimously elected track captain by the team in 1943; after he had served as Prime Minister and had been recognized as a Third World spokesman, he still regarded the track captaincy as one of his proudest achievements. Sports clearly trained Manley in the politics of small, aggressively competitive groups.

As a student, Manley also had to cope with the legendary aura of his father, the Rhodes Scholar. Manley dealt with this by affecting a posture of slightly stylish indolence, as a result of which he won no academic prizes during his seven years at Jamaica College.[13] It was better not to try too hard, because if he tried and failed, Manley joked later, as the son of the West Indies' greatest legal mind, he'd almost have to shoot his brains out.

The school system of Manley's day was wildly inappropriate to the realities of colonial society. Most textbooks contained 'the usual negative images about colonial peoples.'[14] History and geography focused on Europe, and the school reinforced the rampant authoritarianism of the society while it neglected civic and social values. Manley later commented:

> I can recall my own years at school as providing not one single course or class which was designed to give me an attitude about anything. The structure of the language, the dates of Nelson's victories, the laws of motion and the properties of metal were taught with varying degrees of energy and skill, but no one asked me to consider whether there was any reason why I should be my brother's keeper or he mine. Social responsibility, government, political method or public morality were not even words to be wondered at because the school provided no books in which these questions were examined.[15]

Jamaica College did provide an extensive foundation in the colonizer's literature. 'I have a nice lot of English books for Higher School's [exams]' Manley wrote his father in 1941, '19 of them,' including 'Romeo and Juliet', 'King Lear', 'Merchant of Venice', 'Samson Agonistes', Shelley, Keats, Wordsworth, Coleridge, Jane Austin's *Emma*, Peacock's *Crotchet Castle*, Dickens, Hardy, Thackeray, the Brontë sisters, and Chaucer's 'Gardener's Tale.'[16]

Manley later became a severe critic of the traditional educational system, which survived Jamaica's independence in 1962, but he also noted that it had some saving graces such as preparing a few privileged pupils including himself to compete in English and North American universities. Another benefit was that one of Manley's teachers was 'a marvellous English guy,' B. H. J. King, a history teacher who was not a Marxist, but who brought Marxist insights to the study of history by insisting that his student 'look at the modes of production and the broad economic factors working in history and don't just look at who Sir Horatio Nelson shot up in a battle.' Manley recalled later that a group of students at Jamaica College responded by using King's insights 'to really interpret our own political process.' For Manley, King led naturally to Harold Laski, his most influential professor at the LSE after World War II.

Events in the broader world may have been more educational for Manley than the narrow confines and restricting discipline of the colonial classroom. Manley's secondary education ended with the confrontation with headmaster Hardie described in Chapter 1. Jamaica

College had a curiously paradoxical effect on Manley. On the one hand it exposed him, as victim, to an essentially hierarchical, tyrannical, and brutal system, which could wreak havoc on body, mind and spirit. On the other hand, through the college Manley helped solidify his inherited position as one of the favoured few of a society he would later attempt to reform in the direction of equality and social justice.

In the meantime, Norman Manley and his mostly brown, middle-class supporters had begun their challenge of the colonial political system in Jamaica with the founding of the People's National Party. Alexander Bustamante had become the black masses' hero and had started his bitter struggle with cousin Norman. Far away, Adolf Hitler had conspired with history to bring the world to war. Soon after his dramatic departure from Jamaica College, Michael Manley found himself in the Royal Canadian Air Force.

CHAPTER 6

Canada 1943-5

A Jamaican in the RACF

From October 1943 to May 1945 Manley trained in Canada as a member of the Royal Canadian Air Force (RCAF). This period was significant in several ways. The RCAF required discipline and training in new skills. Bicultural Canada contrasted strikingly with Caribbean Jamaica. Manley's letters home reveal his continuing interest in aesthetics and art, which he viewed first hand in Montreal and New York. They also contain some of Manley's first recorded comments on Jamaican politics.

The war years were important in Jamaica. The 1938 labour protests led to the formation of the PNP under Norman Manley's leadership and to a British Colonial Office investigating body, the Moyne Commission. The commission recommended serious reforms in the British West Indies, most notably the colonial Development and Welfare Act of 1940. Criticism of British colonialism in the West Indies continued, however, and was magnified by wartime strategic, ideological, and political concerns. In response Britain began a process of 'constitutional decolonization' that was designed to 'deflect nationalist agitation into governmental cooperation' and thus 'create a counter interest against mass rebellion.'[1] An important result of this was a new constitution for Jamaica in 1944 under which elections for a local House of Representatives were held in December of that year, on the basis of universal adult suffrage. Britain retained substantial power under the Constitution of 1944, but relinquished it gradually until Jamaica became formally 'independent' in 1962.

The parties organized to contest the 1944 election reflected the bitter colour and class divisions of Jamaican society. The PNP, organized in 1938, was preeminently the party of the urban, brown, nationalistic middle class, with its greatest strength in Kingston. In 1942, Alexander Bustamante (already the head of the Bustamante Industrial Trade Union, BITU) split with the PNP, and in 1943, he formed the Jamaica Labour Party, which he ran dictatorially.[2] The early JLP was unorganized and autocratic, based on Bustamante's personalism and on a core of the BITU organized for political purposes. It agitated for, but was not really of, the active part of the working class, especially in sugar zones.[3] Bustamante, with immense charismatic appeal to the black masses, was becoming the central figure of post-war Jamaican political life,[4] a fact which would rankle the Manleys and lead to a bitter political and personal struggle between them and 'Busta'. Bustamante's appeal to the black masses emphasized a corresponding limitation of the often aloof, lawyer-like, and intellectual style of Norman Manley. The PNP's survival depended on competing with Bustamante for the support of the majority class of blacks, much of which was deeply suspicious of the brown middle class as the supposed inheritor of domination formerly exercised by the white plantocracy. The latter formed its own party, the ironically-named Jamaican Democratic Party (JDP), but a 'white' political party could not survive universal suffrage in an overwhelmingly black society. In addition to the JLP, the PNP, and the JDP, several candidates ran in the 1944 elections as independents.

Bustamante and his JLP won a crashing victory with 41.4 percent of the vote and twenty-two out of thirty-two House of Representatives seats. Independent candidates won 30 percent and five seats. The PNP gained 23.5 percent and five seats, while the JDP and other groups won a mere 5.1 percent and no seats. Munroe interprets the election as a victory for the trade unionist welfarism of the JLP over the left-wing nationalism of the PNP, the laissez-faire capitalism of the JDP, and the political of individualism of non-party candidates.[5] Norman Manley failed to win his St Andrew constituency and thus began the long struggle with Bustamante with a crushing defeat.

From the autumn of 1943 until the spring of 1945 Michael Manley followed the Jamaican scene from the cool distance of Canada. David Coore, Manley's best friend at the time who had won the island's academic scholarship for boys and went to McGill University in Montreal, recalled that on the trip north Jamaicans were lodged in segregated hotels in Miami, then a Southern hick town. Discrimination against blacks, Indians, and Jews in Montreal was palpable, and French Canadians were ruled as a subject people. Brown West Indians such as Coore were tolerated as a kind of curiosity, especially at the university. Coore found that his academic preparation in Jamaica made the university work at McGill quite easy. He felt strange in Canada, despite its partly British heritage, in contrast to his later years in England, which felt like home.[6]

Manley probably had similar feelings. He kept informed about Jamaica through *Public Opinion*, a progressive PNP-leaning weekly founded in 1937 in an attempt to break the media monopoly exercised by the conservative (and increasingly anti-Manley family) *Gleaner*. Manley had gone to Montreal ostensibly to enter McGill, but he stayed for only one day before signing up with the RCAF. He chose the RCAF over the British RAF because the RCAF offered twice as much money for the same duty: 'Maybe I was being very much the trade unionist by instinct,' he later joked.[7] Several other factors figured in his decision: a desire to fight fascism; the romance of air combat after the Battle of Britain; the prospect that Douglas, a student at Columbia University, was going to be drafted into the United States Army (Doug in fact stayed on at Columbia); and Manley's mortification 'at the thought of having to face my son: What did you do in the war, Dad?'[8] Perhaps the spirit of Norman Manley, ever intimidating and inspiring, hung about his son's shoulders, since Norman had been a World War I hero, albeit at the cost of a temporarily shattered psyche.

Young Manley began his RCAF stint on 1 October 1943 at Lachine, Quebec, moving on in late November to pilot training at the Initial Training School (ITS) in Quebec, from which he graduated in the spring of 1944. Then, because of a surplus of pilots, he attended wireless navigation school in Calgary and finally gunnery school at Mont Joli, receiving the final status of wireless operator/air gunner.[9] Although Manley volunteered for the Eastern Front, Canada began demobilizing before he could see combat, and he was deactivated shortly after 8 May 1945, V-E Day.

A fascinating aspect of Michael Manley's Canadian experience is how the temperamental, sometimes rebellious aspiring aesthete and individualist responded to the discipline, rigours, and outright boredom of military training in an environment which starkly contrasted with Jamaica. After a week of service Manley wrote from Lachine that 'this place has made me utterly aloof and self-contained already.' On parade, his mind was 'a peaceful blank in which I am intensely alert.'[10] Manley thought that discipline was going to bother him but it did not, and he contrasted the RCAF regimen with that provided by the Jamaica College headmaster, Hardie: 'That was an impertinent little 'Mussolini' upstart with piglike manners and no sense of morality. To resent his discipline was quite different.'[11]

In early 1944 Manley studied navigation, but noted 'I'm only mediocre at it.' 'I grasp well and am a cinch on theory, but in actual plotting I'm prone to make slips which of course are fatal and never tolerated!' His second progress test didn't go well: he missed two out of seven questions and his aircraft recognition score dropped from ninety to forty-six but 'my Signals continue perfect.'[12] He continued to have problems in late February and was beginning to find the subject matter of pilot training dull: 'Anti-Gas' was 'another trivial boring affair', and 'Administration' was a 'tragedy' and 'drier than Latin'. He failed an exam on the 'law' part of the pilot's course and was sick and tired of 'these bloody trivial little subjects'.[13] Once entranced by the idea of fighting fascism and the romance of flying he was becoming bored by the trivial steps required to attain the grand goals. Impatience with detail was a criticism, later, of Manley, the leader.

In March Manley became more interested in the course once it turned to learning why a plane stalls and 'how to pull out of the resulting spin.' The drama appealed to Manley and he found the theory of flight really fascinating. Assessing the prerequisites of air crew, he stressed 'a clear head and even temper'. With pride he reported that 'I think the thing that will really amaze you, dear Dad, is that Michael, the mechanical moron, actually understands those deep hidden mysteries which when added produce the Internal Combustion Engine.'[14] In April, Manley reported that 'at last the hurdle has been cleared—the course is over—I'm through and a Pilot!' He had nearly failed at the end with a 30 percent in aircraft recognition, but he passed on a supplemental test with a 58 (50 was the pass-mark).[15]

Six months later, however, on the anniversary of his enlistment, Michael admitted that his RCAF career 'has been chequered—definitely not "officer material"!—I've been lazy, indifferent and sometimes a poor companion.'[16] Later he wrote that his academic average was 90, that he had no bad marks on his conduct sheet, but that he shuddered 'to think of what my character assessment will be like.' He stated that his reasons for wanting to get into combat were 'purely selfish and fall into two categories—personal and realistic,' the latter being combat pay and a possible government scholarship after the war.[17] In early 1945, Manley reported that he was known as the 'Flying Fool', that he had lost flying-time due to the flu (and shook down the thermometer to speed his hospital release) and that he had 'sunk to 17th out of 258 graduating' in wireless navigation.[18] Three months later he was demobilized as the war in Europe ground to its bloody close. Michael Manley had been able to adjust to the discipline of the RCAF, but his training career had had its ups and downs.

Canada, Art, Politics

Canada provided Manley with his initial experience of the world outside Jamaica, a not altogether pleasant one. His impressions were naturally largely coloured by his RCAF experiences. He disparaged Canada and the Canadians. Manley's rapidly-drawn conclusion was that Canada was a bitter disappointment: 'one of these days I'll go to a university in a country of real people. New York might do—or England.'[19] The slight note of Anglophilia is interesting in the light of his father's nationalist campaign.

Manley's views were those of a young man alone in a strange land very different from his own in the midst of war. They were aggravated by the fact that his initial experiences were in French Canada. Through British imperialism he would have absorbed vulgar anti-French and perhaps anti-Catholic prejudices. In the benign setting of a Catholic hospital where he was treated for chicken pox, these prejudices came out subtly and humourously. All the nuns spoke French and Michael (for whom words were gold) was thus 'excommunicate'!

> I'm stylishly established in a room to myself with two pictures of the Good Lord Jesus to every wall & a peaches and cream statue of the Virgin, her rather bewildered Husband & an innocently prophetic Babe. Any attempt to swap ideas on dinner with the Sisters rapidly becomes a vivid study in the intensely ridiculous.[20]

Manley's discomfort with Canadians was aggravated by Canadian soldiers' views of the Soviet Union, the communist war-time ally of the democracies.

> It's amazing how few guys around here know anything at all about Russia. They either live in a state of indifference over it or in a fog of mistrust. Actually it's amusing to argue with them & watch them either be shaken or just blank their minds to reason. They admire her contribution to winning the war but are still largely living in the spirit of the old 'bogey' attitude. And they have compulsory education!![21]

Manley's hostility to a different culture increased. In late November 1944, he penned a diatribe against the French and the French Canadians, culminating with these words:

> I'll never forget the French-C. girl who wept because I defiled her porch with the word Communism & then proceeded in a rage to defend it & finally when she prattled rot about her Church. I told her I thought her Church had betrayed Christ and everything he stood for—which is so tragically true! The amusing part was that what hurt her was that I, her 'boyfriend,' could have possibly been monster enough to believe there was some good in Communism. The Church had taught her that the word was synonymous with the Devil.[22]

Cultural conflict and ideological clash had become mixed with personal passion.

Solace came with opportunities for Manley to develop his interest in aesthetics, art and literature. In the earliest surviving letter from Canada, the only one to refer explicitly to the all-important Jamaican topic of race, he mentioned 'a fresh snowfall' that 'leaves the earth sensuous looking despite the dead white.' A few lines later he wrote of one of the guys in the canteen, 'the glowing skinned, coal-black, poker-faced negro': 'there, he's smiled as only a negro can & honestly it was like a sudden ray of sunlight in a dim room.' In the background, a pianist played some of Norman Manley's favourite Chopin.[23] Two days later, lonely, isolated, and missing the wide-ranging family discussions on music, books, and art, Manley wrote his father that 'I need to hear the Rachmaninoff & discuss "the Horse," ' ('The Horse of the Morning' sculpture completed by Edna Manley in 1943 and later one of Michael's most prized possessions).[24] In April 1944 Michael reported seeing an exhibition of Dutch paintings in Montreal—Rembrandt, Hals, de Hooch—'it was a real thrill.' There was also a big Van Gogh collection, 'warm and vital'.[25] A reading of William Saroyan's 'Human Comedy' led to the opinion that Saroyan's good reputation in the United States 'doesn't speak at all well for American aesthetic standards,' a lofty judgment for a nineteen-year-old.[26] It is interesting that, in these letters, Manley was critical of Canada and, much less so, the United States, but not of Britain, against which his father was leading a nationalist campaign.

Jamaican domestic politics, however, was an inevitable concern. In the struggle between Norman Manley and Alexander Bustamante, Michael Manley was understandably an impassioned partisan. He observed that 'with Bustamante, an unprincipled semi-maniac, the other side needs to show all the control & balance it can.'[27] Soon, in 1943 Manley penned this revealing view of Jamaican politics:

> Have you gone 'on the streets' yet? when you do I hope you'll take good care of yourself & not do anything foolish like West and work[28] without suitable protection! From what I remember of the situation in Politics I think your timing (October) of the PNP 'street invasion' is good. The time is ripe now for a 'doughty' blow at Busta. I think the PNP is solid enough to stand the inevitable shock of the conflict. Before it might have gone under because the shock of the conflict might have torn it apart. Now you (may I put we) can take it. And besides the only way to weaken Busta is to get right into his 'den.' Once you can do that & remain intact & sound, you have lowered his prestige immediately. Successful withstanding of the Great B's violence shows him up as more human being than diabolic demi-God.[29]

Michael added this personal note:

> You know, Dad, I am far away but my heart is right with you in the fight—I hope one day to be able to carry on in my own way—if not from a platform—at least from an editorial desk. Well, that's a long way off.

Bustamante was not the only obstacle the PNP faced. In February 1944, Manley referred in his characteristic manner to the 'filthy lucre' by which the white-led, laisse faire capitalist JDP attempted to advance its unpopular political cause.[30] Unfortunately, the surviving letters do not reveal Manley's reaction to Bustamante's great triumph in the December 1944 elections, though these are readily imagined. Michael hitchhiked by land and air from Calgary to Jamaica shortly after the elections, an arduous journey which included a freezing night shivering 'until the rafters shook' in a barn near Omaha.[31] In Jamaica he consoled his father and discussed the political situation, a dismal one for the Manleys.

Manley's Emerging Character

The most interesting thing about Michael Manley's Canadian letters is what they reveal about him as a young man between the ages of eighteen and twenty. His character emerged in his attitudes (described above) toward the RCAF, toward Canada, toward aesthetics and politics. The letters also contain more explicitly personal evidence. Family ties were very important. Manley's relationship to his father was very intense, wide-ranging in its embrace of art and politics. Manley often sought and praised his father's advice and views. The tone of the son's letters was by turns serious and jocular, and expressions of Manley's strength and ego were often offset by self-mocking humour and frank confessions of weakness. Manley's first letter stated that he wanted to come home after only six months, if he got leave, and asked his father's opinion. At the same time, Manley was anxious to show his father that he was tough enough to stroll around in fifteen degree weather in an ordinary sweater.[32] In March 1944 Michael thanked Norman for his advice—'always very sound'—and noted that he was in 'full agreement re "tiptoe" living!' He added that he had come down with chicken pox—the bout that delivered him 'excommunicate' into the hands of the French-speaking nuns—and indicated that Edna had not succeeded in her earlier campaign to get her men to respond to each other's injuries with compassion: 'You might as well just sit back and roar with laughter.'[33] The following October Michael responded to a letter from his father: 'as ever I found it immensely interesting and satisfying.' 'I've often missed our talks . . . on how to annihilate the forces of evil in two brilliant hours.'[34]

Manley's relations with his artist mother at this time appear, judging from available evidence, to have been less intense. He closed one letter with an airy 'Well my dear I must be off now.'[35] Another letter asked his father 'How is mother?' as if he were out of touch. He mentioned writing to 'Auntie Mu', Muriel, but not to Edna.[36] It may well have been that in 1943, with both her sons abroad. Edna had partially withdrawn from family affairs into her intensely private world of art, which included the sculpting of 'Horse of the Morning' that year. Earlier, probably during the war years but before Michael and Doug left Jamaica, she had written to Norman:

> I am working, just done an extraordinary drawing— . . . I'm turning into a damned mystic. For god's sake tell Michael not to bring Doug, because I'm not coming down yet & the three of them & Smith can pack into the house & I'll take Minimus. I shan't interfere with them if they will leave me alone![37]

Manley spent at least three of his precious leaves on family visits, the one home to Jamaica in December 1944, and the other two to see Douglas in New York in November 1943 and April 1944. On the second trip to New York he noted that his older brother had flunked his astronomy midterm, but had received an 'A' in philosophy. Doug wanted 'to do Law'—a hazardous path given the footsteps of their nonpareil father—'but is still unsure through a lack of self-confidence.' 'He is ever so much easier to talk to these days & has much less of a shell—less reserve about himself. I think he is losing his fear of competition & acquiring instead a tremendous will to win at everything in college—in other words he is losing his fear of his ego I guess.'[38] Although Douglas experienced racism in New York and was made to understand he could not even think of applying to a fraternity, his years at Columbia were happier than those at Munro College in Jamaica. He went on to a successful graduate school career in Britain, culminating in a doctorate.[39]

Toward the end of his Canadian stay, Michael seems to have grown more doubtful and reflective. By November 1944 it was clear that Canada was not going to use any new pilots in the war and Manley admitted some confusion: 'Maybe I won't see my way straight till I've taken a peep at the Mountains from Nomdmi.'[40] Earlier he had written a letter to his father which was truly remarkable for one not yet twenty. It was occasioned by a hurricane in Jamaica and perhaps by a sense of the impending defeat of the PNP in the 1944 elections. Manley sympathized with his father's sense of time slipping through his hands. He added that 'Tolstoy says there is no "strategy at a glance" but rather action as the sum of an infinity of infinitesimal wills & the inevitability of time & the historic flow.' This led to a reflection on reality:

> Tell me, how does one achieve a sense of reality? At the moment the tragedy of loss & suffering is the reality at home—and a grim one. Yet in a Historic sense there may be a truer reality in which the Hurricane becomes a blessing—so which is real—is it a tragedy or a blessing? Or is there no reality but only a series of causes & effects which only seem real relative to individuals or groups as they are affected personally? My God, we live a contradictory life!

As if these thoughts brought him to a wilderness of uncertainty which he wished to avoid, Manley went on to express contempt for the search for the meaning of abstractions by 'many spineless intellectuals' and 'puerile butterfly intellectuals'. He also denounced 'debauched city-slickers' and preferred those who lay 'a cornerstone in an edifice of national unity' or cultivate an acre of land.[41]

The young Michael Manley's Canadian letters culminated with a missive to his father containing a lengthy rumination on 'a remarkable eye-opener', war-time sex morals:

> It seems to take very little to kick people over the traces of self-restraint into licentiousness and self abandon. For one thing it shows how near the surface the whole thing must be & how universally so. For another, just how little guts we, mankind have. I can see adultery when the union is empty; (as you said); but when people admit (or claim) their union as a thing of glory & truth & still find animal pressures a thing at no price to be denied—then where indeed does truth begin or end?
>
> And the 'war wives'. There is something frightening in that—in spite of all attempts at cynicism. An astonishingly widespread betrayal of faith is taking place there. Presumably the husbands fight on in ignorance—there is going to be a rude awakening shortly after armistice day!
>
> Of course one must ask oneself what right one has to be horrified or anything else. 'The mote' observed, beware 'the beam'. But I think honest observation of what one conceives as wrong in others can help towards a clearer understanding of the moral problems involved in one's own code—however vague and variable. One thing I'm becoming aware of is the need for elasticity, in conscience, in one's approach to the subject. An elasticity without which one is incapable of meeting the ever changing 'values' as

> growth comes. Without it you seem to become easily confused & sometimes totally lost. And yet even with it, it is strange the pang each relinquishment can cause—that presumably is the price of conscience, without which 'elasticity' is disaster![42]

The letter must be seen in its proper context: the war years, a young man struggling to find 'one's own code—however vague and variable', the universality of the sex drive and of sexual hypocrisy. Read sympathetically and in this context, several phrases stand out: the contrast between self-restraint and self-abandon, as if there were no middle ground: 'animal pressures', with its Victorian connotation; the singling out of 'war wives' as if there were no 'war husbands' resorting to adultery, prostitutes, and rape; the need for 'elasticity' and the need to temper it with conscience.

The Canadian experience, albeit relatively brief, was significant and revealed much about the personality of Manley in early manhood.[43] Manley came from a family of powerful parental personalities that could inspire but also intimidate. Geographical distance required the independence which Norman and Edna Manley preached, but which propinquity would at times make it difficult for a young man to practise. Bilingual Canada provided Manley with experiences in a different culture that would inevitably test notions ingrained by a colonial Jamaican childhood. Canada and the RCAF tested Michael personally, and sometimes he found himself wanting, forced to admit a boredom with essential details, selfishness, laziness, and rage at a French Canadian girlfriend. He struggled honestly with the ethics of sex. Perhaps most significant there was the glimpse of a Tolstoyan vision of the infinity of historical causes and the infinitesimal role of the individual in grand historical dramas. Such a vision might serve a future trade unionist and national leader well.

CHAPTER 7

London 1945-51

I see the mighty city through a mist—
The strident trains that speed the goaded mass.
The poles and spires and towers vapor-kissed,
The fortressed port through which the great ships pass,
The tides, the wharves, the dens I contemplate,
Are sweet like wanton loves because I hate.

Claude McKay
'The White City'[1]

London, West Indians, Africans

New York, not London, was the 'white city' of Jamaican poet Claude McKay (1890–1948), and 'hate' is too strong a word for Michael Manley's reaction to London, but McKay's imagery does capture something of the subtle brutality Manley experienced in London from 1945 to 1951 and his love-hate relationship with the metropolitan world. Those years brought Manley into direct contact with the imperial centre against which his father and the People's National Party were waging a nationalist campaign for independence. During this period, Manley studied at the LSE, a seedbed for many nationalistic Third World leaders of the post–1945 era. The young Jamaican was deeply influenced by the scholar-politician Harold Laski. Manley continued to be absorbed by art, and to a lesser extent, issues pertaining to science and religion. From afar, he continued to be much concerned about the ongoing saga of Jamaican politics, as his father led the opposition to Bustamante's slowly waning dominance. On a personal level, Manley grappled to understand himself, attempted to demythologize his parents, and entered into the first of his four marriages. The London years were the final stage of young Manley's preparation for entry into the maelstrom of Jamaican life in the '50s.

Singham has argued that West Indians are a 'marginal people' because they are the descendants of ancestors transplanted from abroad and because of the generally marginal economic and geopolitical role of the West Indies. The very marginality of West Indians allowed them to adapt fairly easily to British society as migrants.

In England, faced with a hostile white population, . . . West Indians for the first time were able to become unequivocally proud of being black. This is largely because for the first time, meeting Africans and gaining a knowledge of the larger world, they realized, that they belong to a culturally and numerically important group of the world's peoples.[2]

The atmosphere that both West Indians and Africans faced was reflected in the experience of Kwame Nkrumah. Nkrumah, who arrived in London in mid–1945 after ten years in the United States, stayed in the English capital until November 1947, when he left to join his native Ghana's struggle for liberation from Britain. Nkrumah wrote:

> I quickly felt at home in England. And the longer there, the more I felt how easily my roots could take a hold. Nobody bothered about what you were doing and there was nothing to stop you getting on your feet and denouncing the whole of the British Empire.

Despite the advantages, racism was a daily problem. Many landladies, whom Nkrumah called 'dragons', objected to taking in 'coloured' lodgers. Occasionally blacks and browns ended up in trouble with 'a girl picked up in Piccadilly or the Tottenham Court Road'. Much racial antagonism and many fights 'after a few drinks' occurred. Nkrumah recorded this scene involving himself and the West Indian communist, George Padmore:

> One day Padmore and I were sitting in one of these cafes discussing some problem or other. All of a sudden I became conscious that we were being carefully scrutinized. I raised my eyes and there, peering at us with most profound interest, was a small girl. Suddenly she screamed with excitement as she said: 'Mummy! It talks!' The poor mother was crimson with shame as she came across and tried to snatch her awkward child away. But we laughed and told her not to mind.

Nkrumah found the West African Students Union hostel chilly and formal, but his own office at 94 Gray's Inn Road became a rendezvous for all African and West Indian students and their friends. But Nkrumah was disappointed to observe that the British Labour Party's colonial policy differed little from that of the Tories.[3]

Manley and England

Michael Manley arrived in late 1945 in a London scarred by World War II. He lived for a while in a room with a gaping bomb hole.[4] London also bore evidence of a different type of war to come: the complex anti-imperalist campaign—violent in some colonies, peaceful in others like Jamaica—directed against the British Empire. Douglas Manley, who did graduate work in Britain while his brother Michael completed his undergraduate studies, recalled that being a university student in London during the late '40s was like being on the youth front of an anti-colonial war.[5] Some years later, Michael Manley recalled what it was like in Britain at the end of World War II:

> 1945 saw the Socialists in real power in Britain for the first time. It also saw Britain economically on its knees after a long and bitter war. Civilian production was at a minimum. There was a world food shortage. Most British industries were harnessed to arms production. For seven years there had been no normal capital investment, no replacement of worn-out plant. Inevitably there was bound to be a period of acute shortage both in Britain and the world.[6]

Shortly after his arrival, Manley observed that 'there is a curious spirit growing in England. The mood is twilight as England moves towards the settled plumpness of Middle Age. Innumerable people that I have met are wanting to leave England & go to the colonies.' As part of the curious spirit in England there was a 'Christian Reform movement who say Hitler was the son of man or God—or both.'[7]

Manley's reactions to England were governed by his personality, by postwar chaos, and by the tensions inherent in racism and colonialism. Very light-skinned, he nonetheless 'witnessed and experienced an immense amount of colour prejudice' in which 'small Jamaicans, West Indians and Africans' were snubbed in restaurants, refused lodgings, and 'Jim-Crowed' out of jobs.[8] Not all the English were racists, but—as Manley wrote in 1955—'one of the great trials in England for a coloured man is the uncertainty about his welcome. He can never assume anything about the attitude of the English native towards himself . . .' Racism, in Manley's view, resulted from history, inadequate education, and the 'misconceptions fostered by cinema, the comics, and inferior literature . . . [in which] white heroes do battle with wild and simple black tribesmen.'[9] When Michael, Douglas and their friend Dossie Carberry visited Parliament and met some of its members, they were lectured by an egotistical politician—a 'cooperative bigwig'—who told 'the three little savages from down under just how to set their island right in five easy minutes of co-operating.'[10]

Manley's Anglophobia deepened with time. Concerning the prospects of an English edition of George Campbell's poetry, Manley observed that 'half the English people are dumb & a quarter terrified of anything positive—the remainder would get him.'[11] Difficult domestic conditions contributed to his views: he had established a household with his future first wife, Jacqueline Kamellard, which soon came to include two small children, was living on a 'potato diet' and walking ten miles a day looking at over-priced flats. The strain of these conditions erupted in a July 1946 letter to his father.

> I have learned to hate the atmosphere of this country with its stupid narrow outlook on love—its foul hypocrisy that allows legal discrimination against coloured people while parading the divine impartiality of British justice—the ghastly, smug impartiality of its accents—its raised eye-brows when a Jamaican throws back his head & laughs—its sickly rotten faces with their frozen death-masks of respectability. . . .
>
> *Eh bien*—every man to his tastes but for me—back to the South & Jamaica

> where a man can feel that no convention restricts the normal pulsing of the blood in the veins. They can keep their blasted 'Revolution' & their Constitution too—& if we in Jam. can't achieve peaceful Socialism by evolutionary methods without totally stagnating as human beings, then I'll personally start an Anarchist insurrection just to keep the ball rolling.[12]

Manley went on to apologize to his father for releasing the strain he had been under: 'And I know it's not nearly as bad as that & that they are a great people with a remarkably controlled & hence, sustained, vitality & I know that they will be mainly instrumental in the peaceful dissemination & acceptance of Socialism.' Nonetheless, he longed for 'my mountains & blue sea & white horses—my blazing sun & people of my own colour & temperament.' Toward the end of his stay in Britain, Manley had concluded that 'our "rulers" are not really a very intelligent lot of people—lacking in vision, courage or an elementary appreciation of consistent thought.'[13]

Art, Science, Religion

Despite these negative views Manley's years in Britain provided him with the chance to further his interest in the arts and to reflect on broad philosophical matters, experiences which helped form a politician notable for his broad interests. In November 1945 Manley wrote to his father that he had listened to a tremendous amount of Beethoven: 'it is from that I've learned the most.' He also learned 'from & about Van Gogh, Cézanne, Velazquez, Brahms, Vermeer, Bach, Purcell, El Greco, Donatello, Angelo & Roman Art, César Franck, Murillo—just those so far.' His study of Beethoven was intense:

> First I listened to the Symphonies & Concertos, the big Sonatas etc. Then I listened to the experimental sonatas & quartets & finally to the Last Quartets. With records and concerts (lunch-hour ones too) you can get to hear a lot of Beethoven. I think that much of Beethoven is invalid, bad if stood buy itself . . . But again if you have got even a breath of the meaning of the Quartets then the symphonies or elements thereof become valid because they themselves are unconscious statements of the very nature of growth.

Manley ended a remarkable letter by listing careers he might follow: (1) an editorial writer on economics and politics, (2) a 'critic and appreciator' of art and music, (3) a politician, if he could free himself of inhibitions about speaking.[14] The fear of public speaking was intense. When David Coore, then a law student at Oxford, was married in London, Manley served as best man; he absolutely refused to make the customary remarks to the wedding party of about twenty-five, and Dudley Thompson (another Jamaican law student and minister in Manley's government later) had to make them.[15]

Like the best of undergraduates, Manley became absorbed in the sweeping issues of the day, in the immediate aftermath of Hiroshima and Nagasaki. 'Science', he wrote in December, 1945, 'has created a state of amorality for itself when in truth it is immoral. Art has tried to do likewise when in truth it should be purely moral.' Manley distinguished the technical aspects of science from its moral effects on society, but nonetheless regarded science as 'Lucifer—the whole—the end—method, reason, push—and spiritual decay & filthy bloody mess.' He thought there was no way to put science in its place, but hoped that 'if Art and religion can combine on the right road they can touch people & so politicians & administrators[,] so counteracting the *influence* of Science.'[16] Of the three—art, science, religion—it was clearly art that engaged Manley the most. After he had been in London for over a year, he and Coore arranged a ten-minute segment on the BBC's West Indian hour in which they discussed the 'Jamaican Art Movement in Its Historical Context'. They covered Jamaica's contribution to art, drew parallels between art and Jamaica's national political movement, and with the Dutch art movement of the seventeenth century after Dutch independence from Spain. As Manley admitted, 'all that in ten minutes by two fledglings is a tall order!!'[17]

LSE and Laski

Manley's artistic and moral concerns coincided with his somewhat difficult adjustment to formal course work at the LSE. As at Jamaica College, an institutional setting such as LSE produced frustrations in the mentally restless young man. Because of his shyness he generally avoided seminars and relied on lectures and readings to pass his courses. His initial course of study was economics, the British Constitution and constitutional history, economic history since the Industrial Revolution, and logic.[18] Young Manley's developing personality expressed itself in his attitude to course work: 'taken in small & regular doses it provides a kind of ballast against the wilder flings & plunges of my highly volatile interests!'[19] In the spring of 1946 his frustration with formal course work increased. Advanced economics was too technical and divorced from his personal realities. He wanted to spend two years doing the history of art and philosophy so as to become an art critic 'eventually'.[20] During his first year in London he had 'spent more time at the Royal Albert and Wigmore Halls listening to music, and at the National and Tate Galleries looking at art than with the pages of Mr Benham's exceptionally conservative introduction to the discipline of economics.'[21]

Soon Manley withdrew from LSE and moved to St Agnes, Cornwall, in the region where Edna Manley had grown up. 'An improbable six months' in the remote Cornish village had three results: 'the birth of my first child, my daughter, Rachel'; a credit in Latin; and a conviction on the part of the villagers that Manley was a spy from 'foreign parts', the last as a result of 'long jogs at dusk in a black turtle-neck sweater past the deserted shafts of ancient tin mines'.[22]

In Cornwall, Manley's first priority was Latin, and he studied Horace and Livy. He also kept up with politics and economics, but art and philosophy had to await his return to London. In July 1947 he wrote that 'I'm pretty well out of the sky now & becoming more & more interested in Economics & the Social Sciences generally.'[23] He decided to return to LSE and was readmitted in the autumn as a second-year student, largely because of his Latin exam. He outlined his projected course work: analytical and applied economics, economic history, social philosophy, statistical methods, government (constitutional history, comparative constitutionalism, administration), and French and German translation ('as luck would have it can't do Spanish').[24]

In his second year at LSE Manley came under the influence of Harold Laski, whose impact on him was second only to that of his parents. Born in 1893 (the same year as Norman Manley) and brought up in an orthodox Jewish household, as a youth Laski read Beatrice and Sidney Webb and was moved by the struggles of the Scottish miners. He joined the socialist Fabian Society at Oxford, from which he graduated in 1914. He taught at McGill (1914–16), Harvard (1916–20) and Yale (1919–20).

His socialism created trouble with American university authorities, and he learned from his experience in the United States that 'liberty has no meaning save in the context of equality.'[25] Laski then joined LSE, which he served for thirty years, while lecturing in such places as the USSR, Ireland, France, Spain, and pre-Hitler Germany. In 1935, Max Ascoli said of him that 'no other political scientist is giving such a stimulus to the development of political thought. . . .'[26] His intellectual concerns included combining individual freedom with social order, worker freedom in the face of growing demands by the state, and the marriage of socialism and democracy. Laski was active in trade unionism and the left wing of the British Labour Party. He was regarded by Labour's right wing as a near-communist and by the communists as a social democrat.[27] Laski served on the Labour Party's National Executive from 1937 to 1949 and became increasingly pessimistic about achieving socialism through constitutional means, though he continued to urge that approach in Britain and the United States.[28] For some a model of the scholar-politician, Laski's academic production may have suffered from his political involvement.[29]

Manley recalled Laski's lectures as being enthusiastically received by the students. Manley himself became a partisan of Laski and foe of the followers of the conservative professors in the ascendancy at LSE. Laski's erudition and showmanship may well have provided Manley with elements of a stage manner, although one obviously adapted for the Jamaican political arena. In January 1948 Manley submitted his first essay to Laski, one intriguingly entitled 'Democracy and Demagogy'. In it, Manley tried to prove that democracy is a function of 'a certain necessary standard of living (to affect every member of society) and of everyone in the society possessing a degree of economic bargaining power.' He worried about Laski's reaction: 'God knows what he'll think of it & frankly I'm a trifle nervous!'[30] Manley was to give a talk to his discussion group on the nature and evolution of the Jamaican Constitution and to present a critical analysis of Rousseau's *Social Contract*. Laski's influence on Manley was very deep. He later wrote of LSE that 'moving from Laski to Hayek on the same day was both a stimulating experience and a guarantee that the attentive student would be forced to think.'[31] Nearly forty years after his university days, Manley still regarded Laski as the preeminent authority on democratic socialism, the creation of which became Michael Manley's controversial mission in Jamaica in the '70s.[32]

Stimulated by the academic ideological wars at LSE, Manley's thoughts turned frequently to politics. With a strong bent toward idealism, Manley soon learned a sad truth about practical politics, as he wrote to his politician father: 'What blurred edges the game has & how seldom your issues bite cleanly into the public consciousness & carve a distinct body of reaction.' He went on to defend himself from the 'slightly reproachful finger' his father shook at him for his 'alleged contempt' for the British experiment in socialism.

> My quarrel is with History as it is repeating itself in the reactions of the men who voted Labour in, to socialism in practice. . . . To the masses of people who voted Labour for selfish reasons & who are now squawking to hell & beyond because they are asked to work for planned prosperity—instead of sitting down to reap its fruits on padded backsides. . . . The workers in every industry save steel & coal & transport are fighting every socialist move tooth & nail. They will lose the fight though[,] for an enlightened (politically) Middle Class with a solid section of Labour are enough to keep the Labourites in for . . . "20 years".'
>
> As to Socialism itself, I still maintain that my contention as to its 'impurity' in England is true.[33]

Manley noted the lack of support for socialism in many small trade unions which voted Labour out of habit more than out of conviction. As in Britain, a divided labour movement in Jamaica became one of the country's highest and most tragic obstacles to socialism, as Manley himself would experience, first as trade unionist and later as Prime Minister.

An issue which divided Jamaican workers, the rural and urban masses, and the middle class, was the bogey of 'communism'. Bustamante shrewdly and successfully confused democratic socialism and communism in many Jamaicans' minds, helped by the *Gleaner* which relentlessly smeared the PNP red. At one point, possibly July 1948, Manley sent his father a clipping from the London *Sunday Dispatch* which referred to 'the communist-influenced PNP'. He noted that the idea of a communist-influenced PNP was taking root in England 'as the effects of *Gleaner* propaganda . . . seep through the British press.' Manley went on to say

> . . . quite apart from the lack of truth in it all, can we afford to have that take hold with the world in its present state over just that issue. For us, Whitehall is still the ultimate source of power. They must o.k. full self-government when it comes & Federation likewise. If you win in 1950, you will share power with a governor under Whitehall's instructions—& Whitehall's attitude to Communism is now only second (in antipathy) to Washington's.
>
> Well, we are not Communist and I don't think can afford to be branded as such. And Busta is surely going to play the card for all its worth.'

Manley asked his father for advice on how he could help counter *Gleaner* propaganda in England and the imminent appearance of Bustamante in the imperial capital. 'I've seen your judgment proved right & that of other PNP men wrong, so often that I am nervous of advices that are not authorised by you.' Finally Michael confessed that 'I myself am a baby in all this. I have no experience and am not essentially a very level-headed person though I'm trying to learn.'[34]

Possibly in response to the above on 13 July 1948 Norman Manley sent his son a long letter instructing him on political matters. One matter of concern was the London branch of the Caribbean Labour Congress (CLC), whose secretary was Richard Hart, a communist member of the PNP. Norman Manley wrote that Hart was doing an excellent job and would not send secret instructions to the CLC; still, he did not like the PNP-CLC connection. On another matter, Norman Manley opposed the idea of an anti-Bustamante demonstration in London as 'stupid':

> After all, he is where he is by election under our own Constitution. Our fight against him is here and not elsewhere. People would regard it as vicious if he pursued me when I went to America in a similar way.

The elder Manley termed his cousin and foe Bustamante a 'first class showman', but argued that although his attacks on the PNP were harmful, they were not as damaging as it seemed.

> Something useful might be done in England if we had any group there influential enough to make some reply to his charges that the PNP is dominated by Communists. The truth is . . . the Press is not particularly concerned with the truth about these matters and it will probably prove very hard to get any number of papers to publish even letters refuting the charge.

Norman approved of Michael's decisions not to be involved with the CLC in London and to stick to his studies. The first duty of students was to equip themselves for effective service when they returned home. In addition they should take an active interest in the 'living affairs' of the country where they studied: activity in the British Labour Party or a trade union would be more valuable than trying to influence affairs in Jamaica from four thousand miles across the sea.

> The West Indian scene is intricate, confused, and confusing. Even we who are out here find it hard to follow the ramifications of local affairs in other Islands where things shift and change almost overnight.[35]

One of the primary political challenges facing Jamaica was (and is) poverty and its hoped-for solution via 'development'. In January 1948 Michael attended a lecture in London by the West Indian economist Arthur Lewis on industrialization in the West Indies. Lewis was a social democrat, member of the executive committee of the Fabian Society in London for many years, and eventual winner of the Nobel Prize in economics. His address advocated what would later become known as the 'Puerto Rican model', industrialization promoted by generous incentives to foreign capital. To the young Manley, Lewis's address was a cold, irrefutable analysis, which two-thirds of the audience, including himself, accepted. He disdained the one-third of the audience that strenuously opposed Lewis's analysis; these 'neo-socialists', Manley said, based their case on the 'false assumption' that industrialization can be achieved without foreign capital. They accused Lewis of a sell-out to 'monopoly capitalism'.[36] Much later, Manley would argue that Lewis had sold Norman Manley on the 'Puerto Rican Model',[37] but long before the elder Manley came to power in 1955, his son had been sold the model by Lewis. At the time, political issues were preeminent in the Manleys' plans, but economic problems would bedevil their political careers and their island homeland.

Another concern was the United States. As the British lion aged, the American eagle soared, eventually replacing the lion as the new metropolitan power over Jamaica.[38] In 1946, Manley showed an understandable concern with the role of the United States in the post-war world. 'What the hell gives in America with these strikes—don't they know there is a World waiting on their production or is it that they aren't mature enough to care.' 'I see', Manley went on, 'that American foreign policy now vies with that of Russia for supremacy in the field of enigma.' 'I was appalled to see America approving the sale of 547 aircraft to [Spanish fascist dictator Francisco] Franco this morning—doubtless [Secretary of State James F.] Byrnes will deliver a scathing commentary on Fascistic remnants in Southern Europe tomorrow!'[39]

Manley's main political activity in London was as a leader of the West Indian Students Union at LSE. He regarded it as his most useful deed of those years.[40] He began to work with the student union in his third year at LSE (1948–49) and 'worked like a dog' on it until he left England in late 1951. A member of the union's executive body, he refused to accept its chairmanship, partly because of his shyness about making speeches. In 1948 Manley led a deputation of students which successfully claimed about £3 each in income tax which was wrongfully deducted from their pay: 'Now I see', he wrote home, 'what Laski means when he talks of the will of minorities to fight for their rights as being fundamental to Democracy.'[41] Manley was also a principal organizer of a strike protesting living conditions at the Hans Crescent students hostel, which was provided for West Indian Students by the British Colonial Office. He participated in a large demonstration when the British government surrendered to South African racism and exiled Seretes Khama and his white wife, Ruth, from the homeland Khama served as chief. Manley overcame his shyness to lecture students in 1950 on the virtues of a West Indian Federation, destined to be a fatal issue in Jamaican politics.[42] At the LSE Manley also met and formed lasting friendships with such future leaders as Forbes Burnham of British Guiana (independent Guyana) and Errol Barrow of Barbados.

Around 1948, Barrow and his wife Caroline arranged a memorable evening for Michael and Jacqueline with Paul Robeson. All-American football star, actor, playwright, scholar, singer, political activist, champion of racial justice, Robeson had long frequented London's West African Student's Union, a hotbed of anti-imperialism. He understood that the oppression of his fellow American blacks was linked to that of Africans and West Indians, and was a hero to the Left and the oppressed.[43] Manley grew up hearing Robeson records and later joked that Robeson's 'Joshua Fit the Battle of Jericho' was 'responsible' for the Joshua mystique which attached itself to him in the '60s. When Michael was in the RCAF and Douglas Manley was a Columbia University student, they had been thrilled by Robeson's 'Othello' in New York City, which Manley recalled as one of the greatest experiences in his life. When Robeson arrived at the Barrows' flat Manley had 'the extraordinary feeling that the room contracted' and that singer/activist's physique and personality expanded to fill it. The conversation that night ranged widely over history and international politics, but Manley was most impressed by Robeson's passionate and tender comments about Marian Anderson. The great contralto, Robeson believed, had such beauty of soul that she thought that if she sang more beautifully than anyone else, racial prejudice would simply blow away.[44]

Personal Realities

In addition to the development of his attitude toward Britain, his reflections on art, science, and religion, his formal coursework and exposure to Harold Laski, and his politics, the years 1945 to 1951 were also years of intense personal experience for Manley. Early on, he wrote of finding his own personal realities.

> I know that I shall live more or less by the pen . . . There are certain realities of my own that I'm beginning to find a door to in things like the [Beethoven] Quartets, Blake, Lawrence, Taoism, the Geeta [sic] & the Upanishads. Christ, the Old Testament. There are further doors to it, negatively and by inference from Absence, in Society. Picasso & Modern Art, the way of Science, the Church . . . Parallel to & commensurate with

> the growth of this 'world' is a need to put it into worlds alongside of the Society in question . . . [45]

In January 1946, he added these insights:

> The one thing I knew, still know & will always know is the relation between Action & growth. I talked a lot of Absolutes [in previous letters] & I'm sure I didn't know what I talked about! I suppose my answer to myself should be 'So it goes, O facile one!
>
> It's taken me my 21 years to make a most amusing little discovery about myself. I wish I knew why, but as it is I instinctively react away from any advice of any sort & will do anything to avoid following it. I run the whole gamut of rejection from forgetting to convincing myself of its invalidity. Often, I've discovered thinking back, sound advice will divert me from an intended action similar to the one [being suggested].
>
> Why? Where aren't I free and so what am I rejecting unconsciously . . . It's not like the strong ruthless action of a sure person—on the contrary it's purely vacillatory & limited to advise as such.[46]

Such introspections probably resulted from Manley's stage of life and his relatively restricted circle of friendships amidst the strange sea of English faces. Both he and Douglas made very few, if any, English friends during their years in Britain. Their close friendships were with fellow Jamaicans (such as David Coore and Dudley Thompson), other West Indians, and family members. When Michael first arrived in England, he stayed with his English aunt, Nora Swithenbank, and her flatmate, Margaret Gray, until he could find his own quarters. He also saw his Aunt Vera—'She's been grand terribly nice to us all'—but they had 'the inevitable argument about Jamaican politics & the PNP.' Aunt Vera and Michael 'agreed to differ & parted with real respect for probably the first time in our mutual association or certainly since the Hardy [sic, for Hardie] fracas!'[47]

Important to Michael's personal growth was a more adult relationship with those almost overwhelming personalities, his parents. His friend, Michael Smith—a future anthropologist of considerable repute—helped Manley 'bust up the "myth" of you two.'

> I'm beginning to see what I must accept & what I must reject of my 'attitude' to you people. The truth being that I must accept you people & reject my 'creation' of you & learn to distinguish between the two things. Mother helped me see that in a letter the other day—I was very, very grateful.[48]

The dynamic is revealing: Michael had tended to mythologize his parents, but his mother and his friend Smith had helped bring his image of them closer to reality. In rare letters to his mother, Manley adopted the rather remote 'My dear Edna' as a form of address (he never addressed his father as 'My dear Norman'). Since he could not give Edna a Christmas gift in 1945, he sent her a movement by movement analysis of Beethoven's B Flat Quartet (Opus 130), featuring such comments as '2nd Movement. Completely objectivised intensified awareness in tightly spinning & dancing forms.'[49] The tone of his letters to his father was often jocular: 'You have been withdrawn from the bottom of the bottomless pit on my opprobrium and reconsigned to your former status on this Earth before God & man. Amen.'[50]

In London, Manley also first experienced the joys and sorrows of marriage and children. His first wife, Jacqueline Kamellard (shortened from Kamellardski), the daughter of a Polish Jew and a woman from the south of France, had been raised in England and was seven or eight years older than Michael. They had met in Jamaica, when she was married to a Jamaican, Robert Verity. They renewed their acquaintance in England, when Jacqueline and Verity's marriage was falling apart, and married in 1946. Jacqueline had a daughter, Anita, by her former marriage, who lived with her mother and Manley. The little family grew to four with the birth of Rachel Manley on 3 July 1947.[51]

Manley remembers these years as rough, but fun. He received £90 per quarter from his father and an RCAF demobilization scholarship. During school vacations Michael, Douglas and their friend, Jamaican quarter-and half-mile champion Arthur Wint, worked for the post office and prepared tax returns. Money tended to run out in the third month of every quarter, and Manley pawned his typewriter repeatedly to tide them over. There were other problems:

> We were kicked out on a week's notice with no reason given! We suspect colour but think the two decrepit old ladies beneath us a more plausible cause of the trouble—noise & the fact that I sat in the bathroom chatting to J. the first night she came home from Lowestoft[?] while she had her bath.[52]

Looking for a new place to live took all day, while time and money ebbed. In October 1946 Manley reported that his time was taken up by 'the triple drudgeries of house work, income tax & Latin'. Jacqueline, Michael and Douglas and his wife Carmen—a beautiful woman born in Panama of Jamaican parents—shared a flat for a while, but eventually separated because of different interests, friends, and lifestyles.[53] Michael and Jacqueline received food parcels from Jamaica. By early 1948, with Michael back at his studies, Jacqueline got no help from him with the housework and 'has to do Rachel, Anita, meals for five [including Jacqueline's mother] and keep the house clean & the nappies washed all by herself. It's a hell of a job (13 hours of steady drudgery in fact).'[54]

Perhaps such toil contributed to the end of Manley's first marriage in 1951, after a prolonged crisis during which he developed his first ulcer, Rachel had been sent to stay with Edna and Norman, while Michael and Jacqueline made a last, futile attempt at reconciliation. While his personal life thus took a dramatic turn downward, Manley completed his undergraduate education with a less than outstanding bachelor's degree in economics. As consolation and reward, he spent much of his last year in England following the exploits of the visiting West Indies cricket team. He also began his career in journalism, which was his major preoccupation during the next few years.

CHAPTER 8

'The Root of the Matter' 1952-5

Budding Journalist

In 1951, Manley began to learn the newspaper trade at *The London Observer*. In mid-year, he sent home this account of an editorial conference at the *Observer*:

> A Bikini bathsuit beauty contest must be covered with the paper's best photographer standing by—must have a sex angle on page two. For page three, take a crack at that American female—what's her name?—you know, the one who claims to be the best tailored woman in the world. And remember to get her impressions of British clothes. I (the editor) will do a leader on Persia and Oil and get Miss A. to look into the Festivals Hotel racket—good social angle if she can get at the facts—but for God's sake tell her to be sure that the facts *are* dead right. And so on, and so on.[1]

From London he sent home an occasional column entitled 'London Periscope' to *Public Opinion*, to which he would make a major contribution in future years.

Manley returned to Jamaica in December 1951. From early 1952 until late 1955, he wrote an opinion column called 'The Root of the Matter' for *Public Opinion*, a journal established in 1937 by O. T. Fairclough and H. P. Jacobs as a nationalist, progressive, anti-establishment weekly. It leaned toward the PNP, but efforts were made to keep it from becoming an organ that merely preached to the converted. Manley's salary was twelve pounds a week and he wrote editorials, covered stories, worked on features, handled lay-out, and managed the sports and feature pages.[2] 'Root of the Matter' and his trade unionism (see the next chapter) renewed Manley's involvement in Jamaican affairs after his long absence in Canada and England. 'Root of the Matter' provides a detailed picture of Michael Manley's views as he approached and passed, on 10 December 1954, his thirtieth year. The columns reveal Manley's personal values, his attitude toward journalism; his feelings about Jamaica and Jamaicans; his concerns about British imperialism, racism, West Indian nationalism, and federation; Jamaican politics and economics; and the international scene.

Values and Journalism

Among Manley's personal values were his high regard for individualism: 'The sanctity of the individual, the belief in his infinite value, is the basis of our civilization.'[3] Manley also professed the value of knowledge which he said was 'the one condition in which responsible action is possible[4]. This wedding of action and thought would also characterize his subsequent career.

Manley's columns also reflect a strong personal interest in heroic figures. Probably this contemplation of such figures served partly to prepare for his own political leadership. Characteristic was Manley's reference to the unsuccessful United States Democratic presidential candidate as 'the great Adlai Stevenson'.[5] More relevant to Jamaica was Manley's nationalistic, anti-colonial view of Winston Churchill:

> As he bows off the stage one must give pause to remember his indomitable spirit, his unique feel for the spoken word, his grasp of history, his blindness to the social revolution around him, his failure to realize why freedom should have a significant meaning for people in the British Empire. All these things are Churchill who is, perhaps, the last of the great Nineteenth Century Liberals.[6]

As a Caribbean nationalist, Manley also looked for heroes closer to home. Puerto Rico's four-times governor Luis Muñoz Marín was the architect of the controversial 'Puerto Rican model' which was emulated by Norman Manley as Chief Minister in Jamaica. To Manley, in late 1955, Muñoz was 'the one-time Greenwich Village intellectual who became Governor of Puerto Rico and the nearest thing to a general symbol of popular aspirations throughout the Caribbean.' Manley wrote of Muñoz's 'iron political integrity' and said his political 'philosophy' was 'Never tell the people a lie.' Manley added that 'Puerto Rican achievement under Muñoz has been stupendous.'[7] Under changed circumstances in the '70s, Manley would become a severe critic of the 'Puerto Rican model'.

Manley was on surer ground in selecting the Jamaican writer Roger Mais as a hero of the island's cultural and national development: 'It is given to few men to become a symbol in their lifetime for the complex strands and forces of a great movement.' 'Roger burned with a sense of the injustice which proliferated around him in Jamaica for most of his life . . .' and was 'never content to bemoan the fate of working class Jamaica from the smug security of a middle class verandah.'[8]

Manley's Caribbean pantheon included women such as Beryl McBurnie, the organizer of a Trinidadian dance troupe which made a highly successful tour of Jamaica in 1955. Manley remembered having met McBurnie in 1951 on a dull, bleak, late winter London afternoon. To him, she was a dedicated nationalist 'determined to create a dance form which drew its basic forms from within the West Indies by studying our own folk dancing and customs.' Manley testified to her deep courage, the significance of her vision, and the role she played as a contributor to and symptom of the emergent West Indies.[9]

As a progressive journalist, Manley frequently criticized the *Gleaner*, Jamaica's conservative press monolith. His interest was professional (as a journalist himself) and political (as a partisan of the PNP, led by his father and frequently attacked by the *Gleaner*). Manley characterized the *Gleaner* as an 'ancient, and somewhat unwieldy, Colossus' edited by 'hoary monarchs of all they survey.'[10] A lengthy column explained his own journalistic ethics and exposed the *Gleaner*'s alleged lack of such. Manley's basic rule came from 'late Editor Scott' of the *Manchester Guardian:* 'Comment is free but facts are sacred.' Manley accused the *Gleaner* of publishing a false story that his father and fellow PNP leaders Noel Newton Nethersole and Wills O. Isaacs were going to seek election to safe country seats instead of reelection to their Kingston Corporate Area seats. To correct such an apparent abuse of the press, Manley relied on voluntary self-regulation and public awareness:

> Freedom of the Press must be preserved at all times. And it would be disastrous for a Government to attempt to curb the worst excesses of newspapers. Action on those lines can only lead to totalitarianism in the long run. . . . Thus it is up the the profession itself and the general public to create a mental climate in which no newspaper can survive once it is openly associated with the peddling of deliberate lies.[11]

The *Gleaner* did not change its ways, however. In 1954, Manley criticized it for functioning as 'Bustamante's "Big Bertha" '; for allotting only two inches to 'a thanksgiving service at the Kingston Parish Church on the occasion of [Norman] Manley's return to active duty' after a serious illness; and for being owned and operated by 'big men' so that it would 'always support the "big man's" party which is the JLP.'[12] Although he acknowledged that 'opinion is free,' Manley also attacked individual columnists. He was happy when Peter Simple, 'that arch purveyor of spite', was replaced by 'Thomas Wright' (Morris Cargill), who wrote 'intelligently and amusingly' and 'has occasionally come near to commanding respect for the impartial wit with which he twigs the two major parties.' But Manley soon attacked 'Wright' for his verbal assault on the black leader. Manley presented 'Wright' as a defender of plantocratic interests and referred to Cargill's 'particularly pale pigmentation'.[13] The *Gleaner* was a severe obstacle to the politics of the PNP and the Manleys for decades, to the point of raising significant questions as to its impact on public opinion in Jamaica where healthy journalistic competition and impartiality were seriously lacking.

Another concern of Manley's 'Root of the Matter' was the theme of Jamaica and Jamaicans. Jamaica was 'this troubled but vigorous little Island'.[14] Its massive social problems were symbolized by a man Manley passed on the road:

> He was an old man, small and very bent. He had a stick and tattered clothes and he was resting his forehead against a concrete wall. The sun beat on him relentlessly, its very energy an affront to his hopelessness. About his shoulders there was the slump of despair. In a profound sense

> he was alone; not only because the road was empty, but because he had no future, no past worth speaking of and a present that was intolerable.

Manley drew the lessons that 'a country without a social conscience is a country without a soul' and that the old man was 'a terrible commentary on our society as it has developed through the years.'[15] In December 1952 Manley wrote:

> Kingston's masses are angry. They are angry because they are hungry. They are angry because they have been abused for two long. . . . They are losing what little faith they ever had in the future.[16]

One cause of Jamaica's problems was the 'Malthusian horror of over-population',[17] in response to which a frank film about sex among young people, 'Street Corner,' was shown. The Catholic Church banned the film and the *Gleaner* called for it to be censored. Manley defended the film as 'a plea for care and self-control on the part of young people in sexual relations together with an indication of the horrors of abortion.'[18]

Manley's columns also posited characteristics he felt typical of Jamaicans at large: love of verbal combat (at which he himself was no slouch); political gullibility; 'super-egoistic' public personalities; an abundance of cynicism; and 'our insular Jamaican pride'.[19] A lasting image of the country came from a column written about the time Norman Manley took power as Chief Minister, having defeated Bustamante in 1955. Jamaica, Michael Manley wrote, was a country of 'mountainous problems, staggering in their immensity,' with a hundred and fifty thousand unemployed and fifteen thousand youngsters entering the job market yearly out of a population of one and a half million: 'Jamaica remains a desperately poor country.'[20]

Imperialism, Racism, Nationalism

Manley also wrote much on imperialism, racism, and nationalism. Unsurprisingly, Manley viewed British imperialism with indignation and frustration. 'To my dying day,' he wrote, 'I will never understand why the British, possessed of such political genius at home, are so politically obtuse in their dealings with colonies. They announce great intentions and proceed to erect every obstacle in the path of their achievement.'[21] To Manley the British politicians and public were 'abysmally ignorant' about the colonies. Visiting politicians 'tend to ignore the progressive movement, social welfare field workers and the people as a whole. Thus they learn nothing beyond the generalized prejudices of Jamaica's upper crust—which, like the upper crusts the world over dispenses bigotry with stupidity in equal measure.'[22] Manley's critique extended naturally to the economics of British imperialism: 'the prices we pay for British goods rise faster than those which she pays for ours,' which was 'just another example of Britain exploiting her superior bargaining position without regard to her moral obligations under the system.'[23] In its anti-imperialist campaign, however, the PNP had been handicapped since 1938 by 'the indifference of the many who did not, perhaps could not, understand the relationship between colonialism and their own situation.' This reflected the nature of the beast:

> Colonialism, be it enlightened or repressive, is a state of dependency in which the destinies of a people are wholly beyond their control. In such a state there can be no incentive to effort, for effort is a function of will and will can only be exercised where it is possible to choose from amongst comprehended alternatives. Colonialism permits of no choice.[24]

While Manley criticized British imperialism, one of his ambiguities was considerable admiration for the Royal Family, particularly Princess Margaret, who visited Jamaica in early 1955.[25]

Racism was a byproduct of imperialism. Manley, although light-skinned, passionately opposed the injustice of racism. He wrote of South Africa and the southern United States 'where it is the next best thing to a crime to be coloured.' In Britain racial discrimination grew as 'the coloured population' grew. West Indian and West African workers there competed for jobs and faced racism in trade unions. West Indian and West African students in England, as Manley knew personally, had difficulty in finding rooms. Such problems were rooted in English assumptions of 'Negro' inferiority, which in turn stemmed from 'the way in which British history is taught—generally on the assumption that British imperialism existed to elevate the backward "native" and never as the ruthless system of economic exploitation which it really was.' Until England solved such problems, 'she has no real claim to the allegiance of the coloured part of the Empire.'[26]

Manley fully recognized that nationalism, the inevitable response to imperialism in this century, has its dangers and is not the promised land:

> . . . In its aggressive form nationalism leads to the excesses of Nazism; I appreciate that popular movements have been frustrated in the name of patriotism. Nonetheless it is a stage through which any people must pass if they are ever to achieve stature. It is also an inevitable concomitant of imperialism and the only answer to it.[27]

The first goal of nationalism in a colony like Jamaica was self-government. After that goal was achieved, nationalism could inspire Jamaicans for economic development: 'experience elsewhere in the world has shown that an indispensible ingredient of economic development is popular enthusiasm.'[28] But nationalism also included cultural and social patterns. Under the touch of artists 'the inchoate longing of a people for self-identification may leap into a brilliant, conscious flame of self-awareness and pride.'[29]

Nationalism requires heroes, and Manley supported a suggestion to remove the statue of Dr Louis Quier Bowerbank from Kingston and replace it with a statue of 'the great Gordon:' George William Gordon was the martyred brown hero of the Morant Bay uprising of 1865 and Bowerbank the custos of Kingston who had ordered his arrest.

> Surely the time has come for Jamaica to recognize who its real heroes have been and who have been the treacherous instruments of an alien power, lionized in the past for their services, not to Jamaica but to Britain.[30]

Manley suggested three types of heroes: (1) those who defended the country against external foes, which was not possible in Jamaica since that task fell to England: (2) those, like George William Gordon, who did good social works without challenging the system; and (3) those who, in the twentieth century, challenged the imperialist system.[31] As an 'inveterate lover of athletics',[32] Manley also believed that the world-record Jamaican four times 440-yard runners Arthur Wint, Herb McKenley, George Rhoden, and Leslie Laing were 'amongst the greatest ambassadors this country has ever boasted' who made people around the world aware that Jamaica existed.[33] An example of nationalistic selflessness was his friend McKenley: 'the grandest thing about Herb has been his steadfast refusal over the years to succumb to ceaseless American pressure to take out US citizenship papers. Jamaican to the backbone, he has always preferred to take his chance with our tiny contingent.' Twice McKenley ran the hundred, two hundred, and four hundred metre events and the relays rather than save himself for the four hundred; thus 'he never won a gold medal but earned Jamaica vast piles of place points.'[34] Regional team sports also served excellently to unite the otherwise diverse British West Indians.

Manley's nationalism took the form, still controversial in Jamaica, of support for a strong federation of the Anglophone Caribbean. Other than for sports, Jamaicans had little use for federation. Norman Manley would unwisely use the matter of federation to test his political power in 1962. As Michael Manley wrote in 1952,

> . . . We are a remarkably Federation unconscious people. There is certainly no mass feeling over it one way or the other and even at other levels the interest in it is little more than academic and not always even polite. . . . In the Southern Caribbean talk of Federation does not savour of an unholy and too intimate alliance with aliens. In our case, one gets the feeling that it does—failing which it appears like asking Jamaicans to accept something which is so remote as to be meaningless.'[35]

Manley believed that the only spur to federation was economic need and that federation promised 'incalculable material benefits'.[36] He doubted whether a federation which lacked real power to take on the vast job of development planning for the West Indies as a whole was worthwhile, but—and this became a characteristic Manley posture when facing monumental problems—it was better to start with 'a little Federation than no Federation at all.'[37] After Norman Manley took power in 1955, his son continued to argue for federation but was finally forced to admit that: 'It is a damp squib!'[38] As in all the Anglo-Caribbean territories, federation got bogged down in local politics.

Politics, the JLP and the PNP

Politics was of huge importance to Manley from 1952 to 1955, when his father was first Leader of the Opposition and later, after 2 February 1955, Chief Minister of Jamaica. Personal and family loyalties—and the deep emotions they aroused—were inseparable from politics. Manley's views were understandably deeply partisan, although he was convinced that the truth was also on his side.

Manley described the political evolution in Jamaica as follows. Before 1944 the 'plantocrats' had controlled politics by their hold on the governor, appointed by the Crown. After universal adult suffrage in 1944, the plantocrats 'felt that Bustamante would, through his abuse of power, prevent the growth of a real party system.'

> For they know full well that so long as there is no real party system, democratic politics are ineffective from the point of view of the small men in a country for the simple reason that it is the little man's interest that must be organized and expressed through politics.

The PNP's gains in 1949, wrote Manley, scared the plantocracy into organizing the Farmer's Party in 1953, to destroy the party system. Bustamante, in turn, was an example of the 'personality politics' that in contrast to party politics made effective political action and change impossible.[39] Writing in this context, Manley's comments on politics fell into three broad categories: politics itself, the JLP, and the PNP.

Discussing politics, in general, Manley clarified his deep commitment to 'democracy', as 'a political system . . . infinitely worth preserving.'[40] 'Democracy predicates tolerance and self-control. It is essentially the political system of restraint.'[41] Its foundations included the 'ability to master the complexities of modern government' and 'iron integrity'.[42] During JLP dominance before 1955, Manley saw 'democracy' threatened by 'the failure of the Majority party to take seriously its responsibility for Government that is above suspicion.'[43] Referring to 'the fascist corruption' he believed was rampant in 1952, Manley offered a stern standard which seemed to echo the tone of his English Methodist grandfather Harvey Swithenbank:

> Public morality is indivisible. There are no degrees of probity. A public servant is either completely honest or a danger to society. . . . No country can afford to tolerate anything less than complete honesty from its servants for without honesty there can be no confidence; and confidence is the very basis of those ordered relations in a society which Government makes possible.[44]

Thus spoke the idealist. But Manley also assessed politics realistically.

Six months after his father became Chief Minister Manley acknowledged that politics is a tough, sometimes dirty game, which he likened to a boxing match: one 'soon learns to absorb punishment—or gets out.'[45] When his father was the most powerful politician in Jamaica, Michael Manley, journalist cum trade unionist and eventual politician himself, offered this less than heroic analysis of the breed:

> It is a commonplace of public affairs that the practising politician in a working democracy tends to articulate public needs when they have become chronic. Indeed, one is tempted to the view that politics is the art of making the obvious dramatic. In this respect the successful politician is not unlike a water diviner. Discovering a powerful, subterranean current of public opinion, he gives it voice and is acclaimed a leader of men, a fashioner of events, a shaper of history.[46]

Manley did not exempt his father from that judgment.

A second political theme of Manley's 'Root of the Matter' columns was the JLP. During the early '50s, the two parties converged around a common agenda, as the JLP adopted many PNP ideas and the PNP shed its more radical, Marxist left wing. Nonetheless bitter partisanship continued, and Michael Manley was both its product and its perpetuator. He referred to the JLP as 'that motley crew', 'like a pack of mad dogs', 'Busta's little boys', 'Bustamante and his stupid little gang', and a 'gang of puerile incompetents'.[47] 'The leaders of the JLP,' he wrote in 1955, 'are completely contemptuous of democracy, are basically indifferent to the welfare of the country, and dedicated solely to the end of their own lust for power.'[48] Bustamante had been elected as a 'messiah' by the people in 1944 (and reelected in 1949), but he betrayed them by 'his refusal for nearly seven years, even to recognize that there was an economic problem which required an inspired assault.'[49] In 1953 Manley listed corrupt JLP practices: distributing public housing on a partisan basis; receiving bribery from merchants; selling 'our bauxite' for one shilling a ton; receiving illegal pay for 'farm tickets' to work in 'the promised land to the north'. He concluded: 'I am personally sick of living under the grinning, buck-toothed mask of corruption. I think the time has really come to smash [. . .] through that mask lest we get to the stage when the Party in Power decides that only [it] has a right to be alive.'[50]

The third theme of Manley's articles on politics was the PNP, which he presented in a heroic light. The launching of the PNP was, he wrote, the first organized response 'by the people of Jamaica to the challenge of poverty and imperialism'. He went on to say that:

> a small band of far-sighted men . . . knew that 'the end' is never in sight; that the bastions of privileges are impervious to a strike action and that unlike the days of Jericho, modern walls do not oblige by lying down before the blast of a trumpet.[51]

From heroic beginnings, in Manley's view, the PNP went on to heroic combat for its existence, as in 1947, when the PNP, 'fighting back in self defence on the streets of Kingston, smashed the power of the Jamaica Labour Party in Kingston once and for all.'[52]

In 1952, the PNP faced a decidedly unheroic crisis which led to the expulsion of its left-wing leadership, the 'four H's', Richard Hart, Arthur Henry, Frank Hill and Ken Hill. The four were examined by a PNP tribunal which established, said Manley, two central facts: 'First, a group of men in high positions in the PNP formed a secret Communist "cell" in the Party with the intention of training men for the task of subverting the organization to Communist purposes. . . . Second, these men used, or acquiesced in the use of, a document which preached that the PNP and its leadership were to be "used" only for the time being.' Manley found the evidence decisive. He described the PNP as 'a democratic socialist party pledged to the preservation of free political institutions . . . [and] the preservation of the freedom of the individual.' He said the 'loyalty to its ideals is the first condition of membership' in the PNP, and that the 'PNP is not, and never can be, either wholly or in part a Communist movement.' 'Democratic socialism and planned economic progress for Jamaica,' he wrote, 'are the bases of PNP political loyalty and integrity, the pillars of its existence.'[53] Manley's own initiation into trade unionism was intended in part to compensate for the loss of grassroots organizing ability in the PNP occasioned by the expulsion of the four H's.

The election of 1955 gave Manley a chance to celebrate. 'For all its imperfections as an institution of men and run by men, the PNP has always functioned in terms of ideals and of service.'[54] A week later he triumphantly described the installation of the new PNP government, captained by Norman Manley as Chief Minister.

> Outside the House [of Representatives] some 15,000 people had jammed Duke Street from the House to East Queen Street and beyond. When the eighteen PNP Members, extremely smart in pearl grey suits and matching ties, arrived at ten-thirty to march up to the House, pandemonium broke loose. It was as though ten, indeed sixteen, years of frustration and pent up emotion were released.[55]

Even the triumphant political year of 1955, however, had its bitter aftermath for journalist Michael Manley. The *Gleaner*'s political reporter Ulric Simmonds, chose JLP aide Donald Sangster as 'Man of the Year, 1955' instead of Norman Manley. Michael responded that 'everybody knows perfectly well who the Man of 1955 has been,' but 'it is equally well understood that Political Reporter would choke to death on the attempt to name that person.'[56]

Economics and Trade Unionism

With his developing sense of the interrelatedness of things—one of his major intellectual assets—Manley saw the intimate bonds between politics and economics. He reckoned that 'it is a sound political maxim that people must feel that they are part of a system before they will subscribe to its moral values,' and noted that no one could expect Jamaica's youth to respect a system which 'can give no work to more than a third of its population.'[57] Manley argued that 'the remedy to economic exploitation lies in political power.[58]

In the '50s, the Manleys adopted as their model of economic 'development' that of 'ever progressive Puerto Rico', based on foreign investment attracted by low wages, tax holidays, well-controlled labour unions, and infrastructure provided free by the local government.[59] Manley did have some reservations. He wrote that the Puerto Rican case showed that unemployment in an 'under-developed territory' could not be licked in a day 'or even five years', but at least a start had been made in Puerto Rico.[60] As a novice trade unionist of 'socialist' bent, Manley also realized the intrinsic contradiction between capital and wages under the Puerto Rican model:

> Some of us know that low wages are an inducement to the foreign capital which can do so much to promote . . . development. But it appears that in the enlightened section, there is a further sub-section which . . . has forgotten that our objective in this matter is development and not low wages.[61]

Manley seemed to have less reservations about the tax break aspect of the model. After his father won the 1955 elections, Manley wrote that 'to secure foreign investment with all that it means for the solution of our employment problems, we have to attract it'; tax concessions under the Pioneer Industries Law was one device to do so.[62]

From mid–1953, Manley devoted 'Root of the Matter' more and more to trade unionism, as he developed into a union leader. He dealt at length with the 1953 Stresscrete strike. Stresscrete (Jamaica Concrete Products) was resisting unionization under the leadership of its manager Eddie Hunter, 'an American from South of the Mason-Dixon line'. Manley observed that industrialization in Jamaica depended on 'vigorous cooperation on the part of labour' which would be more forthcoming 'if trade union action has ensured both the appearance and reality of justice for the workers for whose benefit the effort of industrialization is being made.'[63] As the strike dragged on for months over the right of union organization, Manley observed that 'it is only in totalitarian countries like Fascist Spain and Communist Russia that the worker is denied the right to join trade unions of his choosing.'[64]

Another concern was the JLP's economic policies. In late 1953 Manley criticized the JLP government for buying two million yards of textiles abroad per year rather than from the financially-troubled but local Ariguanabo Textile Mills.

> So here we have a so-called 'private enterprise' government calling for industrialization, a textile mill providing part of that industrialization, but that Government withholding the support which could help the Mill to survive.
>
> Can confusion be worse confounded?

However Manley later determined that it was actually 'certain merchants of considerable power' who preferred foreign cloth because there was no regulation on the wholesale mark-up, whereas there was a twelve and a half percent limit on Ariguanabo goods.[65]

Communism, the US, Africa

The final broad focus of Manley's opinion columns between 1952 and 1955 was international affairs. A compelling topic was communism, given its place in Jamaican politics: the PNP schism of 1952 and the constant red-baiting of the PNP by Bustamante, the JLP, and the *Gleaner*. Manley's column of 12 April 1952 treated the subject at length. It deplored a British ban on West Indian inter-island travel by communists, real or suspected. Manley's stance was that freedom of association is one of the principles of democracy and 'the answer to Communism as an internal threat lies in positive, never in negative action.' Britain's socialist government allowed communists freedom of movement in Britain and only reluctantly banned them from some government posts involving military matters because of 'the painful fact that Communists feel themselves in duty bound to transmit military secrets to Russia.' British socialists responded to communism 'by building a society in which Communism would be a redundant ideology.' Manley argued that the socialists achieved this between 1945 and 1951, when the communist vote dropped from one million to one hundred thousand. He compared this to the conservative strategy of the Kuomintang in China, which 'lost' the country. Ironically, he concluded, 'Communism has become the modern opiate of suffering people; to attempt legal proscription is to invest it with a new magic, a fresh martyrdom.'[66]

The overthrow of the Marxist, Cheddi Jagan, and his People's Progressive Party (PPP) government in British Guiana in 1953 enabled Manley to clarify his position on communism. Although all the facts were not available, Manley offered these pronouncements:

> If Jagan was really plotting a specific coup then I condemn him for seeking to lead the people of B. G. out of one imperialism into another. I despise the Communist system and that is an end to the matter.
>
> If, on the other hand, the PPP was seeking to prepare the ideological groundwork for a future Communist State then again I condemn him for a worthless intention which, to my mind, involves a betrayal of colonial nationalism. . . .
>
> I conceive my duty to be to fight for the interests of things that I like, against Imperialism, against Communism and above all clear of the power struggle which is a function of both Imperialism and Communism.[67]

In the last statement was the germ of the non-aligned, Third World-oriented foreign policy Manley would pursue as Prime Minister in the '70s.

Manley viewed communism as a form of imperialism. The communists in Italy and France were being defeated by American power and by 'a congenital nationalism which instinctively rejects the Imperialist overtones of the Communist movement.'[68] Underlying Manley's view of communism was not only the Cold War, but Jamaican politics. Shortly after the expulsion of the leftist 'four H's' from the PNP, Manley noted that:

> the Jamaican political scene has taken on a strange and amusing appearance. Communists call for 'three cheers' for Busta at their meetings; and capitalists leap to defend him against the charge of trifling with the economic fate of this nation.

In Manley's view, Jamaican capitalists and communists shared at least one motive in their support of Bustamante: 'A socialist government would upset both applecarts.'[69]

Another concern of Manley, the journalist was the way in which the United States was replacing Britain as Jamaica's leading metropolis in the 1950s. In 1952 Manley complained about the detention of Grantley Adams, the 'Socialist Prime Minister of Barbados', by United States immigration officials. This was a special insult to West Indians and part of the "red hysteria" which colours every phase of . . . [Americans'] existence.' (The 'detention-happy Americans' had previously detained Norman Manley.) Manley compared United States actions to 'the equally hysterical travel and news restrictions which create the reality of the "Iron Curtain".'[70] Because of racism, Manley considered the United States, like South Africa, a place where 'the dice are loaded against us.'[71] The attempt to impeach President Truman over the arbitration between workers and the steel industry showed Manley that 'the deepest of all American diseases is alive and kicking,' 'that the rights of property are more sacred than the rights of man.'[72] Manley lamented that in their election of 1952, the Americans 'should desert the dictates of reason and fall a prey to hero-worshipping emotionalism' by electing Eisenhower over Stevenson. He felt that the Americans' need for a father figure symbol during the Cold War was a major element in the election.[73]

Africa also particularly interested Manley. He was already a veteran opponent of 'the militant racialism of fascist South Africa'. When violence and rioting erupted in Bechuanaland, Manley urged that 'we' join the Bamangwato tribesmen's struggle against the South African régime by demanding that the Jamaican government and political parties call on Britain to change its South Africa policy.[74] In 1954, Manley defended the PNP suggestion to invite Ethiopian ruler Haile Selassie to Jamaica, a suggestion attacked by the *Gleaner*. Selassie was the living God to

Jamaica's Rastafarians, and Manley argued that the *Gleaner* seemed to have forgotten, 'the great place in history which Selassie holds.'

> It was he and his kingdom who first felt the lash of fascist aggression; he and his [kingdom] who wrote the first chapter in the story of heroism in defence of freedom which stretches across a decade.

'Selassie is an African and an Emperor,' and Manley could think of nothing more normal than that Jamaicans, 'ourselves of African descent', would welcome 'our distant and distinguished kinsman.'[75]

Manley's identification with the Afro-Jamaican majority came early and would later stand him in good stead as a politician of the '70s and '80s. His short but intensive career as journalist and columnist played a valuable role in re-integrating him at an intellectual level into the Jamaica from which he had been mostly absent between 1943 and 1951. A different and deeper integration, which helped Manley overcome the liabilities of his middle-class status, occurred through his long involvement in Jamaican trade unionism.

CHAPTER 9

'Young Boy,' Trade Unionist 1952-62

From 1952 to 1972 Michael Manley worked as a trade unionist, acquiring the nickname 'Young Boy.' From 1955 to 1962 a PNP government led by Norman Manley was in power; from 1962 to 1972 the JLP ruled. Given the close relationship between unions and political parties, control of government strongly influenced Manley's trade unionism. It is thus logical to divide his trade union career into two stages separated by the change of government in 1962, after which Manley was increasingly involved in politics. His work at *Public Opinion* served to re-involve him with Jamaica at the intellectual level; as a trade unionist he came face-to-face with the working majority of the people.

Jamaican trade unionism was a crucial aspect of 'a real distribution of economic and social power within the Jamaican class system' that occurred after the labour protests of 1938. According to Gordon Lewis, this development was the result of racial self-respect and organized political parties (both pioneered by Marcus Garvey); nationalism developed by the PNP and eventually the JLP; and working-class consciousness. The achievements of the '40s, '50s, and '60s bore steep costs. The PNP, in deciding to follow the post-war British Labour Party government, surrendered its early anti-imperialism and collaborated with the British ruling class. Bustamante, to achieve power, was largely responsible for the destruction of working-class unity, with profound and tragic consequences felt to this day.[1] For a short time after 1938 there had been the possibility of a single major political party, the PNP, led by Norman Manley, to advance the nationalist cause against Great Britain, and a single blanket trade union, the Bustamante Industrial Trade Union (BITU), led by its namesake, to advance the cause of labour versus capital. This possibility broke down in 1939 when the PNP formed the Trade Union Advisory Council (TUAC) in an attempt to control Bustamante. Bustamante soon bolted with his BITU, and in 1943 formed a political party, the JLP, to rival the PNP. The TUAC was meanwhile transformed into the Trade Union Council (TUC) which in 1948 became a blanket union in competition with the BITU. The TUC's initial strength was among workers in small Kingston factories, on the railroad, in public-works and the government, and in white-collar jobs. Norman Manley, primarily a politician, was an occasional union advocate who looked upon the TUC as a political ally of the PNP. The TUC's early leadership came from middle-class politicians such as Florizel Glasspole, Noel Nethersole, Wills O. Isaacs, Ken Hill and Richard

Hart, and was deeply divided by the ouster of the left—including Hill and Hart—from the PNP in 1952. Hill and Hart took the TUC with them, and the PNP formed a new union, the National Workers Union (NWU), through which Michael Manley would help achieve considerable benefits for a substantial sector of Jamaica's workers, and rise to power.[2]

To fully appreciate Manley's career as trade unionist, it is necessary to consider the nature of trade unionism in Jamaica. Its most important aspect is that it has bitterly divided the organized sectors of the working class,[3] for reasons indicated above. Notwithstanding occasional episodes of inter-union solidarity, much money and effort that could have advanced the interests of labour versus capital was spent instead on inter-union rivalries; relatively easily organized trades have been fought over, while the more difficult to organize, more marginal, poorer workers have been neglected. A second characteristic, following from the first, was 'political unionism', or the systematic commitment to mobilize votes and funds for a party in return for some control over economic and social policy, as was the case of the NWU with the PNP and the BITU with the JLP.[4] One of the vagaries of 'political unionism' was that a union might be more militant when its affiliated political party was out of power, since it ran no risk of embarrassing the party as government. On the other hand, when the party with which it was affiliated was in power, it might expect a more sympathetic hearing in some, but not all cases. In the first years of his trade union activity (1952–4), Manley represented a union allied with a party out of power; from 1955 to 1962, the party with which his union was affiliated was in power, and the government headed by his father; from 1962 until 1972, Manley as trade unionist again faced a political environment dominated by the hostile JLP.

It is necessary, finally, to consider the broad historical and social context within which Jamaican trade unionism developed in the late colonial era between 1938 and 1962. Manley, in his autobiographical account of his trade union years, *A Voice at the Workplace*, argued that the Jamaican work environment was created by colonialism, slavery, and capitalism.[5] Jamaica's traditional economy was dominated by authoritarianism, paternalism, brutal exploitation of labour, and a strict class/caste system. Bustamante and his creation, the early BITU, represented the authoritarian tradition in trade union form. The BITU was built around a few men at the top with almost no local organization or leadership. Its first application for membership in the Inter-American Regional Labour Organization (ORIT) of the International Confederation of Free Trade Unions (ICFTU) was refused on the basis that the BITU was undemocratic. It was reported in the '50s that there was a 'growing belief or suspicion among workers that Bustamante has sold out to the employers.'[6] The leaders of the PNP and the NWU clearly represented a more 'democratic' thrust than Bustamante, the JLP, and the BITU, but they were also products of a history dominated by

Edna and Norman Manley at Drumblair.

Edna with Michael, 1930's.

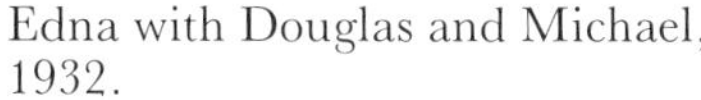

Edna with Douglas and Michael, 1932.

Manley (left) with members of the 99 Squadron, Wireless Gunners in Tuxedo, Canada, in 1944. Manley was then serving in the Royal Canadian Air Force.

The Twelfth Constitutional Convention.

Touring constituency in the 1970's.

Addressing a public meeting in rural Jamaica in the 1970's.

Participating in the Labour Day project in the 1970's
in East Central Kingston.

colonialism, slavery and capitalism, with its deep-seated traditions of authoritarianism and paternalism. Many Jamaican workers in 1952 were ill-equipped to understand the objectives and purposes of either 'democracy' or trade-unionism. Thus leaders like the Manleys faced the delicate task of attempting to build 'democracy' in a late colonial environment hostile to it, one which sometimes even encouraged them, in the name of efficiency, to act in an authoritarian manner. This conflict between a colonial, authoritarian tradition and the struggle to democratize Jamaica remains one of the most enigmatic aspects of Michael Manley's career.

First Steps

Manley described the society in which he began his trade union career as one where 'class relations were stark in their intolerance: There was no subtlety, and little mobility, because a man's class was stamped on his skin as much as on his clothes. To middle-class eyes, the working classes were an opaque mass—without individuality and without rights—because they were without humanity.'[7] In Jamaican society of the early '50s the rule of law conflicted with the laws of capitalism, and thus labour law favoured the capitalist. Among capitalists and workers alike, attitudes toward work, justice, and cooperation had been perversely shaped by colonialism and by slavery. 'Discipline', in the minds of workers, conjured up 'the wishes of the slave master', while employers treated their employees with imperious disdain and subjected them to daily indignities. The workplace of 1952, Manley wrote, was like a 'primitive cave'.[8]

Some progress had occurred. The BITU had helped establish the right to protest and had claimed a larger share of the economy. The PNP and the TUC had challenged the idea that 'the free enterprise system existed by divine right as part of the natural order of things.' However, Manley believed that even the PNP and the TUC 'had not, up to that time, challenged the deeper inner equation involving the relationship between those who manage and those who obey within the economic process itself.'[9]

Manley's introduction to trade unionism was unplanned. He was working at *Public Opinion* and starting a local PNP group in a nearly destitute enclave in his father's generally middle-class Kingston constituency, Barbican. Michael was soon asked to deliver speeches to political and trade union educational meetings. Still very shy in public, he had to overcome a tremendous fear of public speaking. Before his first speech, Edna Manley prescribed a strong drink to ease his panic. Manley described the event in these terms:

> I very nearly never made it. In the end, having eaten nothing since the night before, and bolstered by a half tumbler of neat rum I arrived at the Patriot Group around 8.30 p.m. in a state that bordered upon abject terror. My speech was distinguished mainly for the fact that it had not occurred

> to me that you have to work out an ending. After I had said all the things for which I had elaborately prepared myself for at least two weeks, I suddenly realized that I did not know how to bring it all to a close.
>
> I remember stumbling through a series of sentences each more desperately disconnected from the main theme than the one before, as I searched for a climax. None came and in the end I astonished my audience by sitting down in the middle of a sentence which didn't seem to begin in any particular place and was certainly proceeding in no discernible direction. The applause that followed sometime later stemmed, I suspect, more from sympathy than admiration.[10]

A rather inauspicious beginning for the spellbinding orator-to-be, but Manley made over a hundred speeches in 1952, and was elected to the National Executive Committee and the Central Executive Committee of the PNP. In the autumn of 1952 he was asked to become a member of the NWU's negotiating team attempting to win over striking employees at the Ariguanabo Textile Mill. In March 1953 the task of representing the Ariguanabo workers fell upon Manley. Within a month, he was also involved in a minor capacity in the NWU's first major action, organizing the bauxite workers of ALCAN (The Aluminum Company of Canada). These early activities convinced him that trade unionism 'suited my egalitarianism, my commitment to social justice and my instinct for activism.'[11]

The Sugar Struggle

Manley's first major union activity was in the sugar industry. He became a full-time NWU employee in June 1953 at a salary of eight pounds a week, with the use of a union Hillman Minx, and a weekly travel and subsistence allowance of seven pounds.[12] In September he was appointed 'Sugar Supervisor' of the NWU, although no sugar workers were members of the union. By December he was criticizing Bustamante's plan for a £20,000 unemployment fund for sugar workers. Given the number of unemployed and the length of the 'dull time' after the harvest, the plan would give each worker about eight pence a week, a 'vote-catching device', not a serious solution.[13] Manley's job was to make inroads into a labour sector long dominated by Bustamante and the BITU; although he consistently maintained that his trade unionism was separate from his politics, success in sugar organizing had the political bonus of broadening the support of the predominantly urban PNP. Sugar was a vital, though chronically troubled sector of the economy, employing some fifty to fifty-five thousand workers during the harvest season, and half that number during the 'dull time'.

Sugar had a deep and sinister reputation in Jamaica and the Caribbean, for its associations with slavery, injustice and national dependency on unpredictable foreign markets. Moreover, the early unionization of the industry had aggravated the monumental schism between Bustamante and Norman Manley; a 1941 sugar workers' agreement negotiated for the BITU by Manley when Bustamante was incarcerated, was later renounced by the BITU's dictatorial chief. In 1950 the TUC, before its split from the PNP two years later, began a struggle to gain the right to organize workers in the BITU-dominated sector; after a 'long, brutal, and sometimes bloody strike' the TUC succeeded in gaining the right to represent sugar workers at two estates but was defeated at twenty or so others that composed the sugar-growing sector. The TUC also gained the right to participate in the general bargaining system.

Manley organized the NWU campaign to represent sugar workers using four general tactics. The first was to avoid attacking Bustamante or the JLP, both of which were highly revered by many sugar workers; the NWU's ties with the PNP were de-emphasized for similar reasons. The second tactic was to concentrate on four estates where the NWU already had significant support. A third tactic involved the use of meetings to get the NWU message to the workers. This required 'an enormous investment in time . . . in little villages all over the countryside.' Manley had his union car equipped with a sound system to attract workers to meetings. The lover of Beethoven 'personally chose all the calypsos, rock-and-roll numbers, the "yank" tunes and the other popular music which were in vogue.' After 'dazzling displays of dancing' by the young boys and girls of the region, serious speech-making would ensue. The message was always the same: 'Here are the problems, here is a chance to tackle them yourselves through democratic organization.'

The fourth tactic was Manley's own personal involvement, which had a profound influence on his life and career. Political work had previously brought him close to urban slum life in Jamaica, with, as he wrote, 'its grim and endlessly repeated tragedy of poor housing, children and parents crowded into single rooms, and whole families undermined if not shattered by unemployment and the general conditions of shanty town life.' But before the sugar campaign, he had not personally witnessed the endemic poverty of the rural plantation system, a situation he set out to remedy.

> For months I walked up and down the estates getting to know the field gangs, the cane cutters, the cane loaders, the grass weeders, the tractor drivers and sidemen, the trench men and, indeed, all this gallant company of men and women who were the heirs to and sufferers of the plantation system. . . . This was an invaluable experience for me because it provided the living evidence of conditions which cannot be fully grasped from any written account, however vivid.

His fiancée, his future second wife, Thelma Verity, often accompanied Manley during their two-year engagement as he launched himself amongst often hostile crowds around the island. Thelma later recalled the 'terrible sadness of the really awful plight of the workers' which often reduced her to tears upon returning home. She recalled that Manley 'did some pretty revolutionary things,' but it did not come easy. Very shy, he was flung into unionism very unprepared. He occasionally vomited before making speeches. Eventually, however, Manley developed a flair for the performing, theatrical side of trade unionism. As Thelma Manley noted:

> There was a great deal of sex appeal in it. . . . He had this marvelous old sort of suede jacket that he brought up from England. I use to turn the collar and patch the things, and keep it in one piece and all that. . . . And you know he's tall and imposing and when he was young he used to sort of get the attention of the audience, he would stand up at the mike when he was ready to speak, he'd wear his jacket, I mean it could be ninety degrees, he wouldn't say anything, he'd just slowly start to take it off. It sounds like the strip tease, well it really wasn't, but it had the same effect . . . on the female part of his audience: 'Lord, Young Boy!' I would stand at the back, squirming quietly, but knowing that it was necessary.[14]

Manley recalled that 'their collective experience of suffering and exploitation had made [the sugar workers] into a strange and wonderful kind of élite amongst the sufferers of the land.' He found it stimulating and rewarding to spend a morning with a field gang listening to the conversation, 'a fascinating mixture of worldly wisdom, durable personal philosophy and sometimes incredible prejudice'. The sugar workers dubbed Manley 'Young Boy' despite his thirty years, and the name stuck; Manley suspected that 'the name symbolized to those who used it the idea of a new start with a young leader;' perhaps, also, it was intended as an affectionate, subtle reminder to the thirty-year-old city 'boy' that there were many with deeper experience of the cane field and sugar factory than he.

Manley remembers the sugar-organizing time as 'great days', a memory shared by Thelma Manley. She recalls that they 'never had any money and he worked like a brute. Anything extra we had to give it away.' When they were married in 1955, they had virtually no money and no furniture; Edna Manley gave them the 'Horse of the Morning' sculpture in gleaming red wood, three paintings and a bed. There was a terrible lack of privacy in Michael's and Thelma's home, as people were constantly coming and going on union business and other matters.[15]

Despite the good memories, 1954 was 'a savage year' of NWU-BITU struggle to represent the sugar workers.[16] The NWU won the right to represent workers on three estates including New Yarmouth in the heart of Bustamante's South Clarendon constituency. The advent of the PNP government in February 1955 improved the NWU's prospects. By 1956, the NWU had defeated the TUC at the Worthy Park and Bennett estates, thus becoming the only opposition to the BITU. In 1957, the failure of a BITU strike advanced the NWU cause further.

Nonetheless, the years after 1954 were mixed. On 18 June 1958,

Manley wrote to the Ministry of Labour alleging that 'either on the decision of individual Estates or, possibly worse as a matter of SMA [Sugar Manufacturers' Association] policy, or both, . . . an attempt is being made to destroy the NWU in the Sugar Industry.' His letter detailed three cases to illustrate the point. The SMA denied the allegations, to which Manley responded that the SMA had missed the point: the tendency of the SMA and its members to prejudge incidents involving workers or NWU representatives and take unilateral action before joint SMA-NWU discussion or investigation.[17]

Sometimes, pragmatically, the NWU and BITU joined together, and ancient rivalries were moderated. One such occasion occurred in early 1959, when Manley—by then Island Supervisor of the NWU—met with Bustamante and the Island Supervisor of the BITU, Hugh Shearer. Busta served an 'extremely pleasant champagne at ten in the morning' and an agreement was reached for a joint strike at three estates. During the strike Shearer and Manley shared the platform and produced moments in which, temporarily, endemic divisions of the working class were put aside. At the Moneymusk Estate of the British firm, Tate and Lyle, as Manley described it, Shearer 'made, as always, a clever, tough, hard-hitting speech full of sarcasm . . . where advocacy was well reinforced by the skilful selection of facts and figures.' Manley's description of his own speech suggests much about his style:

> For a long moment I stood silent surrounded by that magnetic force a crowd exerts when it is tense with expectation. I heard a voice that sounded like mine but seemed to come from someone else going through the preliminaries 'Mr. Chairman, Mr. Island Supervisor of the BITU, officers and delegates of the BITU and NWU . . .' and then I paused feeling for that inspiration which if it comes, will express the inner essence of the moment. Suddenly, I said: 'Brothers and Sisters of the BITU; Brothers and Sisters of the NWU, we have a common struggle. We have been divided . . . but now, we are joined . . . and the word is love.' Never before or since have I experienced anything like what took place. The simple phrase, 'the word is love,' which I had not planned but seemed, rather, to have drawn from the emotion of the crowd itself, struck some deep chord of longing and need. It was as if the place exploded. It was quite a long time before I could make my speech and when I made it I knew that it was an anti-climax because I had said it all already: 'the word is love.' I knew that to the crowd this meant that the workers were to be united again. . . . [18]

Manley concluded that this experience represented a 'perhaps unexpressed, semiconscious recognition that no one trade union can bring about the revolution in the circumstances of the working man that is so ardently desired, so incompletely understood and so terribly needed.'

In 1959 a commission of inquiry into the sugar industry was presided over by H. Carl Goldenberg, a Canadian lawyer and later senator. The NWU team before the commission—Carl Hall, Research Officer of the

NWU, Wes Wainwright, Assistant Island Supervisor, and Manley—assembled massive data, which showed that, among other things, the real wage increases received by sugar workers since 1938 amounted to the value of one pack of cigarettes per week. For Manley the preparations for the hearings was 'one of the great intellectual experiences of my life'. The Goldenberg Commission found that the Sugar Manufacturers' Association had consistently underrepresented its profits, and awarded a large wage increase to the workers. It also established permanent guidelines for the reporting of industry finances to unions.

The Bauxite Story

The bauxite industry was a second major area of Manley's trade unionism. The mining of bauxite, the source of aluminium, was the most dynamic sector of the post-World War II Jamaican economy. The NWU soon gained a dominant position representing the bauxite workers. As described by Manley in *Voice at the Workplace*[19] there were three major problems during his tenure as trade unionist: the effort to raise wages; the Vickers case involving internal union control and discipline; and the Alpart Case, involving the BITU's challenge to NWU dominance in the field (to be discussed in Chapter 12).

In November 1952 the NWU launched a campaign to raise the wages of unskilled bauxite workers to four shillings (equivalent then to one US dollar) an hour. Jamaican workers were receiving one shilling four pence; unskilled aluminium workers in the United States and Canada received the equivalent of twelve shillings. The NWU's position was that bauxite was part of an international industry geared to the world market; thus the 'bauxite wage should not be tied to ordinary wage rates in Jamaica, but should bear some relation, however slight for a start, to American and Canadian rates.'[20] As the bauxite wage dispute headed for arbitration in 1953, Manley was aware that the arbitrator's decision 'is going to be of significance, not only to Jamaica, but to every part of the underdeveloped world where vital materials are being exploited by foreign capital.'[21] The NWU's case was argued by Norman Manley and characterized, in this context, by his son in these terms:

> Generally conceded to be the greatest Jamaican advocate of his time, some say the greatest Caribbean advocate of all time, he brought to the case an unusual mind in the sense that he combined to a rare degree the analyst's gifts of logic, simplification and incisiveness with the humanist gifts of breadth of interest and compassion.[22]

Norman Manley argued that the NWU's wage claim would merely allow the worker to fulfil his minimal duties to home and family with enough left over for a pack of cigarettes a day. The English arbitrator selected by the Bustamante government was unmoved and awarded a settlement of one shilling ten pence, a major defeat for the NWU.

Michael Manley characteristically searched for the silver linings of a dark cloud. From the defeat of 1953, he drew three lessons. The first, taught by his father, was 'when all seems lost, act!' He was immediately dispatched to explain the setback to the workers. The second lesson Manley described as 'one of the turning points of my life': the 'loyalty and courage in adversity' of 'those people whom we so inappropriately describe as the masses.' His belief in them had up to then been the product of faith; 'now it was confirmed by experience,' as the workers and Manley consoled each other and pledged to continue the struggle. The third lesson was learned while Manley waited in a bauxite company office. Manley discovered on a personnel officer's desk comparative wage figures that showed the companies had lied about the top wages paid to unskilled labour in Jamaica.

> It was one of the greatest shocks I had ever had for up to then I had genuinely believed that honesty could be taken for granted at the sorts of levels that were involved in a great wage arbitration.[23]

A by-product of the bauxite struggle was the value of international labour cooperation. Capitalists had learned that lesson long before, but the labour movement in general had been 'traditionally insular in outlook'. To change that, Kenneth Sterling, originally a PNP organizer, was hired by the United Steelworkers of America to organize the new bauxite workers. The American steelworkers' union also funded the effort.[24] The NWU's international labour efforts were opposed by the JLP government, which temporarily prevented Nick Zonarich of the American United Steelworkers and Charlie Millard of the Canadian United Steelworkers from entering Jamaica. Manley called the government's actions 'a naked piece of fascism' and 'quite incredibly stupid'. Millard was a member of the International Executive Board of the ICFTU, 'perhaps the most important of all possible bulwarks against Communism'.[25]

Following the setback of 1953, the bauxite wage issue was fought for some twenty years. Eventually, in 1971, the NWU and the workers received the historic goal of a US dollar hourly wage (by then worth twelve shillings, or one Jamaican dollar). Over the 1952–71 period, the workers received increases of four hundred percent in real terms, or an average of twenty percent per year.[26] These gains were made despite serious problems such as the Vickers incident.

The Vickers conflict in 1958 was Manley's 'unhappiest union memory'. The incident was rooted in the impersonal, non-paternalistic worker-manager relations that often existed between the black bauxite workers and the white North American managers. By 1958, workers' feelings were running high at the Kirkvine facility of ALCAN. Some workers felt the NWU was taking too soft a line towards management. John Vickers—characterized by Manley as 'bright but unstable'—emerged as leader of a movement in opposition to the impersonality of labour relations at Kirkvine and the alleged racism of management.

After a worker died when a high pressure valve burst, Vickers and a senior ALCAN manager clashed. Before NWU headquarters was notified, Vickers and his supporters had shut down the plant. Manley thought the wildcat strike a disaster since it breached a contract and lacked focus. As Manley stalled for time, Vickers began to attack both Manley and the union. In response, Manley began to arrange for a resumption of work and an inquiry into the matter. Vickers lost his job, but he had been a hero to the younger and more racially-conscious workers, and they resented the NWU's action for years. Of Vickers, Manley wrote:

> In the end he paid the inevitable price of disloyalty. Having exposed himself by going far beyond the limits of legitimate union action and being guilty of a flagrant breach of contract he would have been well advised to remain loyal to his own organization. When he cut himself off from that final bastion there was nothing left that could save him.[27]

According to some close to him, loyalty is one of Manley's most deeply held values, and the question remains: in the event of a clash between 'loyalty' and other values such as 'independence' or 'democracy', which is the higher call?

The NWU and the PNP

Another sort of conflict involved the close relationship between the NWU and the PNP, headed, respectively, by Michael Manley and by Norman Washington Manley. The Manleys took pains to define the separate spheres of union and party, a situation accentuated during the PNP ascendancy from 1955 to 1962. During the NWU's fifth annual conference in 1957, for example, both Norman Manley and NWU President Thossy Kelly acknowledged that conflicts existed between the NWU and the PNP government. The elder Manley urged the union to 'prosecute the workers' interests even if it temporarily embarrasses the Government.'[28] Some workers supported the NWU because they hoped its favourable ties with the PNP government would advance their own interests. If the union did not push the workers' cause hard enough it would suffer loss of its members' confidence. If it pushed the cause too hard, it would embarrass the government whose influence and favour it needed.

In 1958 the NWU leadership was said to be separating from the PNP leadership. Knowles reported that the NWU was democratically organized (in contrast to the BITU), and that it was building local leadership based on industrial rather than geographical units. He reckoned that the NWU might well become the 'model union in the West Indies'. Knowles also noted the close personal relationship of father and son that connected union and party, 'a fact which causes some to conclude that the Manley family is not far different from Bustamante in their control of a union-political party.' Knowles added:

> Norman Manley has never attempted to be a union officer. Michael Manley, on the other hand, has refused to run for political office or to hold office in the People's National [P]arty. Instead he has devoted his energies entirely to building a strong, well-run union.[29]

The Manleys maintained the image that the NWU and the PNP were separate, but cooperating institutions, with interests that sometimes converged and sometimes diverged. This ambivalence—rooted in the differences between union and party—often found public expression. In 1959, for example, Norman Manley said that the PNP had been returned to power in that year because of NWU support. At the same time, he said, NWU efforts in the sugar and banana industries and on the docks 'took a completely independent line unaffected by political considerations and solely devoted to the interest of the workers.'[30]

Behind the scenes there was more union-party cooperation than appeared in public. The activities in 1961 of one Beckford who tried to organize the new United Negro Workers' Union, apparently with strong racial overtones, prompted such cooperation between the Manleys. In January Norman Manley wrote his son that Beckford could constitute 'a menace to the Unions as well as to Government' and asked for Michael's advice.[31] Six days later, the elder Manley again wrote his trade unionist son:

> Cabinet decided yesterday that I should arrange to see both Unions [NWU and BITU] about the activities of Beckford who . . . has recently set up an office in Frome.
>
> The purpose of the discussion would be to suggest on a confidential level that the Unions should concert action against Beckford and the disruptions which another Sugar Workers Union would cause.[32]

In early February, the Chief Minister of Jamaica again wrote his son:

> I believe that both Unions are going to be in considerable trouble with WISCO [the West Indies Sugar Company] in Vere and Frome. The Vere situation is serious. I have invited Sir Alexander to consider telling Shearer to come and speak to you about Vere and the activities of the racist union. I told him that I am sure an approach would meet with a friendly reception.[33]

These letters illustrate a much closer relationship between the Norman Manley as political chieftain and Michael Manley as trade unionist than the Manleys liked to admit publicly. They also demonstrated the shared desire to avoid further fragmentation of the union movement that would be caused by the creation of new, competing unions; to accept the rival BITU, and to cooperate with it.

Against this background, it was important to maintain the image of NWU 'independence'. The NWU issued 'warnings' to and 'criticisms' of the PNP that usually stopped short of being genuinely critical. At the NWU's fifth annual meeting in 1957, for example, Michael Manley, as NWU Second Vice President, warned the PNP government against creating excessive monopolies.[34] Similarly, in 1961 NWU President Kelly mildly criticized the PNP government concerning 'very grave' unemployment problems.[35] Although Manley claimed trade unionism as his own, the close NWU-PNP and son-father connections mitigated his independence. In some minds, the connections symbolized a Manley union-party dynasty similar to the one-man empire of Bustamante.

Assessing Manley's Trade Unionism

Manley would continue to be absorbed by trade union activities for another ten years, until 1972, but after 1962 he would go into politics. Although there were future trade union battles still to come, it is convenient to assess the nature of Manley's trade unionism at this point.

Gordon Lewis, writing in the late '60s, expressed a common view when he observed that 'Jamaican unions have come to be run by the political parties as vote-catching annexes instead of being controlled by the workers, [which] has inevitably taken the edge off [their] militancy.' He observed that 'union leader names like Manley and Shearer are respectable Establishment names in the Jamaican situation today.'[36] Another observer set Manley's trade unionism in the context of his relations with his father. Trade unionism provided an opportunity, wrote George Eaton, for the younger Manley to exercise his leanings toward advocacy and egalitarianism; it also provided a career relatively unmarked by Norman Manley's footsteps, which Michael had been anxious to avoid. But trade unionism led irresistibly to politics.[37]

In 1970, Terry Smith gave Manley major credit for making the NWU into a major rival of the BITU. According to Smith,

> Mr. Manley served a hard from-the-bottom-up apprenticeship, . . . because he arrived at a period teeming with suspicion of the personality cult, and being the son of the Party leader, Norman Washington Manley, he started out with a built-in strike against him.

Smith considered Manley's lasting contribution to be a new level of sophistication in West Indian trade unionism.[38]

Manley's own thoughts reveal a mixture of pride in accomplishment and recognition of the limits of trade unionism in a poor, underdeveloped, dependent capitalist country like Jamaica. He wrote that as of 1952 'my own analysis of society had been inhibited by the socialist doctrine which I had unquestioningly accepted up to that point.' His trade union experience added a 'humanist and individualist focus concerned with the human equation within society as distinct from a more general and structural focus' to be found in 'socialist doctrine'.[39] However, in the late '60s, Manley had already voiced considerable dissatisfaction with the state of the Jamaican trade union movement, indicting it for destroying labour unity, for squandering resources on irrelevant, internal struggles, and for engaging in a 'sordid scramble for power'.[40] In 1974, Manley offered this assessment to an appreciative audience:

> The trade union born in the capitalist equation . . . has so surrendered its own moral purposes to the capitalist system that it is as venal in its appetite as the very management with which it contends. (*applause*) And I say this from the depths of my own experience—because I knock no man where I am not knocking myself—one of the things I had to learn as a trade unionist was the very fact that what I was doing was inadequate. Because to get more wages is fine; to get better fringe benefits is fine. To get grievance procedures is fine. But when I have them all, I still am essentially writing a prescription for a person in an inferior relationship that I am protecting with rights, rather than writing an equation in which to engage a human being by accepting their creative responsibility. (*applause*)[41]

In 1976 Manley argued that although trade unionism can divert some of the product of labour into workers' hands, it cannot by itself fundamentally reorder social and economic relations. Manley's ambivalence was reflected in his statement that the trade union movement was 'an important but marginal' response to a more profound problem, capitalism in the West and in Jamaica.[42]

Manley's trade unionism strengthened his grassroots political base and exposed him directly to the awful realities of life at the lower reaches of Jamaican society. At the same time, trade unionism was revealed to him to be a limited response to the 'problem' of capitalism and perhaps even an obstacle to socialism. With his bent toward idealism and global approaches to fundamental problems, Manley would eventually hope to transcend unionism via politics.

An unfortunate personal tragedy of these years was the collapse of Manley's second marriage with Thelma Manley, in 1960, after five years. Their son Joseph had been born in 1958, but grew up scarcely knowing his extraordinarily busy trade unionist and politician father, except as an occasional visitor bearing cricket equipment, which unlike his father, Joseph never learned to appreciate. Personal differences figured significantly in the divorce, but the extremely frenetic pace of Manley's life inevitably made normal domesticity impossible. Manley thus faced the '60s, in both personal and professional terms, with a mixture of triumphs and blows behind him.

CHAPTER 10

The Political Background 1938-62

Jamaican Politics

Michael Manley must be understood in the context of the political history and political culture of Jamaica. Jamaica's colonial history, formally ended only in 1962, produced such durable legacies as a strongly authoritarian tradition symbolized by an all-powerful governor, the separation of most people from power, and a tendency to regard the British Westminster model of parliamentary democracy as a superior and inevitable form of politics. The main features of the Westminster model are a powerful prime minister, the cabinet system, a supposed distinction between 'policy' and 'administration', a bicameral Parliament with an upper house to brake popular opinion, institutional continuity, and a bias in favour of a two-party system. The dubious value of a system developed in England for the late colonial and neocolonial West Indies naturally led to certain local modifications of the Westminster model: gerrymandering became common and widely regarded as legitimate, and the granting of favours by political patrons to clients became a central and crucial structural adaptation to Caribbean realities.[1] Trinidad's Eric Williams, a critic of the model, nonetheless recognized its positive points: the rule of law, freedom of speech, separation of church and state, and the orderly transfer of power through regular elections.[2] Nonetheless, the Constitution Commission of Trinidad and Tobago noted that 'in reality the Westminster political system has a propensity to become transformed into dictatorship when transplanted in societies without political cultures which support its operative conventions.'[3] One reason for such a transformation would be strong internal antagonisms creating the need for powerful central control: in Trinidad and Guyana such antagonisms are racially based; in Jamaica the antagonism comes from the 'political tribalism' (as it is commonly called), the intensely bitter rivalry of the PNP/NWU and JLP/BITU camps.[4] The system is also inherently contradictory in that 'it advocates national unity but legitimizes, in practice, national divisions,' a contradiction which leading politicians like Trinidad's Williams or Jamaica's Michael Manley reflect.[5] Manley, a fervent practitioner of the system and a critic of its limitations, has observed that the Westminster model is good for defending against tyranny (a situation which preoccupied its Anglo-American founders), but not so good for mobilizing a Third World country for such awesome tasks as national development in the late twentieth century. Indeed, the real question about the Westminster model and other forms of 'democracy' in the Third World, is what do they contribute to the basic task of development, assuring all citizens of basic human needs and a reasonably dignified life?

There is, finally, the question as to the extent to which the system has been internalized by West Indian leaders. Gordon Lewis has reckoned that even radicals such as the Guyanese Marxist Cheddi Jagan 'became the half-willing victim of British colonial indoctrination' who 'could not break out of the restrictive assumptions of the electoral game.'[6] Wendell Bell has also concluded that West Indian leaders have easily accommodated to Western liberal ideas.[7] Singham, however, argues that that accommodation is much more cynical than Bell suggests. The élites latch on to constitutionalism and other ideas as long as they do not unleash tension and insecurity or threaten their positions. But in small societies where élites have frequent contact with their 'enemies', constitutionalism becomes bogged down in procedural questions which take precedence over matters of substance. Tension builds as proceduralism fails to solve fundamental problems, and predisposes people to the emergence of a 'leader who will take charge and get things done.'[8]

Thus features inherent in the Westminster model combined with elements of Jamaican history to create a hero-Messiah complex that, in some ways, is the antithesis of 'democracy' which is supposedly a rational system. Carl Stone has argued that Jamaican voters are highly rational in their perception of, for example, economic suffering, but he has also noted several factors which constrain voter rationality in Jamaica:

1. The level of hopelessness and despair, which leads to overreaction to slogans in the absence of detailed programmes.
2. The low level of information and education, which limits the ability of voters to assess propaganda.
3. The tendency of most voters to vote based on party loyalty, rather than on the state of the country, or policy choices.
4. The tradition of 'personality politics' with its undue emphasis on personality cults, hero worship, and the projection of messianic leadership.[9]

The messianic tradition, which Michael Manley tapped, was deeply rooted in the religious culture of Jamaica and expressed in such popular figures of hope and opposition to oppression as Bedward, Garvey and Bustamante. Norman Manley (who like Bustamante eventually acquired the political name 'Moses') lacked the messianic personality, but the PNP attempted to create one for him to compete with Busta. A 1962 PNP comic book called 'The Man of Destiny' presented a more appealing image to the Jamaican majority: 'From the very beginning young Norman belonged to the soil' and his holidays were spent chopping log-wood giving him an 'undying love for the "bush" ever since.'[10]

The most complete analysis of the hero-messiah political leader in the West Indies is Singham's study of Eric Gairy in Grenada. According to Singham, the hero-messiah leader is characteristic of the late colonial era, and may carry over into the early years of 'independence' or neocolonialism (since there is no sharp break with the past following 'independence'). 'Despite their espousal of egalitarian and democratic ideas, the élites in colonial societies tend to have basically authoritarian personalities,' Singham writes, which develop into the hero-crowd syndrome where the hero 'tries to impose change from the top rather than through a process of genuine political socialization.' Authoritarianism and aggressiveness tend to characterize the West Indian personality. The 'crowd' which is attracted to heroes exerts negligible control over them, but on the other hand its commitment to them may fluctuate widely. Electoral politics may succeed in exploiting popular hostility against the state, the economy, or 'them', but it is difficult to organize the crowd for long-term, difficult tasks. The hero is also circumscribed because 'even after independence the mass hero is limited in his actions, for the vested and strategic interests of the international élites continue to set limits on what can be done.' A political leader in this mold who wishes to challenge the establishment must base his legitimacy on a special relationship with 'the masses'; he will tend to be both authoritarian and paternalistic; national rather than regional in his base, rely on schoolteachers and local leaders, and articulate the grievances of the poor majority. Singham identified two types of heroes, the middle-class hero (of which Eric Williams and Norman Manley could be exemplars) and the trade union hero (Bustamante, Gairy).[11] Michael Manley is a blend of both, but he has a conscious awareness of the dangers of the hero-crowd syndrome and desires to move beyond it.

Other features of the Jamaican political system should be briefly noted. Kuper lists some general features: strong popular involvement, a high rate of participation in elections, the two-party monopolization of electoral support, multi-class support for each party, and the dominant influence of established business and professional groups in determining electoral choices and policies.[12] Others endorse the final point by arguing that the concentration of wealth and power among 'the twenty-one families' (the oligarchy that Jamaican leftists say dominates the economy) intimidated any party because the 'twenty-one' could cripple the economy by decapitalizing it.[13] Gordon Lewis observed that the middle class controls Jamaican politics:

> . . . The general truth of political Jamaica . . . [is] a two party system in which both parties, although necessarily structured on a mass basis, are fundamentally bourgeois in spirit, led by a foreign-educated middle class leadership.[14]

Like ward politics in the States, concrete rewards motivate many party members more than liberal democratic ideals. Common Jamaican political appeals involve the promise of a job in return for a vote, or a handout or a meal in return for harassing the opposition.[15]

The similar class bases of party leaders and followers helps explain the convergence of PNP and JLP policies in the '50s and '60s. In the '60s both parties shared many similarities: leaders of indistinguishable background and ideology; financial dependency on well-heeled business supporters; markedly similar programmes; international alliance with the United Kingdom and the United States; strong emphasis on the party leader; patronage, clientelism, and paternalism to maintain mass support; the strong 'personality basis of Jamaican politics'; and unofficial linking to urban street gangs which 'in return for handouts of food, the occasional job, protection from the police, . . . helped preserve party territories.'[16] Another factor binding the parties together was the 'family' nature of the contending politicians. Bustamante and the Manleys were cousins, and the JLP leader Hugh Shearer (a close friend of Michael Manley) was Bustamante's adopted political 'son'.[17]

Although the parties converged in the '50s and '60s, differences remained. By the early '70s, many Jamaicans saw the JLP positively as a labour party and negatively—after ten years in power—as tainted by corruption and factionalism. The PNP was seen as the party of intellectuals and educated people, but tainted by the red smear and, in some middle-class eyes, too eager to assert 'African' values and hence racist.[18] Taking into account the 1972 elections and earlier trends, Carl Stone demonstrated that in urban Jamaica the PNP tended to get a disproportionately large share of the better off and more educated working-class vote, while the JLP traditionally had a similar advantage among the lower strata of the working class and among big business.[19] Stone also showed that PNP voters in the small peasant, skilled labour, and unskilled labour groups were consistently more 'radical' than their JLP counterparts on issues like the nationalization of the bauxite industry and government acquisition of the idle lands of big land barons. The two parties also had vastly different styles of political mobilization.[20] To understand the political environment that Michael Manley encountered when he became more actively involved in politics after 1962, it is necessary to consider the evolution of the PNP in the quarter century after 1938. Between the 'hero' and the 'crowd' stood a party that, although molded by its leaders, could and did frustrate them and limit their power.

The Early PNP, 1938–62

The People's National Party was founded in 1938, in the aftermath of the labour protests of 1938. It was explicitly a nationalist party, seeking self-government. Many of its early leaders were educated in England, however, and its anti-imperialism was limited by a basic acceptance of English values and institutions. Its primary concern was political independence, and it also championed economic planning and development along the lines of the Puerto Rican model. Its early leadership was from the middle class and its support from the better educated sectors of the urban working classes, a situation which began to change in the '50s with the trade unionism of Michael Manley. Four central issues help illuminate the nature of the PNP: the role of Norman Manley, the party's brand of 'socialism', its internal dynamics, and its relationship to 'democracy'.

Norman Manley was the central figure of the early PNP, although several others made large contributions to it. Norman Manley represented the urban middle class, a class that was gradually filling leadership roles in late colonial society. According to Beverley Manley's critical analysis, Norman Manley affected the early PNP in four major ways: building a democratic organization and insisting that it usually play a role in decision-making; forging a relationship with the British Governor Sir Arthur Richards during World War II 'which blunted the edge of the liberation struggle;' acting as a 'mediator/compromiser' in holding together the PNP's initial middle-class alliance; and committing the party to a two-party democracy, thus shaping 'the institutional form of the liberation process'.[21] In Beverley Manley's view, Norman Manley's role as leader of the PNP could not be separated from his character, a product of the contradictions of colonial society. He was born and raised in modest circumstances, but not personally exposed to poverty or exploitation. His earliest successes (other than athletics) came as the result of his mental powers. 'A moralist, his learning and his instincts combined to develop a strong sense of "right" and "wrong" along with a conviction that reason and organization would prevail over all social ills.'[22] Although intellectually committed to democracy, the elder Manley sometimes succumbed to authoritarian decision-making and allowed himself to be mythologized to compete with the Bustamante charisma.

The early PNP's 'socialism' bore Norman Manley's stamp. First declared in 1940, PNP socialism was in the well-established mold of the British Labour Party. While the reforms its proposed were substantial, from the start PNP socialism was bound by allegiance to the British empire, the Westminster model, and existing economic relations in Jamaica.[23] In 1938, Norman Manley had held over a hundred retainers representing nearly all of big business in Jamaica, though after that date he never appeared in his capacity as a lawyer in any cases involving worker interests.[24] That he continued to represent big business at all—as in the 1951 Vickers case in which he made his name as an international lawyer—must certainly have affected his commitment to socialism whose prime aim was to liquidate capitalism. Norman Manley's socialism, as described by his biographer and associate Philip Sherlock, was compatible with Rooseveltian liberalism, except for a vague call for a 'complete change' in social and economic conditions. Otherwise it promoted equality of opportunity, government-created opportunities to liberate the potential of the individual, Jamaican solutions to Jamaican problems, and parliamentary democracy, and rejected the domination of the individual by the state.[25] The early PNP left criticized Norman Manley's vague concept of 'socialism', his lack of analysis as to how to achieve it in a colony, and his presentation of the struggle against British imperialism as not in conflict with the basic interests of the Empire.[26] Under Bustamante's incessant and effective red smear tactics and the development of the Cold War, 'socialism' soon became a liability in Jamaican politics and international affairs. Thus in 1949, Norman Manley had to reassure his constituents in Eastern St Andrew that if the PNP got into power he had no plans to go beyond the nationalization of Kingston and St Andrew bus transportation, light and power, broadcasting, and the telephone service. Talk about nationalizing land and ordinary businesses and industry was 'absurd and untrue'.[27] The expulsion of the four H's in 1952 had obvious implications for PNP 'socialism'. In its successful 1955 campaign the PNP dropped its call for nationalization for fear of losing the financial backing of the rich and the votes of the middle class.[28] In 1959, Gordon Lewis wrote of 'The Decline and Fall of Manley's Socialism', the 'slow transition to a mild social-reform programme' influenced by Norman Manley's and the PNP's gradually developed intimacy with the local business class and by the need to create a climate of confidence for foreign investment.[29] Surveying the early PNP from his vantage point in the '80s, even Michael Manley admitted that it stood for 'more or less radical change from time to time'.[30]

The internal dynamics of the early PNP also illustrate its character. From its earliest days the party was internally divided into a liberal right wing and a more radical left wing consisting of Marxists and others. According to Richard Hart, a Marxist leader of the PNP left, Norman Manley's decision to declare the PNP a 'socialist party' in 1940 was never discussed with the left; it was completely his idea.[31] (A similar departure from internal party democracy occurred with Norman Manley's decision to hold a referendum on the West Indies Federation in 1961, a macho response to a Bustamante challenge.) Norman Manley admitted in 1940 that he had learned that Hart and Frank Hill were 'theoretic Communists', but he took no action because of the left's small numbers, because it was a useful spur 'for the mild and sometimes sleepy centre (right)', and because 'extremism in youth is a good thing,' dangerous only if 'denied outlet in action' and driven underground. The 'extreme left' also provided the hardest party workers, the best organizers and political educators.[32] The internal dynamics of the PNP was scrutinized by the United States Federal Bureau of Investigation (FBI). FBI director J. Edgar Hoover, a fanatic anticommunist who had framed Marcus Garvey on false charges of Communism (and later attempted the same thing against Martin Luther King) depicted the PNP left wing as including Noel Nethersole, V. L. Arnett, and the Hills, Frank and Ken. Hoover said the right wing included Norman Manley, C. A. McPherson, and Willis O. Isaacs. Hoover documented a heated argument in the party on 19 September 1942, in which Manley was attacked as a 'capitalist'. The left was said to want to elect Nethersole as President of the PNP and Frank Hill as Vice President. The elder Manley was accused by the left of not paying sufficient attention to the party's workings, of devoting most of his time to the Citizen's Emergency Council, of changing from an anti- to a pro-government policy, and of being 'inactive in the announced policy' of the PNP.[33] The four H's disagreed with the centre-right over the PNP policy of not harassing the British during World War II, as Gandhi did in India.

These internal dissensions ultimately culminated in the PNP split of 1952, which Michael Manley described much later as 'a fratricidal struggle . . . decisively won by the centre and right under [Norman] Manley's leadership.'[34] Born of long-standing conflicts, the split was accelerated by the growing middle- and upper-class membership of the PNP and Cold War pressures. According to Kaufman:

> The expulsion of the party's left in 1952 signalled an end to political struggle within the party, a struggle between the social democratic center, and a social democratic and Marxist left. The upper hand enjoyed by the party right and center in the PNP formative years now became a political monopoly. This monopoly secured a doctrinaire imposition of mainstream social democracy in a manner insensitive to the Jamaican reality.[35]

As we have seen, Michael Manley, as a young journalist recently returned from London, endorsed the expulsion of the left and the adoption of moderate social democracy in 1952. Whether Manley really attempted to move to 'democratic socialism' in the '70s is debatable, but if so, he would have had to struggle against a traditiion he helped, in a small way, to entrench in the party. In the '80s. Manley said the 1952 party fracture (and similar events in 1964 and 1965) was personally traumatic, and that when he became party leader he attempted to keep the left, with its energy, creativity, and grassroots organizing ability, in the party.[36]

A final crucial issue involving the early PNP is the nature of its commitment to 'democracy'. According to Beverley Manley, the early PNP was a multi-class alliance in which the interests of the petit bourgeoisie predominated over those of the majority classes of workers and peasants. Up to 1944, the party had no organizational link with either the peasantry or with most of the labour movement. On the other hand the structure of the party and its constitution was 'geared to an intense, internal democratic process,' although it was hard to keep party groups upon which democracy depended meeting consistently. Its advocacy of 'socialism' and universal adult suffrage laid the basis for a future 'potential ideology' for the majority. And, reflecting Jamaica's long, authoritarian colonial history, 'in practice Party decisions were taken largely without the involvement of group members . . . unless the leadership decided otherwise.'[37]

The relationship of the early PNP with the Jamaican majority—peasants, workers, the unemployed, small artisans and merchants—was strained. The black masses historically regarded the brown middle class, prominent in PNP leadership, with suspicion: 'Mr. Manley's upper middle class [origins] generated insecurities and doubts regarding his capacity and ability to successfully arouse the cooperation and support of the broad masses of the Jamaican people.'[38] The 'masses' were suspicious of Norman Manley from the outset because in his legal career he seemed to support 'the exploiters' one day and the exploited the next.[39] Other Manleys were aware of such problems. Michael Manley wrote to his father from London in 1944: 'I wonder if Jamaica's masses are completely ready for the PNP's particular brand of sanity.'[40] In 1965 Edna Manley recorded a conversation among working-class people: 'Three men were saying that the lawyer [in the unsuccessful defense of a murderer] had sold out the case—that all lawyers were PNP and that's why they always sold out the small men's cases!'[41] Another measure of the weakness of mass support for the early PNP was its inability to attract financing from the majority classes, leading the party to rely 'on individual members of the Capitalist Class for funding'.[42] Thus the early PNP's advocacy of 'democracy', while genuine, was somewhat theoretical in view of its tenuous ties with the majority classes which should play a leading role in a genuine democracy.

The early PNP should not be judged only on its own terms, or in relation to ideal democratic institutions. By comparison with the early JLP, it was clearly the more progressive and proto-democratic. The early JLP was 'strictly the electoral arm of the BITU'.[43] Its decline in the early '50s (and the corresponding rise of the PNP, to some extent) was attributed to 'a series of scandals in government' and the lack of a constructive programme.[44] Two early JLP Ministers of Government were convicted and imprisoned for graft and corruption.[45] And in the view of Gordon Lewis, the JLP's dictatorial chief, Bustamante, had 'massive defects': he never understood the democratic ideal; he was always against collective party leadership; 'he has his full share of the colonial mentality;' 'his temperament is messianic, living on mass emotion.'[46] Bustamante may also have diverted the PNP from its nationalist, 'socialist' mission by forcing it to deal with distractions such as his constant red-baiting and the fateful Federation referendum of 1961. Norman Manley staked his government's reputation on the

1961 referendum, which the pro-Federation PNP lost, leading to its narrow defeat in the 1962 general elections.[47]

The PNP in which Michael Manley became increasingly active in the '60s, and led from 1969 on, was thus a complex and unwieldy vehicle. It bore the heavy but enigmatic stamp of Manley's father. It vacillated in its commitment to 'socialism'. It suffered periodically from fratricidal internal conflicts, which did not end in 1952, but recurred in the '60s and '70s. Its devotion to 'democracy', real enough in ideal terms, suffered in practice because of historical legacies and contemporary social and economic inequalities. Even a West Indian hero could find such a party a problematic chariot to ride toward fundamental change.

CHAPTER 11

Joshua Awakened 1962-7

On 6 August 1962 Jamaica became an 'independent' nation. Shortly before, in the 1962 elections, the JLP defeated the PNP by 50 percent to 48.9 percent (winning twenty-six seats to the PNP's nineteen in the House), inaugurating ten years of JLP rule and political wilderness for the PNP. For Jamaica, the '60s were years of some accomplishment and considerable frustration. The economy grew strongly in the context of the world boom of that decade, but most Jamaicans received little from the boom. Violence rooted in poverty and inequality intensified in the political and trade union arenas. In the late '60s a radical cultural critique based on racial and anti-capitalist sentiments produced serious conflicts between alienated protesters and the establishment. During the tumultuous '60s, Michael Manley continued his leadership in trade unionism, but became ever more involved in partisan politics. He was appointed senator by his father in 1962, elected a member of the House of Representatives in 1967, became the President of the PNP in 1969 (replacing his father who died later that year), and began to plan the campaign which would lead to the PNP's resounding victory in 1972 and his own elevation to the office of Prime Minister. In a trade union action, the Jamaica Broadcasting Corporation (JBC) strike of 1964, he acquired the nickname 'Joshua', that symbolized his messianic appeal to many of Jamaica's oppressed. Trade union and political activities were ever more tightly linked in the '60s, and together they helped create 'Joshua'.

The Meaning of Independence

One of the PNP's historic goals was independence, which Bustamante had at first opposed. It was thus ironical that when independence came in August 1962, Bustamante and not Norman Manley became Jamaica's first Prime Minister. Although independence was an undeniably necessary step, it failed to end the enormous problems the island society faced. Thus gloom and doubt surrounded the event, which to many observers appeared almost hollow. Colonialism ended, to be replaced by neocolonialism. When he surveyed the West Indian scene in 1964, the experienced, radical Trinidadian intellectual C. L. R. James reckoned that its underlying reality was a clear drift 'towards reaction internally and a neo-colonialist relation with a great power, preferably the United States.' Whatever form new governments or parties might take, James observed, enduring features of Caribbean society would persist. Politics would be dominated by the facts that the entire population was expatriate, lacking the roots and enduring indigenous culture of formerly colonized Asia and Africa; that the old colonial system survived in that banks, agricultural estates, industries, newspapers, radio and TV, import and export firms were '90 percent in the hands of foreign firms. . . . or what are in essence their local representatives': and the sugar plantation continued to dictate the unhappy fate of many people and whole societies. Thus, 'this new system of independence is only the old colonial system writ large.' James chose as an example of how little politics had changed under independence Norman Manley's statement after the PNP defeats of 1961 and 1962 that the similarity of two parties is necessary for democracy. James argued that the main similarity of the PNP and JLP was that neither intended to challenge the old colonial order.[1]

Observers of varying persuasions have tended to agree. M. G. Smith, a Jamaican social scientist and close friend of the Manleys, described in 1961 the many ways in which élites could maintain control in Jamaica in the face of the majority's strong desire for rapid change.[2] Gordon Lewis noted in 1968 that the 'Manley forces in Jamaica' found themselves 'with a hollow victory in their hands', with their society 'still unnaturally bifurcated between a local political power and an expatriate economic power'.[3] Louis Lindsay argued that

> Jamaica can be used as a case study for illustrating how myths and symbols associated with independence have been manipulated to generate political quietism and frustrate possibilities for meaningful change in Third World countries.[4]

The conservative descendent of the Jamaican plantocracy, Morris Cargill, also believed that independence was a sham, that Jamaica became 'a synthetic country with a synthetic independence, without a balance in the bank'.[5] Toward the end of the decade, Michael Manley declared that 1962 was far less significant to Jamaicans than 1838, when slavery was abolished, or 1938, when labour protests shook the colonial establishment.[6]

The JLP Government, 1962–72

The JLP government of 1962 to 1972 was more dynamic than that of 1944 to 1955. Continuing the convergence of parties that had begun in the early '50s, the party took on much of the PNP's programme. Bustamante went into semi-retirement in 1964, leaving much of the day-to-day management of the government to his Deputy Prime Minister Donald Sangster. Sangster became Prime Minister after the JLP's victory in 1967, but died two months after the election and was replaced by Hugh Shearer. A rising star in the government was Edward Seaga, to become a deeply distrusted nemesis of the Manleys, father and son. Born in Boston of Lebanese-Jamaican parents, Seaga graduated from

Harvard with a bachelor's degree in sociology and studied Jamaican revivalist religions. This and his interest in the Jamaican popular music business, helped Seaga—despite the handicaps of white skin and non-membership in the Jamaican Bustamante-Manley-Shearer political 'family'—to establish deep roots in Jamaican society, particularly among the alienated ghetto youth of West Kingston. He served as Minister of Development and Welfare from 1962 to 1967 and as Minister of Finance and Planning from 1967 to 1972. He was responsible for the JLP's five-year plan of 1963, and brought 'brains' and conventional nationalism to the party. The JLP government adopted a strongly pro-Western and anti-communist foreign policy. It was hostile to revolutionary Cuba, but maintained consular relations with it. Although its foreign policy emphasized ties with the old metropolis, England, and the new one, the United States, the JLP did take the first faltering steps toward regionalism, joining the Latin American group in the United Nations, the Organization of American States, and the Economic Commission for Latin America. The JLP benefitted from the world economic boom of the '60s, but nonetheless had numerous economic troubles. Unemployment increased, food production decreased, inflation reached unprecedented levels, investment sources other than in bauxite dried up, and the business sector chafed under state regulation (and Seaga's economic nationalism and tax collecting). Additionally, the government was increasingly troubled by internal squabbling, as the physical decline of Bustamante opened a vacuum in JLP leadership ranks.[7]

The JLP government from 1962 to 1967 carried out a 'fierce struggle . . . to consolidate its power in a hostile environment' created by the PNP's previous tenure in office. This campaign included the mass sacking of casual labour in public service and pressure on the Jamaica Teachers' Association, the Jamaica Broadcasting Corporation, and the Police Federation, all pro-PNP.[8] The Kingston and St Andrew Corporation (KSAC), metropolitan Kingston's municipal government, had long been dominated by the PNP. The JLP national government dissolved it in 1964, recalling the 'evil habit' of British colonial governments dissolving the elected parochial boards which served as outlets for grassroots democracy.[9] Politics governed the JLP's awarding of government-funded 'Christmas work', a kind of patronage, in PNP-dominated Kingston in 1965. Crowds led by PNP politicians protested and were dispersed by tear-gas.[10] The JLP also engaged in gerrymandering so that in the 1967 elections it increased its seats in parliament although its overall support stayed the same. PNP infighting weakened its ability to resist.[11] C. L. R. James referred, in 1966, to 'the savage ferocity of some of the West Indian rulers today to the populations who have put them in power.'[12] In this atmosphere, Michael Manley as 'Joshua' was awakened.

The Origins of Joshua

Although he professed to be unconcerned with a political career for many years after his return to Jamaica in late 1951, Michael Manley was inevitably drawn to political leadership by his trade unionism, his family connections, and his personality, perhaps in that order. Trade unionism in Jamaica was enmeshed in the political system, though not identical with it and sometimes in conflict with it. But when a trade union leader's father is head of its allied political party and alternatively head of state and leader of the opposition, the ties are bound to be close and personal. All the Manleys were involved in politics. Against her inclination, Edna Manley performed many of the routine tasks of constituency work that Norman Manley did dutifully but without enthusiasm. Following the disastrous Federation referendum of 1961, Norman Manley turned to his sociologist son Douglas to investigate voting patterns.[13]

The twin defeats in 1961 and 1962 severely affected Norman Manley's health and his finances were precarious. The Manleys went along gamely with Jamaica's independence ceremonies on 6 August 1962, but the event had lost much of the joy it should have had for them and for the PNP generally. Norman and Edna were forced into new, more modest quarters, as Edna recorded forlornly in her diary on 3 September 1962: 'We have been in *Regardless* now for four months. Drumblair has been taken down and sold for old lumber. We have lost the election.'[14] Forced to suspend much of her art during the PNP's years in power, Edna Manley had a new creative burst after the political loss of 1962. She concentrated on trying to penetrate to the essence of subjects: 'Nothing really matters except to find a core of simple truth.'[15] Michael recovered from defeat rapidly, as Norman recorded him struggling 'like a lion' in a dangerous street demonstration, apparently one month after the 1962 general elections. Norman Manley also recorded that Michael was 'working extremely well and the news of his getting to grips with the angry young men at the university is excellent.'[16] Norman boosted Michael into politics by naming him to the appointive Senate, the lesser house of Jamaica's bicameral Parliament in April 1962, where his main job was to represent labour.

The lost election and deeper conflicts over ideology, strategy, and tactics caused deep rifts in the PNP. In August 1962, Wills O. Isaacs, the party's first Vice-President, and a long-term member of its right wing, resigned over the way in which Norman Manley made decisions in the party, specifically his selection of the PNP's members of the Senate, including Michael Manley. Isaacs had first found out about the appointments in the press.[17] Also in 1962, plans were announced to return the PNP to 'socialism and youth'. Norman Manley explained that the basic problem for the party was

> how to build our economy so as to create a society which offers the realities of equal opportunity to all our people, and offers an opportunity of decent

> Christian living to every man, woman and child in the land. . . . This is a job for socialism. The basis of socialism is equality among men.[18]

But 'socialism', when forced into concrete policy choices, was divisive in Jamaica and within the PNP as well, as events soon revealed.

In the meantime, Michael Manley became more and more involved in politics via trade unionism. In 1962 the NWU formally affiliated with the PNP, reversing the trend toward separation of union and party in the late '50s.[19] Manley as trade unionist had reached the status of national spokesman on industrial relations. His 1963 address, 'Economic Criteria by Which Arbitrators May Be Guided', attempted to define such things as the fair wage and major criteria for labour arbitration. As a trade unionist he functioned not to end capitalism, but to attempt to harmonize the valid interests of labour and capital.[20] Despite this moderate position, Manley was identified as being part of a group of 'social radicals' including PNP Senators Dudley Thompson and Ken McNeill. They advocated changing outmoded class and colour values, but this was a relatively safe position as opposed to, say, Allan Isaacs' advocacy of land reform.[21] The *Gleaner*'s 'political reporter', Ulric Simmonds, a veteran observer of Jamaican politics, noted a general air of longing in the speeches at the PNP's twenty-fifth anniversary banquet in September 1963.

> This wistfulness, this transference of the father image, was there in full measure in the fulsome praises which speaker after speaker, starting with Mr. Isaacs and going through everyone else but Mr. John Maxwell and Mr. Norman Manley, himself, showered on Mr. Michael Manley, the Island Supervisor of the NWU. . . . The roar after roar with which everything that was said about the young Manley was greeted by the large gathering was to my mind a further indication of the wistfulness that was so noticeable in the speeches of Mr. Isaacs, Mr. Glasspole, and Mrs. Dalton-James.

The political reporter noted Jamaican politics' heavy reliance on charisma, and the irony that without it the country's two-party 'democracy' might never have been created. He wondered whether it was the PNP or the Jamaican people (evidently he excluded the JLP automatically) that was responsible for 'acceptance of the proposition that the great mass of people must be led and not brought into the stream of things.'[22]

The year 1964 was crucial for Michael Manley, involving a momentous strike against the Jamaica Broadcasting Corporation (JBC) and an intense debate in the PNP over 'socialism'. Manley described the JBC strike as a turning point in his life.[23] He explained that the JBC, created in 1957 on the model of the Canadian Broadcasting Corporation, was 'very much my father's brainchild'. Thus the personal blended with the public in what ensued. The JBC 'was supposed to provide an outlet for Jamaican culture, a forum for the discussion of public affairs and an additional means of public education.' It was also intended to be politically independent, something difficult to achieve in any society, and especially in the super-heated politics of Jamaica. Michael Manley maintains that during the PNP tenure of 1955 to 1962, his father's government was often severely attacked by JBC commentators, especially John Maxwell, but Norman Manley resisted PNP pressure to retaliate against the JBC or its workers.[24]

Within a year after it came to power, however, the JLP government began to attack the JBC's independence. A new JBC board was created, under the control of the Minister of Culture and Development, Edward Seaga. To protect themselves, JBC workers turned to the NWU, which had already organized RJR, Jamaica's private-enterprise, light-entertainment radio station. Negotiations over wages, working conditions, leave, and pensions began. In the midst of this Adrian Rodway and George Lee (the NWU delegate of the news room staff, but an avowed JLP member) were fired. The NWU's investigation showed that Rodway was guilty of a technical violation of news reporting procedures, but that Lee was guilty of nothing. The union's plea that the two be reinstated (Rodway receiving a reprimand for his technical lapse) was rejected. The entire matter was saturated with political considerations: the relationship between a government and public broadcasting, and employer-employee relations where the employer is, in effect, the government, and the employee is represented by an anti-government union.

The JBC strike began on 1 February 1964. Since the JBC was not economically vital nor even crucial to the dissemination of news in the island, Manley decided that the strike would have to be waged on the basis of public awareness of the injustice of Rodway's and Lee's firing. Picketing and meals for picketers were supported by a strike fund raised by public contributions, and even some branches of the BITU contributed. JBC commentators boycotted its radio and TV stations. As the strike wore on, some strikers favoured violence, but Manley resisted it except as 'an absolute last resort' to be used 'in the kind of absolute intransigence of situations like that in Southern Africa.' Instead, in early March the strikers began non-violent civil disobedience in the style of Gandhi and Martin Luther King. Manley joined the strikers in lying down and blocking major streets in Kingston with their bodies. Although he really didn't have to join this risky effort, as John Maxwell noted, Manley 'faced the tear gas and the danger of arrest, [and] totally committed himself to a struggle which most of us knew could not be won, but which he realized in the depths of his soul was necessary to fight: win, lose or draw.'[25]

Eventually a board of inquiry was held, with the NWU calling witnesses 'from all over the world'. After ninety-seven days the strike ended with eighty strikers returning to work. Some thirty had been declared redundant and were awarded J$16,000 in severance pay. The board of inquiry found that Adrian Rodway had been justifiably fired, but that George Lee was innocent of all charges and entitled to compensation. The JBC offered him J$150, but was forced to pay him J$4,000 when Norman Manley discovered the basis for a libel suit against the JBC. JBC, Manley said, lost its credibility for many years. In the end, he admitted that 'we lost the JBC strike in the sense of the immediate issues that were involved. . . . We could not win because, short of revolution, you cannot make an elected government do what it does not want to do unless you are able to bring tremendous economic pressure to bear on them.'[26] Manley hastened to add that

> the strike was a unique experience for all Jamaicans, provoking intense controversy, idealistic involvement and contributing to the evolving self-image of a young nation.

It was in the strike that Manley acquired his most enduring nickname. In one demonstration, he faced the JBC building, and stated: 'There are the walls of Jericho!' From the crowd, someone called out to him: 'Joshua'. The name stuck.[27]

Another significant controversy developed in the PNP over 'socialism' in 1964. A new generation had developed on the party's left, organized as the Young Socialist League (YSL). YSL was Marxist-Leninist, neo-Trotskyite, and anti-Stalinist. It wanted the 'rapid and immediate nationalization of everything in Jamaica', a position with some support among the second level of PNP leadership. At its second congress, YSL's Bobby Hill argued that it should not work with either the PNP or the JLP. Trevor Munroe, who would later establish his own trade union and the communist Workers' Party of Jamaica, argued at the time that failing to work with the PNP would cut YSL off from labour. YSL voted to work within the PNP. The lawyer Hugh Small was prominent in its leadership, and he, in turn, wanted to jettison the top leadership of the party. The young socialist left was a threat to much of the older core of the party, which had become comfortable and respectable in recent years. Facing the left and right wings and trying to hold them together were Norman Manley, Glasspole, Wills Isaacs, and Michael Manley who were said to 'realize the country could founder on an "out and out socialist policy".'[28]

The PNP's socialist programme was announced in November. It was the result of a policy advisory committee chaired by David Coore, and including Vernon Arnett, Allan Isaacs, Dudley Thompson, Dr Ivan Lloyd, and Michael Manley as members; Coore, Arnett, Isaacs and ex-officio member P. J. Patterson did most of the work. The programme announced that independence 'has brought no improvement in the vast majority of our population. . . . The political game continues but seems to lose more and more meaning for all except those who are actually engaged in playing it.' The programme highlighted unemployment, income inequality, problems with foreign investment, shortcomings in education and social services. It noted that 'the present conservative posture in which the PNP finds itself is alien to the nature of the party itself' since it 'has always been in essence a revolutionary party.' It went on to spell out a democratic socialist policy for Jamaica based on the premise that 'Socialism is about equality.' Equality did not mean uniformity, dictatorship of any kind (including 'dictatorship of the proletariat'), or equality in misery and poverty. Equality meant a society providing a decent and happy life based on recognized twentieth century material and social standards, a society in which the accident of birth conferred no special advantages; a society in which everyone would have the opportunity to develop their talents to the fullest; a society encouraging everyone's direct participation in the life of the country; and a society in which laws were designed to prevent, rather than protect, inequality, to inhibit the domination of any class or group over others, and to prevent the arbitrary waste and selfish exploitation of Jamaica's human and material resources.[29]

Reaction to the programme was mixed. The *Gleaner*'s political reporter wrote that the PNP's adoption of 'democratic socialism' was bold and honest: referring to the whole history of the PNP, he reckoned that 'the PNP has acted honourably in its political approach to the people of Jamaica.' *Gleaner* columnist 'Thomas Wright' (Cargill), however, criticized the PNP programme on the basis that 'there is . . . no difference whatever between socialism and communism except in the methods used to achieve them.'[30]

Inside the PNP, the 'socialism' controversy was fought out largely on the battle ground of the NWU. Handbills circulated by the 'Unemployed Workers Council' alleged that £50,000 in union funds had been misappropriated. Florizel Glasspole blamed the charges on 'anarchists with headquarters in West Kingston' and 'Communists . . . fed by money from Red China.' NWU president Thossy Kelly and Michael Manley both denied the charges and said that an independent international accounting firm was being brought in to audit NWU's books. During the controversy Hugh Small was dismissed as NWU legal officer for activities and statements 'not in the interests of the union'; Kelly invited Small to address an NWU meeting but he decided it was unsafe to do so. Michael Manley was accused personally of mishandling the funds. He said the talk about him personally amused him: reportedly, 'he said that all the men of the Manley family liked fast cars and some of them like beautiful women' but 'no member of the Manley family has ever meddled with any other man's money.' The younger Manley spoke of 'treachery within our own ranks', and mentioned that two NWU officers had gone to noted columnists with stories about him. It was also alleged that the PNP had given to the NWU £50,000 to make up its funds; Norman Manley denied this as a 'scandalous lie'.[31] In late 1964 and early 1965 PNP and NWU members associated with YSL were either expelled or made to leave YSL. In the late '60s, YSL members supported the 'black power' newspaper *Abeng*, which was highly critical of Michael Manley. As in 1952, Manley cast his lot in 1964–5 with the centre-right of the PNP rather than its left-wing insurgents.

Internal bickering within the PNP and NWU broke out again in 1966, when Dennis Daly, a barrister, was expelled from the PNP and its National Executive Council. He had said the NWU couldn't effectively represent the sugar workers at Frome because of its PNP connection. In violation of the PNP policy of affiliation with the NWU, Daly had attempted to organize the Workers Liberation Union at Frome.[32]

A fundamental problem was the inherent contradiction between trade unionism and 'democratic socialism'. Following a trip to Israel in September 1965, Norman Manley observed: 'I am convinced that what is wrong with most Democratic Socialist countries is that the Trade Union Movement is entirely out of step with the objectives of the Socialist Government in a poor and developing country.'[33] No doubt the elder Manley shared these views with his son, who wrote later:

> The union movement is a product of the capitalist system, and more often than not, represents no more than an attempt by the workers to mitigate its worst effects. Unions, therefore, do not necessarily challenge the system itself. While they may secure an increasing share of national wealth for their members they often act as fundamental props to capitalism because, accepting capitalist values, they divert attention from the fundamental shortcomings and inequities of what is, for all practical purposes, a morally bankrupt way of life.[34]

Given the political importance of the NWU to the PNP, this realization of the conflict between trade unionism and 'socialism' was basic. Equally fundamental, given the tendency to equate socialism with communism in Jamaica and elsewhere, was the likely opposition of even the most 'liberal' and well-educated presidents of the United States, such as John F. Kennedy, to 'democratic socialism'. Norman Manley had been the only statesman to see Kennedy on the date of the Bay of Pigs invasion of Cuba, 19 April 1961, and had been impressed by his self-control and inner strength. He called Kennedy a 'very modern man' 'who has freed himself from old attitudes and preconceptions,' and who 'fully' understood the world, with this exception:

> I'm afraid from a political point of view that not even the most modern of American Presidents can see communism in terms of intelligent modern thinking.[35]

Democratic socialism and the internal and external obstacles to its realization, however, were soon overshadowed by the West Kingston Wars of 1965–7 and the elections of the latter year.

CHAPTER 12

Joshua Prepares 1967-72

The West Kingston Wars and the Election of 1967

Political 'tribalism' between the PNP and the JLP has played a central role in Jamaican politics, and violence has been a tragic element of it. The 1960s were marred by repeated outbreaks of politically-related violence: the 'Henry rebellion' of 1960, the Coral Gardens 'uprising' of 1963, anti-Chinese riots in 1965, political warfare in West Kingston from 1965 to 1967, and the Rodney riots of 1968.[1] Of these events, the West Kingston 'Wars' were mostly deeply and tragically enmeshed in the political system, revealing the credibility gap that existed (and exists) between the constitutional theory of Jamaican 'democracy' and political practice.

The West Kingston 'Wars' began in 1965, with the JLP trying to fight its way onto the revived PNP-dominated KSAC via the West Kingston constituency held by Edward Seaga. Prior to 1962, West Kingston had alternated between the parties, owing to their nearly equal strength there; since 1962 West Kingston has been continuously represented by Seaga, often reelected by enormously one-sided votes. A fatal aspect of West Kingston was the existence of gangs of unemployed youth. Barry Chevannes, a university research fellow and participant observer of the urban youth culture, described how the West Kingston gangs became politicized after 1962:

> What brought it about was the attempt by members of the ruling party [the JLP] to tackle the social dislocation by programmes aimed especially at organising the youths. In fact the constituency representative [Seaga] held the cabinet post of Minister of Community Development and Welfare. Thus it was that the Park gang formed a club called Wellington United and became affiliated to the Youth Development Agency. But if affiliation brought rewards such as greater access to sports gear and equipment, it was not without its costs. Severe pressure was placed on them also to affiliate with the ruling party. . . . so that YDA affiliation was soon regarded as tantamount to JLP affiliation. What this meant for the Park men was that they could no longer visit the Vikings by the seaside, for although the latter were dreadlocks [Rastafarians] they were still favourable to the party which once had a socialist tradition [i.e. the PNP], and moreover as dreadlocks they scorned the subservience to party politics. To make matters worse even the people from Back-o-Wall through which they had to pass to reach the seaside began to refuse Park access, although Back-o-Wall did not really have a gang, as such!

> And so inter-gang hostilities became at the same time inter-party hostilities.[2]

Tensions heightened considerably in June 1966 when the JLP government bulldozed the Back-o-Wall, Ackee Walk and Foreshore Road slums: 'The viciousness with which it was executed led to the belief that it was being used to wipe out opposition, and even the churches raised a voice of protest'.[3]

The JLP's motives for the bulldozing and reconstruction of West Kingston were both political and economic, the latter being to develop the Newport West industrial and shipping area. The bulldozing wiped out five polling divisions in a heavily Rastafarian area of West Kingston which leaned toward the PNP. Dudley Thompson, the PNP's candidate for West Kingston in the forthcoming 1967 elections attempted to capitalize on the resulting anti-JLP sentiment. Gangs connected to the JLP and the PNP began to attack each other with guns and Molotov cocktails. In mid–1966, Norman Manley accused the government of fomenting violence in West Kingston 'to confuse the public mind and to divert attention from the moral shame and wickedness of the conduct of the Government towards the poor squatters'. He referred to a white politician in West Kingston 'whose origins and methods do not relate themselves to our country, its history or background'.[4] For his part Seaga 'named the PNP as the instigators of the violence in the Western section of the city and stated that the aim was the overthrow of the country starting in West Kingston'.[5] The political climate was further enflamed by advertisements in print and on the broadcasting media.

> Hear Senator Michael Manley expose the shocking truth about the Maffesanti Strike and poll in Discovery Bay. RJR 2:15 p.m. Today. We Shall Overcome![6]

Seaga spoke on the JBC on 21 August 1966 on 'The Cancer in West Kingston' which needed to be cut out with a knife.[7]

Violence was deeply rooted in Jamaican history and culture, and political leaders and gangs affiliated with both parties escalated it. There is some evidence that the greater blame belongs with the JLP. The *Gleaner*, generally biased against the PNP, reported that Wills Isaacs of the PNP had asked the JLP for a meeting on the violence and was at first refused, though the meeting subsequently did take place. A *Gleaner* editorial criticized a TV speech by Seaga for including 'unfounded statements' which heightened the hysteria. On 24 August 1966, 'Thomas Wright' (Cargill), a columnist with a long history of antipathy to the PNP, wrote that he was not surprised that people called the JBC 'Seaga's Broadcasting Corporation'.

> As to Mr. Seaga, if he had sat up all night thinking about how to make a

> statement that would be least helpful and least appropriate he couldn't have done better. . . . Mr. Seaga made a number of statements for which he offered not a vestige of proof. . . . If the PNP has been doing all that he says it has been doing, how come the PNP constituency office on Regent Street is so often and so viciously attacked?[8]

On 27 August, the Farquharson Institute of Public Affairs invited both Seaga and Dudley Thompson, the central figures in the 'war', to issue a joint statement denouncing violence. Thompson accepted, but Seaga declined, saying he was the elected Member of Parliament of West Kingston and had made his views known many times.[9] The next day the chairman of the Farquharson Institute, the director of the Jamaica Employers' Federation, and the presidents of the Jamaica Chamber of Commerce and the incorporated Law Society of Jamaica announced that they deplored the violence and the failure of both parties to control it. They specifically demanded that the government inform the public whether Seaga had passed on to the police information he apparently possessed on the importation of arms by the PNP; the announcement tended to support allegations about Seaga's evident penchant for making inflammatory, undocumented statements.[10]

Efforts to organize peace talks involving Thompson and Isaacs of the PNP and Seaga and Sangster of the JLP broke down. In October 1966 the government declared a State of Emergency and mobilized the Jamaican National Reserve. The State of Emergency ended after three weeks, but violence continued sporadically throughout late 1966 and early 1967. In its campaign for the 1967 elections, the PNP promised to end gang warfare and the alleged defiance of the police by JLP officials. In February 1967 Norman Manley organized a 'mission of peace' including Thompson, TUC leader Hopeton Caven, and Ken Hill, whose names had apparently 'been coupled with allegations of violence'. This mission was shot at by hostile forces, according to Norman Manley.[11] Michael Manley later wrote that the 'principal beneficiary' of the West Kingston Wars was Edward Seaga.[12]

Manley's First Campaign

Manley's first campaign for elective office, his bid for a House of Representatives seat in 1967, took place against the background of the West Kingston Wars. His third wife, the former Barbara LeWars, played a major role in persuading him to run. During their courtship and following their marriage in 1966, Manley and Barbara had long talks about his political possibilities, and she helped convince him that he could make an important contribution and overcome the burden (as he saw it) of his father's awesome reputation. According to a PNP poll in the Central Kingston (later East Central Kingston) constituency, Wills Isaacs, an important party figure and vice-president, would probably lose. Party leaders decided to run Manley for the Central Kingston seat and Isaacs for a seat in St Ann. After an all-night talk with

With world leaders including Pierre Trudeau, Harold Wilson, Archbishop Makarios and Indira Gandhi at Commonwealth Heads of Government, 1976.

With President Kaunda in 1976.

With Helmut Schmidt and officer from th
Ministry of Foreign Affairs
at Runaway Bay Conference, Jamaica.

On parade with Fidel Castro, Kurt Waldheim
and Maurice Bishop in Havana, Cuba, 1979.

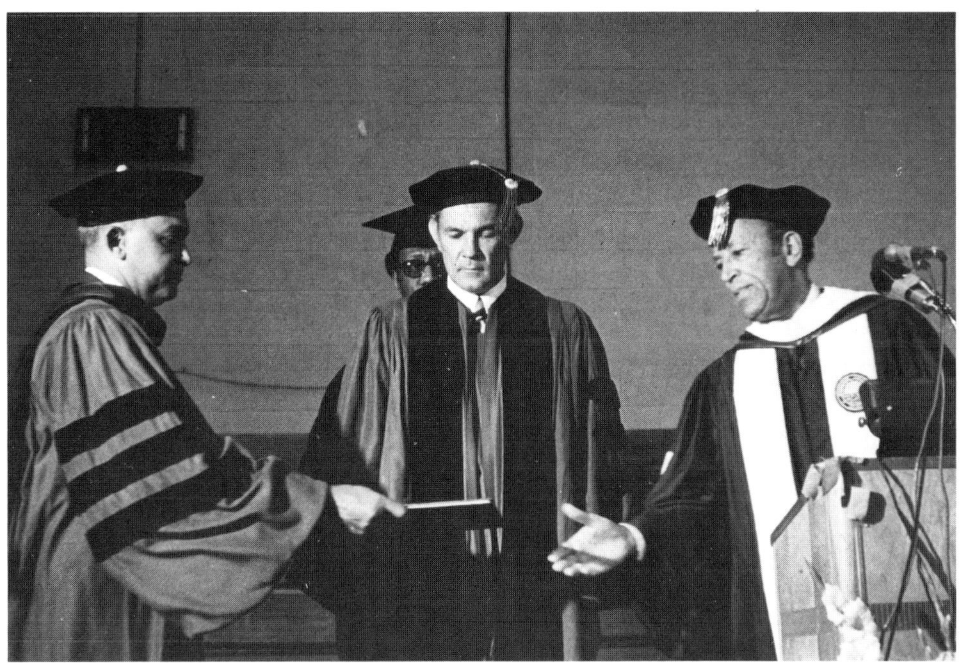

Collecting his Honorary Doctorate of Law at Morehouse College, Atlanta, 1973.

Beverley Manley (pregnant with Natasha), Captain Mark Phillips and Princess Anne with the Prime Minister at Jamaica House, 1973.

The official visit to Mexico in 1975. With the Manleys, President and Mrs Echeverria. *below* with Carlos Perez.

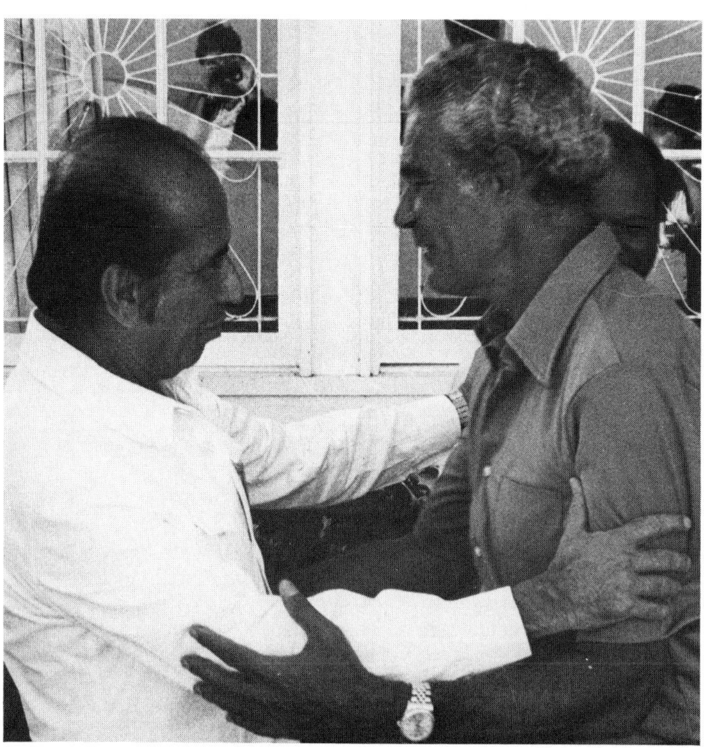

Barbara, Manley decided to accept the party's offer. As he recalls, the decision was made only five days before nomination day and about three weeks before the election.[13]

As a prominent PNP leader Manley was privy to discussions on political violence and its remedy. Naturally enough, he was convinced that the JLP had started the violence and that the PNP retaliated in self-defense. He came under pressure from his own Central Kingston grassroots supporters to show his machismo and supply them with guns. According to Manley, he shamed his followers into pursuing a non-violent path in his constituency, and worked with the police to avoid violence.[14] The PNP made a special effort for Manley, whose opponent was the JLP's E. K. Powell. The campaign was suffused with popular religion, tailor-made for the emerging 'Joshua'. The intellectual, Oxford-trained barrister, Norman Manley, made one broadcast entitled 'Praise the Lord for the Day of Deliverance Has Come.' The PNP claimed the song 'We Shall Overcome' from the United States civil rights movement. In mass meetings, Michael Manley said that rich people had captured the JLP and were behind the violence in lower-class areas, which Busta never would have condoned. Manley was elected to the House of Representatives by the narrow margin of forty-three out of ten thousand votes. Manley might have lost if Powell had been better organized or had he won the unsuccessful court case he brought, contesting the election.[15] Had Manley lost the election, it is unlikely he would have gone on to lead the PNP in 1969 and Jamaica in 1972.

These were hard years for Manley personally. Around 1965, he developed a second ulcer. His marriage to Barbara turned out to be tragically brief. Their daughter, Sarah—Manley's third child—was born in April 1967. In 1968, Barbara died of cancer, an event that almost knocked Manley out of politics. He says that only Carlyle Dunkley's persistent encouragement kept him involved in politics after this tragedy. Even fifteen years later, Manley could hardly speak of Barbara's death.[16]

Joshua Prepares to Lead, 1967–72

Following his election to the House of Representatives, Manley entered into a period of preparation for the leadership of his party and country. Though during the years 1967–72 he continued to be much involved in union affairs, he broadened his interests and sought to reach out to the non-unionized majority of Jamaicans.

During this time a conflict developed which Manley described as the 'most terrible' of his union career. It originated in agreements made by the JLP government in 1967; the Kaiser, Reynolds, and Anaconda corporations were to build a huge aluminium plant on the south coast. Building the plant, to be run by Aluminum Partners of Jamaica (Alpart), would take three years and would employ more than four thousand workers. The BITU decided to make the construction project a battle for control of the bauxite workers, hitherto controlled by the NWU. Pearnel Charles, a young Jamaican graduate of the City College of New York, was chosen to lead the BITU's efforts, and, in Manley's words, 'set about it with commendable energy.'[17] After Manley addressed a lunchtime workers' meeting in the midst of the NWU-BITU struggle he experienced 'one of the ugliest incidents' of his life. According to his account, as he was driving away from the meeting, Manley witnessed Pearnel Charles in an argument with NWU delegates. According to Manley's version, Charles swung first on one of the delegates, Clinton 'Big Cookie' Cooke. Cooke hit Charles, while Manley yelled at him to stop. Charles subsequently drew his gun and started firing, wounding Cooke and another NWU worker, George Whynne. Charles then ran to a waiting BITU car, which sped away while more shots were fired in the midst of tear-gas fired in turn by the police on duty at the work-site.[18] Subsequently, Manley wrote, the BITU 'moved a number of notorious Kingston gunmen on to the site with the connivance of certain of the sub-contractors.' The BITU called a strike at the site and shut it down. As Manley saw it 'the scene had been carefully set to make the strike appear to be a personal rejection of my leadership.' He decided to challenge the strike on the site. After 'an emotional account of how the bauxite workers had fought over the years,' Manley 'started to walk towards the plant gate where the BITU gunmen were waiting with their strike placards.' He plunged through the picket line and found that a 'multitude', even including the BITU group, had followed him, breaking the strike. Manley called this 'perhaps the most dramatic victory of my career'.

In 1968, Manley continued to coordinate union and political affairs closely with his father, as in ALCOA's recruitment of workers, which the company planned to do apolitically, 'based on testing without fear or favour.'[19] True to its history, NWU matters did not always run smoothly. During its sixteenth annual conference in December, 1968, a fight broke out between the rival candidates for third Vice-President, Caswell Rodney, the area supervisor for bauxite, and Fred Clarke, a veteran Kingston organizer. Fists were drawn and chairs and bottles flew, leaving the meeting room—a schoolroom in the Vauxhall Senior Modern School—a shambles.[20] Many have noted the 'tribalism' that existed between the JLP and PNP, but there was also a less obvious mini-'tribalism' existing within such institutions as the PNP and NWU that reduced their effectiveness.

Based in the somewhat turbulent NWU, Manley broadened his appeal by seizing the initiative from the JLP as the 'party or redistribution'. In 1961 Seaga had attacked the PNP for favouring the 'haves' over the 'have-nots'. In June 1968 Manley turned the tables by making his own haves/have nots speech, this time blaming the JLP for the markedly unequal distribution of income in Jamaica. In an address on Roger Mais, Manley argued that the writer's struggle for human rights was waged against colonialism and that 'human rights and poverty are mutually exclusive.' He said that poverty continued to defeat hope and undermine faith, that the classes remained 'essentially myopic about each other', and that in many ways human rights were consciously subverted in Jamaica.[21] Thirty percent unemployment and eighty percent of Jamaican workers earning less than $J20 per week[22], lent credibility to Manley's words.

The Rodney Affair of 1968 crystallized much latent conflict in the society. Walter Rodney, known for his radical black power views, was a young Guyanese historian teaching at Jamaica's branch of UWI. The conservative JLP government refused to readmit Rodney to Jamaica after he returned from a lecture tour in Canada. Students and faculty took to the streets, made some temporary alliances with the working class, and were harshly repressed by the police. Arson and looting resulted in one million pounds (sterling) of property damage and three deaths. On 18 October the PNP walked out of the House of Representatives to protest the JLP administration's actions. Norman Manley—then seventy-five years old and nearing the end of his political career—at first defended the protesters and embraced the concept of black power. But when Shearer broadcast allegations that the government would possibly be overthrown, Norman Manley supported the regime. The Rodney Affair helped organize opposition to the JLP, especially in the form of the *Abeng* group and the Sixth Form Association, both of which contributed members to the PNP Youth Organization (PNPYO) in the 1970s. The Rodney matter reinforced the 'anti-intellectual' image of the JLP. It also frightened the middle classes into believing in the necessity to overthrow the conservative JLP in order to prevent revolution. Finally, it legitimized 'black power' and brought it within the PNP embrace as a potential source of support for Michael Manley, although he had yet to win the support of some black power advocates.[23]

In February 1969 Manley succeeded his father as President of the PNP, becoming only the second leader of the party to date (1989). He defeated moderate lawyer Vivian Blake by the surprisingly wide margin of 376 to 155 votes. In his acceptance speech, Manley emphasized equality, social justice, the problems of youth, crime, victimization, unemployment, the 'scandal of the electoral system', corruption in high places, the need for integrity in public life, and the 'arrogance of the leadership of the Jamaica Labour Party'.[24] The *Gleaner*'s political reporter revealed that there were PNP supporters who did not want a Manley to succeed another Manley as party president. But the political reporter argued that Norman Manley was 'a pragmatist who could not really care whether or not the party had selected Mr Blake over his son,' and perhaps believed—as the political reporter did—that Michael Manley was best equipped to lead the PNP during difficult times because of his union background and his charisma.[25] The radical black power intellectuals who clustered around the journal *Abeng* and who claimed to speak for Jamaica's poor majority of 'sufferers' reacted differently. *Abeng* called the choice between Manley and Blake a choice between a 'Black Dog and Monkey', meaning that there was not much difference between them. *Abeng* editorialized that

> among sufferers the feeling is that the new Leader of Her Majesty's Loyal Opposition is a danger to them. He represents the people who own Jamaica and who therefore control its politics.[26]

Eventually Manley won the confidence of some *Abeng* participants, but others retained deep suspicion that he represented the 'brown establishment' and the PNP-JLP political hegemony.

One of Manley's first acts as PNP President was to call for an independent crime commission to investigate violence and law and order in Jamaica. Prime Minister Shearer responded by calling on Manley to 'stop preaching civil disobedience and support the police.' Shearer accused the PNP of meeting with subversive groups and said it should stop making charges of police brutality. The PNP accused the JLP of starting political violence back in 1944. The JLP responded by exposing 'the PNP record of violence.'[27] This sad ritual would be repeated many times, while violence escalated during subsequent years.

In his budget speech of 1969—his commentary on the government's annual budget—Manley elaborated his views as the new Leader of the Opposition. He opened with the theme of the disappointments following Jamaica's independence. While admitting the success of tourism, high-rise apartment construction, and public acquisition of some corporate shares, Manley emphasized high unemployment, the 'unparalleled' alienation of youth, and the collapse of rural life. He said that the role of the politician must be not only to expose misery but also to elaborate strategies to banish it. He listed five elements of a just society: the search for equality, which widespread unemployment denied: national independence, which meant wresting the 'commanding heights of the economy' from foreign control; the preservation of democracy, threatened by gerrymandering and the exclusion of three hundred thousand eligible adults from voting; the politics of participation; and a sense of national identity. Manley, himself fair-skinned and privileged, noted that

> it is a tragedy of our history that the masses are predominantly black and the privileged classes predominantly fair-skinned. . . . We call on Jamaica to assault the economic system that perpetuates disadvantages and so feeds the delusion that race is the enemy, when poverty is the true obstacle to overcome.

Manley identified the credibility of Jamaica's system as the biggest political problem, and announced that he was opposed to violent revolution or any form of political violence.[28]

On 2 September 1969 Manley suffered an enormous loss when

Norman Washington Manley died. Manley saw his father frequently during his last days, and believes that he died more of boredom than any specific medical malady.[29] At the time Manley spoke to the PNP of 'almost unendurable sadness...because we have lost our ultimate and definitive moral inspiration.'[30] A controversial figure in Jamaican and West Indian history, Norman Manley nonetheless was one of its giants. His biographer, Philip Sherlock, assessed Norman Manley's contribution thus:

> Philosopher and statesman, he had a unity of vision that gave coherence to his many activities. He was excited and inspired by the task of creating a nation, and in doing this he taught us all, who were his countrymen, how to link thought with action, vision with one's everyday work, how to feed idealism into the nation's affairs. He demonstrated how power could be used to promise good government; and he insisted that if power were not linked with morality it would corrupt and destroy the society. It was this union of power and morality, of authority with personal integrity, that was most vital in a democracy, for democracy was a sham unless it involved the use of power to protect and promote the common good.[31]

Invoking the memory of Norman Manley, Michael addressed the PNP's annual conference in October 1969 for the first time as the party's President. He made economic nationalism the centrepiece of the PNP's policy, while reemphasizing many of the themes made in his budget speech earlier that year. He announced that Jamaica's poor were his 'true constituency'.[32] At the end of the year the *Gleaner*'s political reporter selected Manley as his 'political man of the year' over such rivals as Shearer, Seaga, Blake and JLP leader Robert Lightbourne. The political reporter chose Manley because of his superb and persuasive verbal skills ('even when he is saying the most illogical of things'), his revival of the PNP, his dissociating the party from 'nationalization for nationalization's sake', his giving Jamaican politics new interest and new horizons, and his defusing of the 'black power' discontent. The political reporter also attempted to answer the question of who Manley was: 'It is a very strange thing but a lot of people don't know.' He identified Manley's trade unionism as crucial. The political reporter also noted his family's influence:

> There is little doubt that he inherited in greater degree than his brother, Douglas, an intense family pride. His father was of course his hero: his mother, his heroine. . . . Even from [his days in the RCAF] he was deliberately trying to pattern his public life on his father's.

The political reporter added this personal sketch:

> Mr. Michael Manley is definitely tough, a fighter, a hard worker and an egotist. He believes in himself. He believes in his ability. He is intellectually aware. And like Canada's Pierre Trudeau and Jamaica's Hugh Shearer, he

> is a romantic figure, although thrice married. He has therefore all the attributes of leadership in the modern world.[33]

'Leadership in the modern world' involved a variety of experiences, as Manley's personal correspondence files reveal. In January 1970 he received a letter from an unblushing admirer of 'free enterprise', who sent him two speeches, including one by Ronald Reagan on the 'hope, value, and goodness' of free enterprise. She imagined that 'already . . . you are a little like Reagan,' a judgment that would not stand the test of time. In March, Manley wrote to a party worker in Falmouth about a woman in Duanville who had written for help: 'It might be a good idea if you paid her a special visit before the election telling her that I had written to you about her matter.' Also in March, he had to conjure up a reply to the following from a man in August Town:

> Mr. Michael Manley
> Sir,
> Send the folowing words to Mr. P—— G—— that Black Supremacy now call up on him to erect an alter in that church where he is as one of the Church of England church and send Mr. J. N. H—— to St. Ann Browns Town to build one in the church of England there hopeing that you will take theis words as laws from the governor general now in charge or else you can't have a seat.

(Manley's reply: 'Thank you for your letter. . . . The points you made have been noted and I thank you for your interest.')[34]

As a new national leader, Manley's views on foreign policy had become important. In a personal and confidential letter in November 1969 he outlined his views. He regarded the creation of a foreign policy as perhaps 'the most important consequence of independence'. He believed that Jamaica should have normal diplomatic relations with Cuba, and he rejected a simplistic good guys/bad guys approach to international relations. Manley made clear that 'I do not accept Castro's method of political organization nor even all of his concepts of economic organization.' On the other hand Cuba represented 'one of the most exciting areas of experimentation in the world today' in such areas as education, youth, and the use of culture to create an egalitarian society. Manley admitted to being 'an unreconstructed and unrepentant Federationist who feels that the Jamaican electorate were bamboozled into a stupid, but I hope not fatal, mistake when they voted against Federation. . . .' He wrote that Jamaica's role in the Commonwealth Caribbean had been 'obstructionist, ambivalent, and confusing.' In the OAS Jamaica should be a force for stronger dealings in trade with the metropolitan countries and a voice for 'democratic Freedom in Latin America and against American domination.' He supported Cuba's reentry into the OAS, 'provided Cuba ceased trying to export her revolution to other countries.' Finally he said that he agreed with Norman Manley's view that 'Jamaica is of the West,' but he found

Bustamante's offer of military bases to the United States shortly after independence 'unnecessary and inimical to Jamaica's interest.'[35]

Manley's 1970 article in the prestigious journal *Foreign Affairs* addressed primarily the economics of international relations. He urged that economics was more important than ideology in international affairs. He argued that the common features of the Third World were nationalism; the adverse terms of trade between it and the industrialized, developed countries; the need for foreign capital, which conflicted with nationalism and independence; and a diplomacy based on regional economic concerns. He stated that Jamaica needed to overcome its insular tradition and to restructure its economy, in the context of a coordinated pan-Caribbean (and eventually Third World) economic strategy. The choice facing Jamaica, Manley concluded, was between 'a low road of self-imposed, insular impotence and a high road of adventure in Caribbean regionalism leading on to the wider possibilities of third-world strength.'[36] In contrast to his private letter in 1969, Manley's *Foreign Affairs* article contained no reference to Cuba or the USSR, but the issues of adverse terms of trade and Third World solidarity were potentially as controversial.

In Jamaica, foreign policy is a significant part of domestic politics. Black pride, 'black power', and Rastafarianism gave many Jamaicans an intense interest in Africa. Both parties attempted to capitalize on that. Manley was in touch with Rastafarian groups, including the Ethiopian Orthodox Church.[37] Fortuitously, Douglas Manley was in Ethiopia with the United Nations, whence he wrote in February 1970 that 'Our local Rastas have been in touch again. Apparently Shearer's visit caused a slight flurry of activity and then nothing.' Douglas sent Manley information on agricultural schemes and community development in Ethiopia. Michael wrote back that 'a Rasta group that I get on very well with' had asked him to inquire about a professor of Amharic who could be sent to Jamaica at the expense of the Ethiopian government.[38]

In 1971, the PNP intensified its election plans. By January, a public relations firm had prepared constituents' letters for candidates. The PNP's parliamentary candidates were presented in the National Arena on February 28 to a jubilant crowd of fifteen thousand, many of whom waved brooms to sweep the JLP out of power. The arena was decorated with signs such as 'Change or Perish.' 'Change, Baby, Change,' and 'Change Time Now.' Fifteen minutes of cheering elapsed before Manley could be introduced, during which the singing of 'Michael, Row the Boat to Shore' swept the arena. In his charge to the candidates, Manley urged them to build a society based on economic and social justice; to bring to that task political and economic integrity; to work to end victimization and violence both criminal and political; to work for a just distribution of wealth and the development of all citizens; to teach young people the value and dignity of work, and to provide them with training; to know their constituents well and explain the party's ideals, plans and programmes to them; and, finally, 'to strive to be efficient in yourselves.'[39] One of the candidates was Douglas Manley, who was to run in South Manchester, a strong JLP area after 1959. Earlier, when Douglas was still in Ethiopia, his younger brother had written him that 'I wanted to clinch you as the Candidate for the seat because the offer to run in a difficult constituency is a brave gambit for a Manley and therefore long term good plugs.'[40]

The Campaign of 1972

Elections were set by the Shearer government for 29 February 1972. In his campaign Manley used terms from the Old Testament that reflected Rastafarian speech and sensitivity toward the 'sufferers'. Songs like 'Beat down Babylon,' 'Small Axe', and 'Must Get a Beating' expressed the outrage of many poor people toward the 'oppressors', loosely identified as the JLP and associated groups such as 'capitalists'. A walking stick supposedly given to Manley by Haile Selassie, the Ethiopian Emperor (and Rastafarian god), became known as 'Joshua's Rod' and featured prominently in the campaign, as Olive Senior described: 'Thousands of Jamaicans came to believe that the Rod was imbued with supernatural powers, and everywhere [Manley] appeared people wanted to touch this potent source of power, a few ascribing to it healing properties.'[41] For some audiences, Manley argued that 'black power means . . . working towards an equitable distribution, both of goods and ownership of industries.'[42] For others, especially church groups and leaders, Manley's and the PNP's stand against a national lottery promoted by the JLP, was attractive. Also important were large amounts of resources from interest groups, including the middle classes and capitalists, alienated by JLP policies.[43] The slumping economy also played a major role in driving people from the JLP to the PNP.

Still, 'Joshua' Manley was the dominating image of 1972. For Carl Stone, Manley fused the giants of the past, 'the angry populist and almost charismatic style of Bustamante [and] the image of competence, respectability and educational sophistication of the earlier Manley.'[44] Stone referred to Manley's 'cultivated posture of humility, sincerity and God-fearing piety' and reckoned that Manley was a 'phenomenal actor' who 'stimulated an intense level of mass excitement, involvement and millenarian expectations for change.'[45]

The PNP won the election with fifty-six percent of the vote, capturing thirty-seven of the fifty-three House of Representatives seats. The PNP won majorities from all urban classes, especially the better off working class and the middle and upper classes.[46] Afterwards, Hector Wynter, Chairman of the central executive of the JLP (and later *Gleaner* editor), accused some unnamed PNP leaders of 'terrorism' in marginal constituencies where JLP candidates lost preliminary counts, such as in Northeast Westmoreland, where a JLP winning margin of thirty-two votes became a deficit of forty-four votes on recount. Manley, as new

Prime Minister, directed the Commissioner of Police to enforce law and order at centres where recounts were in progress.[47] One of the disputed elections was that of Douglas Manley in South Manchester, whose election was overturned after a lengthy court process, causing great embarrassment to the Manleys.[48] In an editorial in early March, the *Gleaner*, relatively neutral at this point, observed that 'in almost every constituency, adherents of the People's National Party . . . have been making it impossible for Presiding Officers and Returning Officers [election officials] to proceed with the orderly counting of the votes.' 'We do not think we have ever seen an election as confused as this week's election.' The election was held under the auspices of the sitting JLP government, and the *Gleaner* stopped short of questioning the overall validity of the election.[49]

There were some hidden costs and some contradictions underlying the landslide victory of the PNP and Michael Manley in 1972. Lacey has implied that principle was a liability that needed to be put aside:

> Only when the PNP retreated from Fabian preoccupation with party programmes and intra-party democracy, concentrating instead on fighting political warfare 'Jamaican style', did it begin, at the end of the [1960s], to build the foundations for success in the seventies.

Lacey also noted 'the contradiction of the PNP as the most developed political party in Jamaica having to respond to the personalism of the JLP by creating a bigger and better personalism of its own (that of 'Joshua' Manley), with consequent tensions in terms of its belief in promoting intra-party democracy.'[50]

For the moment, however, and for most PNP supporters and many independent voters, such concerns were lost in widespread euphoria attending the election of Manley and the PNP to power. On the night of the election, commuters attempting to traverse Kingston were stopped at every corner and offered toasts of rum by joyous celebrants: John Maxwell said had he accepted every toast he would have been a dead man.[51] JLP hard core supporters were shellshocked by the magnitude of the defeat and by the disarray in the party that followed. Much of the rest of the country looked forward to at least a change in the tone of government under the intelligent, charismatic, idealistic, pragmatic, public-relations conscious, forty-seven year old new Prime Minister.

CHAPTER 13

Populism 1972-3

> And at that time shall Michael stand up, the great prince which standeth for the children of thy people: and there shall be a time of trouble, such as never was since there was a nation even to that same time: and at that time thy people shall be delivered, every one that shall be found written in the book.
>
> Daniel, 12:1.

Context

A proper appreciation of Michael Manley as Jamaica's Prime Minister from 1972 to 1980—of his strengths and weaknesses, achievements and failures—requires that he be viewed in historical perspective. A brilliant article by Carl Stone, 'The Long Road From 1938', written in 1978, serves the need well.[1] Stone noted that Jamaicans tended either to exaggerate the achievements of the post–1938 era or cynically disregard them depending on 'the particular biases and myths which colour [their] view of social reality.' Stone found much that was positive during the era leading up to Michael Manley's administrations. In 1938 perhaps five percent of Jamaicans enjoyed power and comfort: in ensuing decades this increased to thirty percent 'within reach of the opportunities of modern living'. Peasant farmers and unionized workers gained in material and non-material ways. The State, formerly controlled by England and the local plantocracy, came under middle-class control. The once rigid class structure became more complex and flexible. Anglophilism was largely replaced by Jamaican nationalism. Economic dependence on a declining United Kingdom was replaced by economic dependence on the more dynamic United States. Urbanization increased common amenities once regarded as élite luxuries. Health and nutrition improved.[2] The standard of living was enhanced for most of the working population.

However, when Manley took office, serious problems remained. Economic growth had increased inequalities of wealth and income.[3] There was high unemployment, urban overcrowding, and agricultural stagnation. Popular aspirations were greater than the opportunities available in Jamaica and through massive migration abroad. Skilled and semi-skilled workers left the island. A new form of discontent developed in the form of unemployed ghetto youth prone to alienation, crime, and violence. As Stone noted:

> The lumpen culture espouses unbridled sexuality and violence, mastery of

> the gun, hostility to all symbols and figures of authority, class and racial militancy (blending Rastafarian black supremacist notions with misunderstood Marxist class aggression), unrestrained individualism, egocentric behaviour, and a disdain for work, particularly manual work.

By the 1970s, political unionism was being replaced by 'business unionism', with its emphasis on material benefits, and unions increasingly saw their interests as divorced from those of party politics and politicians.

Stone argued that Manley inherited a tradition of political leadership which capitalized on widespread alienation. This tradition was begun by Bustamante and, in Stone's view, continued by Michael Manley. However, from the Bustamante era to that of Michael Manley, the base of political parties changed. In the early years after 1938, parties were supported by vital groups like unions and the Jamaica Agricultural Society. Taking their place as the social base of the parties were, in Stone's words, 'a disproportionate number of pimps, hustlers, lumpen types, rip-off artists, the unemployed, the least productive and those most disconnected grew from the main centres of productive activity in the society.' Thus the parties less adept at generating ideas, work discipline and collective action as the State itself grew in the '60s and '70s.

To understand Manley, Stone urged, it was necessary to understand his 'followership'. The collective character of his followers set limits on what Manley could do and how he viewed his options and alternatives. Manley's approach to 'socialism' was influenced by lumpen pressures as Bustamante's approach had been influenced by the 'working class' in 1938. Stone noted that

> the price of genuine attempts at democracy is cumulative people influence on the character of political organisations, including both [the people's] strengths and weaknesses. The personality of the Jamaican people is stamped indelibly on our political parties, trade unions, and mass organizations.

There were other influences on Manley. Much of the upper reaches of the PNP, including its members of parliament and Parish Councillors, was middle class, as was much of the state bureaucracy. Within the PNP and the government, 'popular mass' and middle class often pulled in opposite directions, despite the party rhetoric of a 'class alliance'. The private sector, churches, the JLP, the army and police, women's groups and foreign governments also set limits on what could be done.

Michael Kaufman has argued that Jamaica's parliamentary institutions were widely supported, but perpetuated the unproductive parliamentary proceduralism characteristic of the Westminster model in its Caribbean setting. Jamaica's socialists wanted to extend democracy beyond the limited Westminster practice. In 1976, they would achieve a plurality of votes for 'socialism'. Nevertheless, throughout the '70s many traditional attitudes, ideas, and practices persisted. One of the most important of these was the separation of the masses from power. Another was the patriarchal divisions between men and women. There was social fragmentation and alienation born of capitalism in a Third World context. The country was dominated by capitalism, but capitalism also existed awkwardly with subsistence agriculture and street hustling. Those oppressed by the system were divided by heterogenous interests, especially political party 'tribalism'. The local capitalist class, although underdeveloped by First World standards, had relatively significant economic power to reward or to punish the state. Externally, Jamaica's economic manoeuverability was limited by dependency on foreign markets, technology, raw materials, energy and capital. Finally, the 1970s witnessed a world economic crisis, marked by increases in the cost of energy, by stagflation, by trade patterns increasingly unfavourable to Third World countries, and by raising Third World indebtedness, all of which had disastrous effects on most of the countries of the Caribbean and Latin America and most of all on their poor.[4]

Popular aspirations for a better life were a compelling aspect of the context in which Manley operated for eight years as Prime Minister. The desires of people in his own poor, urban constituency of East Central Kingston illustrate this. A thirty-two year old fisherman reported: 'I would like to be the owner of a boat, engine, and a net so I would be able to purchase a house.' An unemployed thirty-three year old man reported his goal: 'super market and Liquor Store'. A jobless forty-year-old male wrote:

> I would [like] to live on the land and farm and cattel raising. I would like to help the economy of my contry and also help my family. Its been a long time since the contry be in deffecoute and I would like to work along with the government.

An unemployed man of forty-six expressed a common concern: 'I would like to get a house for me and family.' A fifty-three year old reported his goals: 'Farming. Goat. Hogs. House'.[5]

The latent fears of the upper and middle classes were also part of the setting in which Manley took power. These anxieties were aggravated by the general clamour of the black masses and by such expressions as Barry Reckord's drama 'In the Beautiful Caribbean'. A BBC play aired in England, 'In the Beautiful Caribbean' raised the possibility of a Castro-style revolution in Jamaica. It also argued that black middle-class rule was worse than colonialism.[6]

1972: Populist Beginnings

In a 1972 interview Tony Verity asked Manley whether 'some form of socialism specially adapted to the West Indies conditions' could solve Jamaica's serious economic problems. Manley replied:

> . . . I totally distrust these cliché words like socialism, capitalism and this, that and the other. . . . I am one who believes in equality and the thrust towards an equal society, but methodologically I am a pragmatist.[7]

Confronted later with the label 'populist' attached to the early years of his administration. Manley acknowledged the validity of such a label because of 'the sheer extent of our popularity' and because his government invested heavily in social programmes and fighting poverty. But he also doubted the populist label because the government's efforts were 'an extension of a philosophical position, not simply an attempt to win popularity and power.'[8] 'An extension of a philosophical position' the government's policies may have been, but when the PNP attempted to define its philosophy, it caused enormous internal conflict in the party and was put aside for years.[9]

Although 'populism' is much-debated and varies in specific historical and national settings, many of its general features seem to fit the Manley approach to politics at least until 'democratic socialism' was defined by the party in 1979. Populist features in the Manley government included the use of revolutionary rhetoric; an appeal to 'the masses', especially the urban 'masses', and the expression of their concerns; a leftism independent of, and sometimes in conflict with, communism; identification of local power holders (the 'oligarchy', the 'twenty-one families', the *Gleaner*) as the enemy; a tendency to bypass established institutions; pragmatism and eclecticism in ideology and policy; progressive nationalism (as opposed to the traditional, parochial nationalism of the JLP); all cemented together by the charisma of the leader.[10] None of this is to deny the seriousness of Manley's politics (which may be why he disliked the 'populist' tag). Ideological labels only loosely approximate social reality, and 'populism' and 'socialism' are not necessarily mutually exclusive. Indeed, 'populist socialism' is the most likely label to describe politics in the Caribbean setting.

During the spring 1972 honeymoon of the new Manley administration, ideology could be forgotten in the midst of what the *Gleaner* reported as 'island-wide jubilation' over Manley's victory.[11] Big businessmen drank whiskey to celebrate the victory, while 'the masses' smoked 'herb'.[12] Reflecting the ephemeral post-election euphoria, the *Gleaner* reported that 'the nation as a whole is in good heart,' and 'the nation is healthy.'[13] Its political reporter wrote that 'Mr. Michael Manley has proven himself to be a master of organization, to be a persuasive and charismatic leader' who had run 'a truly magnificent campaign', and 'above all he is a realist.'[14] Earlier, the *Gleaner* had reported that the new government's policy seemed to be based on 'sound social thinking for the good of Jamaica', although it gave insufficient attention to necessary resources.[15] It certainly did not hurt Manley's image that one of his first acts as Prime Minister was to recommend legislation requiring that parliamentarians and their families report their assets.[16] Soon Manley was seen in locales such as Barnes

Gully and the intersection of South Camp Road and Tower Street in Kingston shovelling rubbish, challenging old prejudices against manual labour.[17]

At first, the sharpest criticism of Manley came from the left, and it may have touched Manley's sense of guilt about being a privileged Jamaican, and thus helped move him leftward. The *Abeng* group reckoned that 'the Manley government is more sophisticated and clever [than the Shearer government]. It is likely to foster groundless hopes for a longer period than Shearer ever could.' The group identified a clique running things in the government: Mayer Matalon (Chairman of the Bauxite Commission), Eli Matalon (Mayor of Kingston and Minister of State for Education), Moses Matalon (consultant, Technical Advisory Committee), Leslie Ashenheim (committee to review the salaries of members of parliament), Morris Cargill (banana board), John Pringle (ambassador extraordinary), Peter Rousseau (Deputy Chairman, JBC), Pat Rousseau (Deputy Chairman, Bauxite Commission) and Rabbi Hooker (JBC Board). According to the Abeng Group, Manley had surrounded himself with 'these arch-enemies of the people'. The progressives appointed to government boards – such as Hugh Small and D. K. Duncan – only made matters worse by masking the conservative background of most appointees.[18]

In private, Manley began to cope with the numerous contending interests now anxious for his ear. A member of Jamaica's tiny upper class wrote to 'my dear Michael' hoping that the Exchange Control Order imposed by the previous government with PNP concurrence, in response to an international monetary emergency, could be lifted so that Jamaica's economy could be 'open'.[19] The United States Ambassador, Vincent Roulet – later to be declared *persona non grata* by the Manley government – sent the new Prime Minister a drawing of 'Joshua'.[20] On Labour Day, pursuing a national unity that would transcend political tribalism, Manley made a courtesy call on Bustamante. In July, 1972, a prominent member of Jamaica's artistic and academic communities wrote urging him to keep his commitment to a 'philosophy' to save Jamaica from 'that crass pedestrianism that has plagued us for far too long.'[21] In August, a well-placed American friend advised him that 'the attitude of OPIC [the US Overseas Private Investment Corporation] is even more strongly in favour of investment in Jamaica today than a year ago because of the *improvement* in leadership.'[22] Dick Gregory, the American entertainer-activist, offered enthusiastic support:

> Your message last evening seemed more like an extended period of transcendental meditation. . . . It seemed as though I were listening to Gandhi, Dr. [Martin Luther] King, Malcolm X, Thoreau, all of the great theologians and statesmen of history rolled into one contemporary spokesman for truth, dignity and justice.[23]

A very positive article in *Time* earned its author a personal thanks by Manley who termed the article 'most kind' and 'perceptive'.[24]

The Manley government introduced a large number of major programmes in 1972. These included the promotion of the New International Economic Order (NIEO, an initiative to reduce the international gap between the rich 'North' and the poor 'South'); the Special Employment Programme (Impact Programme, or 'Crash Programme.' aimed at the hard-core unemployed); a skill training programme; the Workers' Bank; JAMAL, a literacy programme; a voting age lowered to eighteen; community health aides; Operation GROW; Project Land Lease; Civil Service reclassification; and a public housing programme.[25] As the year progressed, the political honeymoon ended. An early turning point was Manley's delivery of the annual budget debate speech on 1 May. The speech assumed an activist state involved in many areas of national life, but this was consistent with Jamaica's previous pattern of government. It steered clear of ideological issues. It dealt overwhelmingly with domestic issues. The speech identified the four principles upon which the government's policies were founded as equality, social justice, self-reliance, and discipline. It listed as its highest priorities employment, housing and nutrition for the majority. It announced the necessity for restructuring government, but mostly by reorganizing ministries. The speech announced programmes for addressing inequities facing women and youth. To confront Jamaica's 'very grave' economic situation, Manley proposed an economic strategy that would meet the people's basic needs, coordinate all aspects of development, be pragmatic and would value and utilize the private sector. Manley stressed the importance of leading by example and defended himself from critics' ridicule of his shovel-and-broom-wielding.

> When those who are more privileged in the society . . . will never pick up a broom to sweep their own verandah, they create the attitude that to pick up a broom is a mark of inferiority. . . . I do not believe I have a right to lead this country if I am unwilling to show that I respect the dignity of my own person. They mock, they laugh, they treat life like a joke while people are suffering outside. While they are laughing, let them remember the television cameras are upon them and the sufferers are watching them, and so when you see me . . . get into Barnes Gully and pick up a broom and work for them, I am only setting an example.

Manley closed the long speech with a plea to 'the sufferers . . . who are the first and last concern of my heart': 'Be patient, we are working. With God's help your time will come.'[26]

Foreign Policy.

The brief section of the budget speech devoted to foreign affairs belied the importance this aspect of Manley's administration would have. In the first ten years of Jamaica's independence under the JLP, Jamaica's foreign policy had consisted of a 'quiet and constructive role in world affairs'. The JLP had deliberately avoided close relationships with the Third World and had 'rather ambiguous' relations with Latin America, despite the fact that large numbers of Jamaicans lived in Cuba (twenty-eight thousand), Costa Rica, Honduras, and Panama.[27] The Manley PNP foreign policy changed dramatically as a result of domestic Jamaican factors ('black power' and 'African' identification, the intellectual awareness of Third World dependency) and such international factors as the rise of the Organization of Petroleum Exporting Countries (OPEC), changes in the UN, and the US defeat in Vietnam.[28]

Also important was Manley's own personal commitment to an activist foreign policy. Many Jamaicans, used to a parochial view of their nation, had difficulty comprehending why Jamaica should be involved in such affairs as African liberation struggles. To a friend who held such a provincial perspective, Manley explained that African freedom fighters' struggles were 'very desperately our business. Fascist colonial brutality is every man's business.' He continued:

> . . . I have got to put Jamaica into the leadership of the Third World because we are a part of that Third World and have got to seek ourselves so that we may ask the rich countries to help us from a posture of pride rather than mendicancy. . . . Where these large issues are concerned, it is not possible for me to lead Jamaica as if I were a beauty contestant in a popularity contest.[29]

Sometimes it was easier to find comprehension from the former imperial power. Hugh Foot, Lord Caradon, wrote Manley shortly after his election, including his best wishes and a highly appreciative appraisal of Manley's 1970 *Foreign Affairs* article.[30] Some Jamaicans would feel considerable pride in Manley's vigorous, highly-visible foreign policy, but others would find it disquieting.

The disquiet resulted from specific policies such as rapprochement with Cuba, but also from Manley's basic attitudes and assumptions about world affairs. He believed that 'you cannot divorce anything from foreign policy. Every [domestic] move you make must have its balancing foreign policy.'[31] He saw Jamaican foreign policy as offering a choice between a 'subservient alliance with the US' or 'common cause with the Non-aligned Movement and work[ing] for changes in the world's economic system.'

> . . . The point is that now some two-thirds of mankind have been reduced to a peripheral status in political, economic and even social terms. It is the degree, the scale, the ruthlessness, the completeness of the exploitation which marks this imperialism, that sets this process of domination apart from any that have preceded it.[32]

Manley's assumptions were correct, but they ran up against the facts that the power of the United States was enormous and that the Jamaican people bore good will towards that superpower as a source of opportunity and money for many of them. A tremendous problem was that 'for the masses of the [Jamaican] population, values were those disseminated through the stream of propaganda and entertainment that poured out of Washington, New York and California.'[33] Moreover, according to Seymour Hersh, under President Nixon and Henry Kissinger 'There was little room at the top for concern over what were seen as the lesser issues, such as human rights and the problems of the lesser countries in Africa and Latin America. . . .'[34]: Angola, Cuba, Chile, Jamaica; such places were important to the White House only as pawns in the East-West struggle.

In 1972 Manley and the PNP had developed two basic foreign policy objectives: (1) avoiding East-West polarization and helping build up the Non-Aligned Movement: (2) helping develop Third World economic strategies and cooperation. More specific targets included disengaging from slavish obedience to the US-NATO alliance, supporting 'genuine' liberation struggles, and pursuing a close friendship with Cuba.[35]

Manley's intellect, vision, interests, commitments, eloquence and style served him well as an exponent of foreign policy, and he became a major Third World spokesperson in the 1970s. By October 1972 Jamaicans were reading headlines such as 'JAMAICA LEADS WAY AT U.N.,' after that body adopted a Jamaican amendment on 'international terrorism' stressing that it did not apply to the freedom fighters of Southern Africa.[36] On 2 October Manley addressed the Twenty-Seventh Session of the UN General Assembly. He began by criticizing the Assembly as 'a place where torrents of words that burn with urgency and truth fall upon deaf ears . . . closed by the narrowest consideration of national self-interest.' He called for a 'universal moral foundation for the conduct of affairs' as necessary to the survival of 'mankind'. He said that Jamaica was 'totally and bitterly opposed to the racist regimes of South Africa and Southern Rhodesia and to the continuing Portuguese tyranny in Angola, Mozambique and Guinea (Bissau).' He supported the right of armed resistance when the peaceful political process was exhausted but disapproved of violence against innocent third parties.[37]

The most controversial aspect of Manley's foreign policy was rapprochement with Cuba, which began in a relatively uncontroversial way. At a meeting of the Commonwealth Caribbean countries in October 1972, Jamaica, Barbados, and Trinidad and Tobago decided to establish diplomatic relations with Cuba. This occurred against the background of a gradual diminishing of the United States efforts to isolate Cuba in the hemisphere.[38] Months after Jamaica established diplomatic ties with Cuba, the *Gleaner*, that barometer of conservatism, supported Manley's policy.[39] The pragmatic Cuba of the 1970s, engaging in realpolitik with several not even remotely revolutionary Latin American regimes, moderated by its close ties with the USSR, was not the radical, revolution-exporting Cuba of the 1960s.[40] The United States' rapprochement with the People's Republic of China seemed to signal greater realism in its own foreign policy, but later it became clear that this realism did not embrace such countries as Chile, Angola, Cuba, and Jamaica.

Opposition

Despite the recognition that Manley brought to Jamaica by his growing international reputation and the relatively moderate nature of his government's domestic reforms (which scarcely went beyond institutions established by social democracy in Europe), opposition began to surface soon after his election. In July Manley wrote that 'the *Gleaner* decided to be totally cynical and malicious about the whole effort and it is clear to me that the old Seaga/Shearer/Simmonds axis is at work again to try to tear us down.'[41] To another correspondent Manley wrote of Jamaica as 'a country so given to vicious gossip.'[42] In September a friend wrote him that her Chinese grocer, a PNP supporter for many years, had reported a rumour in the Chinese community that Jamaica was going communist. The 'logic' behind the rumour was that Jamaica appeared to be 'catering' to Tanzania, and Tanzania had close ties with Red China.[43]

In mid-October Manley wrote of being under great pressure and of having 'another attack of diverticulitis'.[44] In early December he had to denounce some spurious pamphlets with the titles 'Behold the Red Rod' and 'Jamaica in Political Upheaval/State of Emergency Soon to Be Declared'. He regarded the pamphlets as seditious and libelous and they made him very angry. He dealt with this red smear by denying that Jamaica was either communist or totalitarian and stating that 'it is not intended by me that Jamaica should ever become so politically inclined.'[45] Manley also countered the red-baiting by meeting with the Evangelical Association of Jamaica which announced that Jamaica's democratic system was not in danger.[46] As 1972 came to an end, Manley had to apologize to a friend and political supporter:

> I am very sorry I lost my temper last night. The truth is that I am desperately over tired and I am going to have to find a way to lead a more sane life next year. This is always going to be my biggest problem; that when I get beyond a certain point of fatigue it manifests itself in explosiveness.[47]

Manley was fortified against fatigue and opposition to his policies by happy personal events during his first year as Prime Minister. Beverley Anderson became his fourth wife. She was a popular radio personality, one of the few to emphasize Jamaican music over foreign imports on her programme. She was to become a controversial and outspoken member of the PNP left and women's movement, but her and Michael's marriage added to the glamorous aura of the early Manley administration. Later in the year, Manley was gladdened by the reading of his daughter Rachel's first book of poems: 'I have never been prouder of anything in my life.' he wrote her.[48]

1973: Domestic Reforms and Opposition

In 1973, Manley's government deepened and extended its pragmatist, populist reform programmes. New programmes included a cultural training centre, subsidies on flour and condensed milk, uniforms for primary school children, free secondary and university education, the National Youth Service, the revision of an act restricting rents, equal pay for women and the establishment of a women's bureau in the government, Jamaica Nutrition Holdings, an increase in property taxes, a key role in founding the International Bauxite Association, the Home Guard to deal with escalating crime, and a prominent role in the Non-Aligned Movement. Despite the relatively mild reformism of these programmes, they generated increasing opposition.[49]

Manley and his government faced a delicate situation in 1973. In a JBC news analysis, Rex Nettleford, a sympathetic observer and Manley friend, warned of the necessity to guard against extremes, the view that everything was either good or bad. He noted that many Jamaicans were unemployed and thus lacked 'the underpinnings of a value framework that can make them genuinely feel that this country belongs to them. . . .'[50] Later Manley himself warned against the dangers of extremism in Jamaica. One extreme was represented in 'the impatience of some who have suffered centuries-old social injustice. . . . who seem to wish to destroy everything in our society, even the hope that is now being engendered'; the other extreme was represented by the attitude that said 'hands off my privileges'. In his Independence Day speech on 6 August 1973, Manley noted that his government 'had a duty to control, direct, and manage change so that it would come with order and intelligent purpose.'[51] Finding a disciplined, intelligent, middle course to serious change would be difficult.

Of the many reforms launched in 1972 and 1973, the one that first led to strong opposition was taxes on wealth rather than that 'mercurial category known as income'.[52] For Adam Kuper, an adviser to the PNP government, the land tax imposed on 2 May 1973 was an example of how PNP reforms and policy-making occurred. Even the preceding JLP government worried about the level of agricultural production and thus started to register idle lands. The reform of May 1973 was a tax on lands not in full production. It was not a radical action, but the *Gleaner* editorialized about its uncertain effects. The Chamber of Commerce called on the government to renounce the tax. Edward Seaga said it was the beginning of a vast land-acquisition spree that would be aimed ultimately at the foreign-owned bauxite companies, an allegation sure to arouse concern in the United States and Canada. Opposition mounted. The government wavered. Administrative details and the purpose of the tax were unclear. On 17 May Manley suggested the tax would pay for new free educational services, but later denied this. The government held meetings to reassure business, instituted a new tax collection schedule, and started an appeal system for 'special cases', effectively emasculating the land tax policy. Those who might have benefited from the tax were left unorganized and thus incapable of bringing pressure to bear. In Kuper's view, a similar process operated on such reforms as extending free secondary education and taxing foreigners' work permits. Original proposals were whittled down, organized middle-class opinion mobilized, and ambiguities in the proposals were attacked. The fervour of debate was enhanced by the fear that individual reforms, innocuous in themselves, might lead to a more general radicalization culminating in 'Castroism'.[53]

Only a few of the more controversial domestic reforms of 1973 can be mentioned here. Many involved the problem of alienated youth, an explosive situation within the PNP, the government, and the larger society, which contributed to the pervasive problem of 'tribalism'. Educational reforms reorganized access to public secondary and university education and provided free tuition at both levels. Prior to the reform, of the nine thousand entering secondary school students, some two thousand received free tuition. Forty to fifty thousand eleven-year-olds competed for the two thousand scholarships: the remaining places went to the children of parents who could afford to pay. The system favoured the affluent, not necessarily more intelligent students. After the reform, all nine thousand places went to the best students, regardless of parents' financial status (private school still existed for the well-off).[54] Middle-class children had advantages under the system, but their edge was cut, and some of their parents were disturbed by the way in which the government attempted to make the educational system more equitable. For example a proposal was put forward for a National Youth Service, which would engage all scholarship students in two years' work for the country, and middle-class children would therefore mix with ghetto and peasant youth; this increased middle-class opposition. Similarly, the Special Employment Programme directed at the hard-core unemployed, was largely wasteful and generated opposition from the 'oligarchy', the middle classes and even some union members; Manley ruefully recorded: 'It was not our finest hour!'[55]

Indiscipline and excessive, often pseudo-Marxist, rhetoric by youthful government supporters added fuel to the opposition. Crime increased and 'robbery with rape had become the new curse.'[56] Efforts to combat crime by community organization—no more radical than 'Neighborhood Watch' programmes in the US—were seen by the opposition as proof of 'Cuban' or 'communist' tendencies, as were efforts to overcome victimization and tribalization by community organization transcending the bitter, enervating, Jamaica-destroying PNP-JLP rivalries. In the autumn Manley warned, in his speech to the PNP annual conference, that 'a one-party state' had been built in Tivoli Gardens, the centre of Seaga's West Kingston constituency. When the Ministry of Housing had tried to place a few families there that had earlier been bulldozed out 'by political tyranny', a JLP group was organized to keep them out.[57]

Foreign Policy: Confrontations and New Directions

Rising domestic tensions were aggravated by international confrontations. The United States Ambassador to Jamaica from 1969 to 1973 was Vincent de Roulet, an 'unreliable amateur'[58] who 'all but out-uglied the original Ugly American.'[59] In 1973, before an astounded US Senate subcommittee De Roulet testified the abuses of multinational corporations and how he had intervened at the highest levels of domestic politics in Jamaica before the 1972 election. His central concern was the bauxite companies, which he believed had been irresponsible corporate citizens; he said that he browbeat them into improving their image by supporting social programmes. De Roulet also tried to influence the election of 1972. He felt Manley was 'much more of a socialist dedicated to an egalitarian society' than Shearer and thus a threat to United States interests. De Roulet devised a deal to keep Manley from being elected: boosting then-Prime Minister Shearer by arranging an appointment with President Nixon, a highly unusual event in United States-Jamaican relations. De Roulet characterized Shearer thus:

> In the whole time I dealt with him . . . I never had to listen to a long speech about his little people with withering bellies and the back-to-Africa crap. He was a hard-boiled businessman. He knew where his bread was buttered.[60]

In his bargaining for a presidential appointment de Roulet played on the fact that several of Nixon's 'closest supporters [had] extensive real/estate holdings in Jamaica' and noted that Shearer was anti-Communist, a well-dressed and courteous 'Negro'.[61] The presidential appointment was granted and a tour of the CIA was added. De Roulet said 'it gave us a chance to talk about Cuba a hell of a lot and . . . to get a hook into him. You know, it makes these guys a Big Thing in the Third World.'[62]

The US Agency for International Development (AID) granted twenty million (US) dollars to the Shearer government just before the election. Years later, it was revealed that ALCOA had contributed to both parties—twenty-five thousand dollars to the JLP in 1971, and twenty thousand dollars to the PNP in 1972.[63] In his Senate testimony de Roulet asserted that he had gotten both Shearer and Manley to agree not to raise the issue of nationalization of bauxite. Both denied that they had made any deals with Roulet, and the governmental and corporate influence of the United States was unable to prevent Manley's election in 1972. The *Gleaner*, while noting the difficulty of proving deals between de Roulet and Jamaican leaders, expressed concern at the possible United States interference in Jamaican politics. Robert Hurwitch, US Deputy Assistant Secretary of State for Latin American Affairs, denied any intervention in Jamaican affairs from 1969 on.[64]

After Manley took office, de Roulet lost his interest in high-level machinations, probably because he realized his leverage with Manley was limited. De Roulet returned to his former habit of discouraging Jamaicans from applying for visas to the United States. In his conversations with George Crile after his ambassadorial term, de Roulet described Jamaicans as 'the most spoiled race of people in the world', as 'idiots', and as suffering from a 'rampant inferiority complex'. According to *Time,* the American ambassador referred to Jamaicans as 'niggers'.[65] Before an audience of Americans and Jamaicans he said Prime Minister Manley had an ambivalent personality attributable to Manley's mixed racial origins. Nonetheless, Manley and de Roulet were pictured in a friendly pose in July 1973 when an AID grant of four million (US) dollars to help finance forestry development was announced.[66] Shortly thereafter, as the revelations of his Senate testimony surfaced, de Roulet was declared *persona non grata* by the Jamaican government. Subsequently, after leaving office, de Roulet referred to Manley as 'an emotional yo-yo' and 'latent alcoholic'.[67] The de Roulet episode cast a sad light on United States handling of Jamaican affairs: an American sociology professor wrote to Manley apologizing for de Roulet's 'boorishness and ignorance'.[68] But the Jamaican government's problems with the United States were far from over.

Jamaica's foreign policy in the '70s—upon which Manley placed such a personal, committed and controversial stamp—was bound to aggravate Jamaican-United States relations. Manley recognized this and repeatedly attempted to reassure foreign and domestic audiences that new directions in Jamaican foreign policy—toward Cuba, the 'socialist bloc', and the Non-Aligned Movement—did not have to threaten Jamaica's traditional relations with the United States and its allies. In a speech to the Board of Governors of the Inter-American Development Bank (IDB) in May 1973, Manley said his government had no dogmatic ideological commitment except to social justice, equality, and freedom. Manley assessed the IDB—a regional pillar of the post-war global economic structure created by the United States and its allies to further their interests—as 'a most imaginative and effective example of fruitful economic cooperation between developed and developing countries'. He also urged the IDB to put people at the centre of the development process; and he criticized the political strings attached to IDB funding. But the approach was not fundamentally threatening to United States interests.[69]

In other ways Manley's foreign policy did challenge Yankee concerns. The United States had long preferred bilateral relations with countries of Latin America and the Caribbean and feared independent regional efforts. Manley made an official visit to Venezuela and expressed his desire to break neocolonial attitudes which led Third World countries to orient themselves to metropolitan powers and turned Third World nations into mutually suspicious rivals. He enunciated principles of Third World diplomatic relations which, in practice, the United States was (and is) unable to accept: mutual respect, sovereignty and dignity; non-exploitive economic relations; and preference to trade among Third World countries.[70]

The apprehensions of conservative Jamaicans and the United States increased in September with Manley's trip to the Non-Aligned Movement's summit meeting in Algiers. Manley and Guyana's Forbes Burnham joined Castro in Trinidad for the flight to Africa. Manley and Castro talked for five hours on the flight, sharing childhood reminiscences, 'stories about our families, and much else that was simple and human'. Manley was 'fascinated by stories of the Sierra Maestra'. Thus began 'a genuine and enduring friendship' that Manley included within his closest circle of friendly national leaders: Canada's Pierre Trudeau, Tanzania's Julius Nyerere, Guyana's Burnham, and Barbados's Errol Barrow.[71] Manley's travelling with Castro panicked some Jamaican conservatives, as if he would fall under the Cuban's evil eye. Manley was irritated that people didn't understand that he could be a friend of Castro's without becoming a communist, while at the same time he was an equally good friend of Hugh Shearer's without becoming a conservative champion of capitalism.[72]

At Algiers, Manley's speech on 8 September focused on Third World economic problems and the liberation struggles in southern Africa, for which he volunteered Jamaican resources. The *Gleaner* noted that his speech received 'high acclaim' and added its own editorial support: 'The futile tedium of the Conference was relieved by [Manley's] fresh approach which the participants acclaimed like sinking men clutching at the last straw of hope.'[73] Following the conference Manley made state calls in Zambia and Tanzania. In Tanzania, led by his friend and ally, Nyerere, he was welcomed by a crowd of one hundred thousand. At the University of Dar Es Salaam Manley commented on United States' reaction to Jamaica's recognition of Cuba and the People's Republic of China:

> I could not detect any loud applause emanating from Washington. On the other hand, in fairness to them, I did not detect any boos either. It was received with a discreet and statesmanlike silence.[74]

Back in Jamaica, Manley pronounced the trip 'an almost unbelievable success' in advancing toward three goals: winning support for a Third World-controlled development and solidarity fund; gaining influence for the idea of small-state economic cooperation; and establishing Jamaica's credibility and prestige as a significant, influential Third World voice.[75]

Less fortunate events were also in the winds. Manley wrote to Kenneth Kaunda, President of Zambia, that he 'returned home to find a minority of "establishment Jamaicans" bitterly opposed to my offer of volunteers for the fight in Southern Africa,' although he believed 'the masses' supported his position.[76]

Much more tragic was the overthrow of the democratically-elected socialist government of Salvador Allende in Chile on 11 September by military and civilian forces, with the complicity of the United States. Allende's fate boded ill for any government attempting the democratic path to 'socialism' in the Western Hemisphere.[77] The PNP executive condemned Allende's overthrow by 'the fascist military regime.'[78] Castro, who had cut short a trip to Vietnam and cancelled visits to Iraq, Syria, and Yugoslavia because of the coup, sent Manley his assessment of Chilean events emphasizing Allende's heroism: 'He fought like a lion in the Palácio de la Moneda [presidential palace], resisting during many hours the fascist coup.'[79] In later years, Manley would draw parallels between the Chilean experience under Allende and his own.

At the end of 1973, a *Gleaner* editorial pictured Jamaica 'at the crossroads'. It had been a year of major problems: a food crisis, international and domestic inflation, crime, controversial government tax policies, the furor over Manley's journey with Castro to Algiers, and Manley's pledge of aid to African freedom fighters. A second editorial blamed the country's economic woes on the government, unions, and the private sector.[80] A somewhat different appraisal of conditions in late 1973 was made by Mortimer 'Buddy' Martin, President of Local 3959 (Toronto), United Steelworkers of America, in an open letter to 'Prime Minister Manley and the People of Jamaica' dated 9 December 1973. Although Martin agreed that people were partly right when they said 'Jamaica is no longer a nice place to live or even visit since the British left,' he urged people to rally around Manley to 'help him carry out the good job he is doing.' Jamaica, Martin concluded, was 'the only place on earth where a black man can stand ten feet tall regardless of his position or stature.'[81]

CHAPTER 14

Return To 'Democratic Socialism' 1974

For Michael Manley, his government, and Jamaica, 1974 was a pivotal year. The government adopted many new policies and programmes. It faced enormous economic problems, many of them aggravated by the OPEC increase in oil prices in late 1973. Perhaps the PNP's most successful policy—the bauxite levy—was announced in May. Support from the middle classes eroded, but the government attracted offsetting votes from the lower classes. Manley's first book, *The Politics of Change*, was widely discussed in Jamaica and led to the PNP's controversial rededication to 'democratic socialism' in November, bringing ideological concerns to centre stage. Throughout the year, the Manley government moved leftward, in response to such events as the 'ghetto triumph of Rastafarianism', the success of black power ideology on Jamaica's campuses, nationalism, and the rise of black pride.[1] Michael Manley himself was a major cause of his government's move to the left.

Policies New and Old

In 1974, for the third and final year, Manley's government introduced a large number of new programmes, thereafter, a few important programmes were started, but the government's concerns would lie elsewhere. Economic reforms included loans to farmers, the establishment of sugar cooperatives, the bauxite levy (discussed below), a development venture capital financing company, and the construction of small business facilities. The government also began to acquire essential utilities: the Jamaica Public Service Company (electricity), the Jamaica Omnibus Service Company, and the national merchant marine. Important social reforms included the establishment of a family court, national minimum wage legislation, an increase in National Insurance Scheme pensions, increases in poor relief, the establishment of Agricultural Marketing Corporation outlets in low-income areas, and free education for the handicapped. Two measures dealt with rising crime: the Gun Court and the Suppression of Crime Act.

Among the social reforms adopted in 1974 was a new mental health law. A letter from Manley's daughter, the poet Rachel, to her father captured with a poet's sensitivity and a Manley's concern, the plight of the mentally ill on the streets of Kingston:

> A sore thumb is the basic inhumanity & indignity to those poor & incapable

> ones who roam the streets sightless save to their own blurred vision of themselves, having to share their nakedness—hunger—often masturbating outside shops—laughed at—scorned. They must be sheltered from this indignity which they have not chosen for themselves—
>
> I beg you—I can't bear to think & to know they suffer helplessly & no one cares or cleans or houses or feeds them . . . please Daddy—next time there's a little money can someone help them[.]
>
> Please care yourself—[2]

In his budget speech of May 29, Manley emphasized the new mental health law, the family court, the national minimum wage law, local government tax reform, free uniforms for primary schoolchildren, the furnishing of prime land to sugar workers, and compulsory union recognition by employers.[3] Assessing the first two years of Manley's government, the *Gleaner* placed its emphasis on developing human resources.[4]

Political tribalism and associated violent crime was a major problem. Crime took a large human and economic toll, hurt Jamaica's image abroad and thus the important tourist industry, and tested the state's ability to govern. During the local elections of 1974, Manley condemned political violence, and the *Gleaner* reported a JLP candidate being beaten.[5] Several prominent persons were killed, and Manley wrote that 'people were beginning to lose confidence in the ability of the State to guarantee their personal security.'[6]

In response, Manley declared a war on gunmen with the Prevention of Crime Act and the Gun Court Act. The Gun Court featured prompt and severe punishment, with indeterminate sentences, reviewed after two years. Conditions in the Gun Court's prison were kept deliberately rough.[7] A new Ministry of National Security was formed. Its first chief, Eli Matalon, described 'interlinkages between illegal foreign exchange transactions, the export of ganja, the import and export of cocaine, [and] the import of guns.'[8]

Manley commented that the decision to implement stringent anti-crime measures was difficult for any person 'steeped by training and experience in the tradition of civil liberty,' and he hoped to relax them in time.[9] The government's anti-crime measures were popular with the public and were supported by the *Gleaner* and, officially, by the JLP.[10] The measures were short-term solutions, but they had some impact. Reported gun crimes dropped from an average of 232 for the three months before the Gun Court opened to 119 for the three months after.[11] Long-range solutions would be more difficult, requiring a war on poverty and joint PNP-JLP commitment to end party tribalism.

In addition to new programmes and the battle against violent crime, the Manley government, continued to administer programmes introduced earlier. A *Gleaner* editorial commended Manley for his 'courage and candour' in admitting that JAMAL, the literacy programme, had begun on a great wave of enthusiasm, but subsequently developed problems. The literacy programme was one of the few in which problems were recognized and corrective measures taken.[12] Manley continued to participate in Labour Day projects to demonstrate the dignity of manual labour and its contribution to national development. In 1974 he helped dig foundations for basic schools and paint a children's home dormitory. Governor-General Glasspole helped plant trees. Hugh Shearer, by contrast, took part in a wreath-laying ceremony.

One of the government's most troubled programmes was the Special Employment Programme to reduce hard-core unemployment, which was constantly criticized for its low productivity. Manley defended the programme as misrepresented and honestly misunderstood. Although it employed eleven thousand people in projects all over Jamaica, criticism focused on its ineffective street-cleaning in Kingston. Many of its employees were women, including 'old ladies many of whom had hovered on the verge of starvation for years.' Manley promised that supervisors would weed out those who did not perform an honest day's work.[13] More successful was Project Land Lease, the centrepiece of Operation GROW. Although the project had its shortcomings, the *Gleaner* regarded Land Lease as successful up to late November. The conservative newspaper stated that 'unreserved credit' should go to Manley and those who conceived and implemented the programme to lease idle lands to small farmers.[14]

The Economy

One of the Manley government's biggest failures was in managing the Jamaican economy. Many economic factors were either impossible or difficult for the Jamaican government to control: world recession, the skyrocketing price of oil, unfavourable flows of capital, the end of expansion in bauxite, and the migration of skilled personnel. There is no reason to believe the JLP would have done any better, as its economic statements were contradictory and incomplete. Nonetheless, the sitting government must bear responsibility for what it could control and the Manley Administration's cardinal sin was that it lacked medium- and long-range coherence: 'Any coherence the policies appeared to have was because they hung together in the mind of Manley, and Manley's strength was not economic planning.'[15] Manley recognized the need for planning, but under the tumultuous and unprecedented world economic conditions of the '70s, Jamaican expertise was unequal to the task.

An example of a major factor the government could not control was the price of oil. Privately, Manley observed in January 1974 that 'the oil crisis has brought countries like Jamaica to the verge of total disaster in our balance of payments and to survive, we must renegotiate the prices for our major exports like sugar'.[16] Publicly, Manley called the oil crisis 'the biggest challenge that the Jamaican people have faced in the 20th century'.[17] The oil-sugar equation was but one aspect of steadily worsening terms of trade—increasingly expensive imports, increasingly cheap exports—that plagued not only Jamaica, but most of the Third World throughout the '70s and '80s. It was this basic equation, resulting in the steady impoverishment of the Third World to the short-term benefit of the First (but threatening ultimate collapse of the world economy), that led Manley to become a, perhaps *the* leading advocate of the NIEO. Even the conservative 'Thomas Wright' (Cargill) recognized Manley's contribution:

> Whatever else may be said about Mr. Michael Manley and the present government, the creditable fact is that this government has been the first Jamaican government not only to understand fully the disastrous implications which had been taking place, but to take effective and vigorous action in an attempt to redress the balance.[18]

In general, however, the *Gleaner* and business groups were very critical of the Manley government's economic performance. In June, Douglas Vaz, President of the Jamaica Manufacturer's Association (JMA), expressed alarm about the economy, citing 'a multiplicity of bureaucratic booby traps and stumbling blocks ranging from chaotic licensing and de-licensing arrangements to excessive or misapplied restrictions on credit, fuel, [and] raw materials.'[19] The *Gleaner* editorialized that 'it would almost seem as if the "cards are stacked" against the legitimate businessman in Jamaica today'.[20] An atypical note was sounded in November when Winston Mahfood, the new President of the JMA, and Jim Lord, a financial specialist, both called on the private sector to have more confidence in the economy. Lord even seemed to think that in Jamaica government control of the private sector was less than it was in the United States.[21]

Even before the PNP's recommitment to 'democratic socialism' in November (discussed below), businessmen had begun to leave the island and to withhold investment in response to increased property taxes, the National Youth Service, Jamaica's foreign policy, and the general economic situation. The export of capital—much of it illegal—had begun even before Manley took office; Edward Seaga said it had already reached 'alarming proportions' in 1971.[22] The Manley government made repeated attempts to reassure the private sector, but outspoken anti-capitalists in the PNP nullified such efforts. The PNP failed to manage the economy properly, making major mistakes in the implementation of programmes, the lack of an overall plan, and excessive borrowing.[23] There was one major success, however, which gave the government and Jamaica some breathing room.

Bauxite

Jamaica became one of the world's most important producers of bauxite, the source of aluminium, after World War II. During the first Bustamante government (1944–55), foreign companies exported bauxite for a royalty of ten cents a ton. Under Norman Manley (1955–62), the royalty was increased fourteen-fold, but for several reasons, the result, Michael Manley wrote, was still a 'grossly unsatisfactory state of affairs'.[24] The Jamaican bauxite economy was an enclave integrated into a global network dominated by powerful multinational corporations: Kaiser Aluminum and Chemical Corporation, the Reynolds Metal Company,the Aluminum Corporation of America (ALCOA), the Anaconda Corporation, and the Aluminum Company of Canada (ALCAN). It provided some good jobs for 9,700 people (in 1972) but otherwise produced relatively little benefit for the Jamaican economy and people.

Manley's trade union work made him familiar with the problems of bauxite workers and had given him contacts in the multinationals. In August 1972, the industrial relations officer of the Kaiser Corporation wrote Manley to congratulate him on his election as prime minister: the Kaiser executive asked Manley to remember him to 'other NWU friends' and recalled that 'even though our negotiations were long, tough, and fatiguing at times, I look back on them with enjoyable and pleasant memories and associations.'[25] Less happily, allegations about the political influence of the aluminium giants had figured prominently in the de Roulet affair and the United States Senate's inquiry into the abuses of multinationals.

For Manley, Jamaica's bauxite industry represented an unusual challenge and opportunity. It represented the largest single capital investment, earned the most foreign exchange, and paid the most taxes of any industry on the island. On the other hand, it was entirely foreign-owned and controlled, withheld much agricultural land from use, was slow to train Jamaican managers, and—logically enough—pursued its own corporate goals rather than Jamaica's national interests.[26]

The Manley government set out to redress the undesirable features of the relationship between the foreign aluminium companies and the Jamaican nation. Before negotiating with the companies, in early 1974 the government formed the International Bauxite Association (IBA), which attempted to unify and coordinate the activities of the bauxite-producing nations to offset the advantage of the multinationals in pitting one country against another and in transferring operations. With the IBA in place, the Jamaican government pursued four goals: (1) a drastic increase in revenues from the mining of bauxite and production of alumina, an intermediate product; (2) control of the rate of mining; (3) reacquisition of all lands owned by the aluminium corporations; and (4) bringing the industry under national majority ownership and control.

Anticipating the concerns of the governments of Canada and the United States, Manley visited Prime Minister Pierre Trudeau and Secretary of State Henry Kissinger. Trudeau, a warm personal friend of Manley's, was reassured that Jamaica intended to negotiate fair compensation for any assets it acquired from the multinationals. Manley told Kissinger that 'we would never seek to affect US access to our bauxite through the legitimate channels of its multinational corporation[s],' that the Jamaican action was purely economic in its intentions, not political or strategic.[27] Nine weeks of negotiations with the companies over the revenue issue turned out to be 'an exercise in pure farce': the companies 'were clearly not prepared to deal with a small island that saw itself as clothed in the rights and dignity of sovereignty.'[28] Jamaica unilaterally passed a law in parliament on 15 May 1974 imposing a new bauxite levy.

Reaction to the bauxite levy varied. Most Jamaicans, even conservatives, supported the action, even as most Chileans, even conservatives, had supported Salvador Allende's actions against the copper multinationals, including the same Anaconda which operated in Jamaica.[29] Rex Nettleford commented on the JBC that Manley had promised that the increased revenue would not be squandered but would be placed in a capital development fund. He went on to note that 'Jamaica, as a major bauxite producer, is at the centre of a war of survival in a world which cares little for the fate of poor developing countries despite pronouncements to the contrary.'[30] When a *New York Times* editorial called for 'importing countries' to protect themselves from international cartels, the *Gleaner* reacted with indignation:

> How dare the *New York Times* call on importers in the big nations to protect themselves against producer cartels in the poor nations, when the developed countries are the fathers of the world's strongest cartels which have been bleeding the developing countries for years![31]

Hector Wynter, chairman of the JLP in the '60s and later, as executive editor of the *Gleaner*, a bitter enemy of the Manley government, reported that, as a result of the bauxite levy, Jamaicans in New York saw Manley as a 'modern Robin Hood', a David slaying the multinational Goliaths, and he predicted that 'this bauxite act will make [Manley] a folk hero.'[32] Not all United States opinion was as negative as the *New York Times*. *Vision Letter*, a bi-monthly analysis of Latin American affairs, referred to Manley as a 'moderate leader' facing 'almost insurmountable problems'. It noted that he did not 'charge the companies with exploitation or whip up public sentiment against foreign devils', and reported Manley's argument that 'obligation to the people outweighed the sanctity of contractual agreements.'[33] American columnist Carl Rowan wrote that 'if you love justice and admire old fashioned guts join me in a tip of the cap to Jamaica.'[34] At the end of the year, the *Gleaner*, having had ample opportunity to weigh the bauxite levy, praised it as 'perhaps the greatest achievement in recent years'.[35]

Numerous results flowed from the bauxite levy and related measures. First, the levy significantly increased Jamaican government revenues, from J$24,504,000 in 1973 (the last pre-levy year) to J$168,422,000 in 1974 and eventually J$367,300,000 in 1980.[36] Much of the increase was absorbed by the increased price of imported oil, but Jamaica would have suffered much more from oil prices without the relief provided by the increased bauxite revenues. Second, instead of investing all bauxite revenues in a capital development fund, as Manley promised, much of the revenue was used shortsightedly, under tremendous domestic political pressure, on current expenses. As frequently happens in open political systems, spending for short-term political needs and to relieve human suffering conflicted with the long-term economic requirements of capital accumulation. Third, it appears that the bauxite multinationals tried to punish the Manley government in 1975 and 1976 by cutting back output and laying off local workers; Manley stated baldly that 'bauxite production was systematically reduced as an act of retaliation by the companies.'[37] Reynolds and ALCOA, through Senator Orrin Hatch of Utah, tried to get Congress to reduce aid to Jamaica.[38] Fourth, the Jamaican bauxite strategy put Jamaica on the map for the United States government. It reacted with some restraint, probably because of the Watergate crisis, its own scrutiny of the multinationals, and perhaps lingering embarrassment at the de Roulet revelations and indiscretions. But US economic aid, $13.2 million in 1974, was cut to $4.3 million in 1975; US loans, averaging $16 million in 1974 and 1975, were virtually eliminated for the duration of Manley's government.[39] The bauxite strategy certainly contributed, along with other events, to strained relations with the United States.

Manley considered the bauxite strategy the landmark of his government.[40] For him personally, it united many deeply held convictions. It attempted to alter the unfair economic relationships between the developed North and the underdeveloped South that he regarded as the most crucial world problem. It was a microcosm of Jamaica's main foreign policy goals. With his own deep interest in heroes and heroics, Manley felt that it provided at least a temporary remedy to Jamaicans' lack of a heroic image of themselves. The closing sentence of the American edition of *The Politics of Change* reads:

> Whatever else the future holds, the Jamaican people at the moment when I announced the [bauxite] Production Levy in the sovereign Parliament of our country knew what it was to be citizens of a sovereign, independent nation.[41]

The 'Politics of Change'

Manley's first book, *The Politics of Change; A Jamaican Testament*, was released in Jamaica in late 1973. In 1974, the *Gleaner* performed a public service by serializing the book and offering $50 prizes to the best reviews of it. When the book was released, in a ceremony including Guyana's Prime Minister Forbes Burnham (whom Manley referred to as 'my good friend'), Manley stated that he wrote it to stimulate thought. He said that Jamaicans spent too little time examining first principles and thus tended to base action too much on impulse rather than careful thinking. Lacking time for extensive research, Manley wrote from 'intuitive understanding'.[42] Later he said that he wrote to clarify his thinking, that he had become a bit out of touch during his years as trade unionist and opposition leader, and was 'slowly struggling back to the rediscovery of my own idealism'.[43] Still later, he wrote that 'in *The Politics of Change* I began to sketch the outlines of a personal philosophy of the kind of structural change relevant to the Jamaican context.'[44]

The book is divided into two parts. Part 1, 'A Philosophy of Change', contains chapters on the Jamaican setting, multi-party politics, equality, self-reliance and 'the problem of attitudes,' and 'social justice'. Part II, 'A Strategy of Change', contains chapters on 'the politics of participation', the restructuring of a post-colonial economy, foreign policy, education, and change and basic institutions. The book also contains Manley's account of the first two years of his government. The book was important in Jamaica because it was an extended, systematic presentation of the views of the country's leader, which was fairly widely discussed and debated. The book avoided extended discussion of 'socialism' as premature in the political evolution of the country and Manley's own PNP, but contained many features of the 'democratic socialist ideology' proclaimed by Manley as government policy in late 1974. Although not autobiographical, *The Politics of Change* does provide insights into its author as a man dedicated to the principles of equality, self-reliance, social justice, participatory democracy, and nationalist policies regarding the economy, foreign policy, education, and culture. It reveals a systematic mind, yet one without the time (or inclination) for extended research. It is in the Caribbean tradition of the politician as man of letters, which includes such notables as Eric Williams, Luis Muñoz Marín, Joaquin Balaguer, and Juan Bosch.

Jamaicans' reactions to *The Politics of Change* spanned a broad range. Lance A. Dunkley called it a 'wonderful testament' of 'great foresight & vision' with 'few relevant gaps'.[45] Father Ignatius Saldanha found *The Politics of Change* to be consistent with recent Catholic social doctrine and commended the book for its 'Christian inspiration', while noting some of its 'fallacies', including that of 'ideological dictatorship'.[46] Alvin Nairne, a Marxist-Leninist, attacked the book as 'an ideological defence' of Jamaican bourgeois, 'anti-masses', and non-revolutionary politics which showed the 'big capitalist character' of Manley's government.[47]

More balanced, judicious reviews came from the Reverend B. L. MacLeavy, C. L. Russell, and S. E. Buchanan.[48] In the context of the Watergate scandal in the United States, MacLeavy wrote of *The Politics of Change*'s 'breath of innocence, almost boyish enthusiasm, [and] idealism so refreshingly different from the tired and cynical outpourings of statesmen elsewhere.' 'One is aware of a sincere, compassionate, ardent reformer with an attractive presence.' Manley 'is completely sincere, he is not out for personal gain or aggrandisement. He has a genuine pride in his people and is determined to give them the best he has got.' On the other hand, MacLeavy criticized Manley for a too one-sided view of colonialism. More importantly he doubted his views of the National Youth Service and education because 'Mr. Manley does not hold dialogue with students in schools nor does he 'permit the genuine involvement of students in decision-making'. He acts exactly as other charismatic leaders, putting before his young followers his plans, inspiring them by his vision to enthusiastic response.' C. L. Russell noted that the book might be better titled 'The Politics of Pride', and noted the conflict between Manley's treatment of the evils of colonialism and his statement that 'democracy' is Jamaica's 'natural tendency'; Russell made the good point that colonialism was a far deeper influence than democracy in modern Jamaica. S. E. Buchanan noted that Manley would probably be revered in Jamaican history for trying to change society, though he was bound to fail. The vast majority of Jamaicans' ancestors were brutalized by slavery and since 1944 Jamaicans had been 'lied to, misled, cheated, neglected, embarrassed and frustrated by virtually every established institution in the land and not the least by their own political leaders whose false promises and weak-kneed political activities have left much to be desired.' People now expected miracles from someone nurtured in the same political system and in the trade union movement. 'As a pragmatist,' Buchanan concluded, 'the author is too deeply steeped in party politics Jamaican style . . . to be as sincere as George William Gordon, as persistent as Marcus Garvey, or as daring as Biblical Joshua.'

The Politics of Change contained most of the elements necessary to democratic socialist development including national control of 'the commanding heights of the economy', increased economic self-reliance, a non-aligned foreign policy, and popular mobilization. But when it was written, the PNP leadership did not share Manley's ideology, and thus the book left some controversial points unclear: the relative weight of different sectors of the economy (public and private, big business, small business, cooperatives); whether the interests of the middle class or those of the 'masses' had the higher priority in the event of a conflict between them; and Manley's attitude toward 'imperialism'.[49] These crucial controversies burst into the open with Manley's rededication to 'democratic socialism' as the official ideology of the PNP and the government.

Rededication to 'Democratic Socialism'

Manley started the PNP along the troubled path to ideological definition with a declaration recommiting the party to 'democratic socialism' in the autumn of 1974. (The specific meaning of 'democratic socialism' was unclear until the publication of the document *Principles and Objectives* five years later [!] in 1979.)[50] The PNP had been a nominally 'socialist' party since 1940 (see ch. 10), but 'socialism' had proved a political liability and had provoked serious internal party divisions in 1952 and the mid–1960s, when Manley sided against the left with party centrists. Norman Manley was an intellectual Fabian socialist. Michael Manley grew up in a household where 'socialism' was part of the routine intellectual fare. His exposure to Harold Laski—whom he still regarded in 1985 as the premier ideologue of democratic socialism—combined with a natural inclination toward rationalism and ideas to consolidate his faith in democratic socialism. The events of the '50s and early '60s—trade unionism, the apparent success of the Puerto Rican economic model, pragmatic politics—caused Manley to give less attention to ideology. The renewed ideological concerns of the mid- to late '60s and Manley's contacts with the young lions of the Mona campus of UWI reawakened in him the necessity of ideological definition. As a condition of seeking the PNP presidency, he insisted on ideological differentiation from the JLP. Most of the PNP leaders in 1972 simply wanted to win elections and lacked Manley's interest in ideology. Shortly after the election of that year, however, he charged an executive body which included party moderates David Coore, P. J. Patterson, and Howard Cooke and radicals D. K. Duncan, Anthony Spaulding, and Arnold Bertram with defining the party's ideology. The PNP Youth Organization (PNPYO) and Women's Movement, led by Beverley Manley, took an active and radicalizing interest in the internal party ideological debate, while, in Manley's words, 'the National Workers' Union, the union affiliate, took much less notice.'[51]

The underlying rationale of Jamaican 'democratic socialism' is that neither the Puerto Rican model of dependent capitalism tied to the United States or the Cuban model of revolutionary socialism tied to the Soviet Union has led to the almost universally espoused goals of growth, equity, and democracy. The Puerto Rican model is particularly weak in equity, and the Cuban model is weak in democracy.[52]

In August 1974, a public storm arose when word leaked of a closed PNP meeting debating ideology. The resulting pressure forced the PNP to reaffirm 'democratic socialism' before it had been widely discussed in the party. There was disagreement on several key points such as the predominant class interest in the PNP's 'class alliance', the role of the private sector vis-à-vis the public sector in the economy, and 'imperialism'. Many PNP moderates were unhappy, but Manley supported the move toward 'democratic socialism' and his position in the party and the nation was unassailable; the moderates could only grumble, drag their feet, or subtly sabotage the leftward drift of party and government.[53]

Manley reaffirmed the PNP's commitment to 'democratic socialism' (which it had never formally renounced) in the party's annual conference in September. He linked the reaffirmation to structural defects in Jamaica's tribalized two-party democracy: 'We borrowed a system which was based on the notion that the public good is promoted by conflict and the proximate instruments of conflict were the P.N.P. and the J.L.P.' Manley explained why Jamaicans were ashamed of their political parties.

> I think that so long as politics is seen as the rather undignified contest between groups that are struggling for nothing more dignified than power, so long as that is true, then so long is shame the proper attitude to apply to the process.[54]

The adoption of 'democratic socialism' was intended to restore moral purpose to the political process. Under capitalism, 'we are all scrabbling like crabs in a barrel to get to the top of the barrel, [and] in the end you get your élitist pyramid and you have defeated the whole purpose to begin with, which was human happiness.' In contrast, 'the Socialist begins with the notion of being part of a community where everybody is committed to the accomplishing of the community's needs.'[55]

'Democratic socialism' was extremely controversial. The *Gleaner* cartoonist Leandro pictured Manley and Dudley Thompson forcibly dragging Jamaica into socialism.[56] The PNP might have been able to weather outside criticism if it were internally united around 'democratic socialism'. To translate the ideology from the pleasing, but meaningless 'Socialism is Love' to a genuine ideology required serious political education in the party. This was attempted in late 1974, based on the pamphlet 'Democratic Socialism, the Jamaican Model'. The 'Jamaican model' was said to be based on 'at least' four basic principles: '1. The Democratic Political Process. 2. The Christian principles of brotherhood and equality. 3. The ideals of equal opportunity and equal rights. 4. Determination to prevent the exploitation of our people'. Several other 'basic principles' were also included in 'Democratic Socialism: The Jamaican Model': the right to own private property; the responsibility of every Jamaican to contribute to the development and welfare of the nation; the belief that Jamaica would flourish best under a mixed economy with an honourable role for private business in partnership with the public sector; faith that the individual and family are the basic units of society; the idea that people create governments to provide for their welfare; and rejection of capitalism as the basis for the future of Jamaica.[57]

Attempts at the political education of party members based on 'Democratic Socialism: The Jamaican Model!' foundered almost immediately. One of the first to alert Manley to difficulties was his brother Douglas, a Senator and Minister of Youth and Community Development. Douglas distributed the pamphlet to his party groups and immediately a quarrel developed between younger and older party members over class relationships and the mixed economy.[58] Divisions in the party forced it to call off the political education drive and definitive formulation of 'democratic socialism' until 1979, meaning that no consensus existed within the party on its ideology for most of the time it was in power. Disunited and ideologically confused, the party rallied for elections, attempted with less success to deal with the economy, and ultimately foundered on the divisive issue of how to deal with the International Monetary Fund. Having proposed the importance of ideology to the political process, Manley had been dealt a bitter blow by his own political party, which itself reflected the state of political development in Jamaica.

CHAPTER 15

Polarization and Turning Points 1975

By 1975, following the recommitment to 'democratic socialism' as the policy of his party and government, Manley had moved left, but was still a moderating influence in the PNP.

> His basic approach, learned in years of trade union organizing, was to conciliate between different views and tendencies within the party. The power he exercised was not of iron-hand discipline, but of the incredible force of his personality, the loyalty he inspired, and his commitment to impose his own political will where he thought it was necessary.[1]

A less sympathetic view, from a leftist university intellectual who on occasion served the government, held that Manley 'merely perpetuates in more sophisticated form the traditional colonialist notion that when the royal governor mingles with his subjects, he teaches them that the great can also be humble and the powerful meek and lowly.' This view dismissed Manley's commitment to physical work projects: 'a kareba-dressed middle-class politician gathering stones in the bush presents but a moment of comic relief in the life of the suffering peasant.'[2]

The year 1975 marked the beginning of wholesale polarization in Jamaica, in which professionals, businessmen, and skilled workers lined up against the Manley government, while support increased from the urban and rural poor. While few new reforms were introduced by the government, existing reforms were extended and with them, support for serious reform of the society. The economy continued to decline, as government efforts at stabilization produced few tangible results.

Two *New York Post* articles by Max Lerner illuminate the situation Manley faced early in 1975.[3] Lerner found Manley 'more self-assured and confident' than he was during the 1972 campaign, 'but still worried about sensible issues'. Lerner believed that Manley's strengths were his ability to handle ideas, his articulateness, and his charisma; his weakness: 'rhetoric that flows on and repeats his familiar phrases.' The *Post* reporter respected Manley's earnestness and integrity. He summarized Manley's brand of socialism as consisting of three essential features: (1) popular access to equality, (2) state control of the 'commanding heights of the economy', and (3) Manley's own contribution, worker participation in economic decision-making. Lerner characterized the heart of Manley's thinking as Jamaica's need as a Third World nation to move away from 'metropolitan psychology'. He noted that Manley occupied both the centre and left of the PNP; Manley responded, 'You have made my day'. The newspaperman volunteered what his big worries would be if he were in Manley's place: managerial skills, crime, alienated youth, and the work ethic. Manley replied that he worried desperately about them all, and stated that 'If I died tomorrow . . . you would find "managerial skills" written on my heart.'

The year 1975 was one of increasing polarization and important turning points for Manley's government, as illustrated by foreign affairs, the media, issues raised in Manley's second book, ideological concerns, the struggle for women's equality, and political violence.

Foreign Affairs

Although Jamaica's warming relations with Cuba—highlighted by Manley's trip there in 1975—was controversial and tended to polarize Jamaican public opinion, other aspects of Manley's foreign policy were highly praised. Manley's trip to Venezuela in April was called an 'unqualified success' by the *Gleaner*.[4] Similarly, Jamaica's hosting of the Twentieth Commonwealth Heads of Government Meeting in Kingston resulted in a front-page *Gleaner* editorial praising the government for 'one of the most successful presentations ever staged in modern Jamaica.'[5] In his speech to the conference, Manley dramatized his central theme of North-South economic inequalities: 'A housewife in America or Europe may be annoyed by the price of her lumps of sugar, but the increment which irritates her may be saving the children of a million sugar workers from malnutrition, if not starvation.'[6] A third venture praised was Manley's agreement with the Mexican government to establish a merchant shipping line which would be twenty-five percent Mexican- and seventy-five percent Jamaican-owned.[7] Despite such successes, relations with Cuba raised great difficulties for Manley, his party, and his government.

In July Manley led a delegation to Cuba, where he was awarded the National Order of José Martí by Fidel Castro. In his presentation speech Castro recalled that early in the twentieth century Jamaican cane cutters in Cuba 'lived in veritable ghettoes just like the Haitian immigrants and many of them spent the best years of their lives in the most degrading and inhuman forms of labour.' Fidel recalled that during his trip with Manley to Algiers in 1973,

> I learnt how while we were fighting in the Sierra Maestra, day after day you followed the events in Cuba with unflagging interest and of the extraordinary delight with which you received the news of the victory of our just cause on the first of January, 1959.[8]

In his response Manley expressed Jamaica's solidarity with Cuba and all 'formerly oppressed' Third World peoples engaged in the struggle for development 'in their several ways and by their separate paths'.[9] In their joint communiqué, Castro and Manley expressed broad sympathy for each other's policies. Castro said Cuba appreciated Jamaica's diplomatic recognition in 1972. Manley said Jamaica admired the Cuban struggle. The two agreed broadly on Latin America, Africa, Asia, and the Non-Aligned Movement. They concurred on economic collaboration and established a Jamaican-Cuban Mixed Commission to deal with common problems.[10] Years later Manley said of this visit, 'we were tremendously impressed with what they had achieved in health, education, and mobilisation, particularly of the young.'[11]

Upon Manley's return from Cuba one of the most notorious events of his first administration occurred, the 'five flights to Miami' speech. Following a rally in Montego Bay, Manley was widely quoted as saying that 'Jamaica has no room for millionaires. For any one who wants to become a millionaire, we have five flights a day to Miami.' Manley later said that he was the victim of calculated lies to make it appear that he was inviting those who disagreed with his policies to leave Jamaica. What he actually said, Manley insisted, was that no developing country can afford a get-rich-quick mentality, and that, since Jamaica was an open country, those who did want to get rich quickly were free to try elsewhere.[12] Some people believed that the trip to Cuba and the media coverage of the 'five flights' speech constituted a fatal turning point, particularly in Manley's relations with the private sector.[13]

The warming relations with Cuba troubled Jamaican moderates and conservatives, members of the JLP, the upper and middle classes, and some Christians, especially fundamentalists (including many poor people) worried about 'Godless communism'. In 1975, however, attitudes about Cuba had not yet hardened. Before Manley's trip, the *Gleaner* carried a very positive story highlighting two Jamaicans who had lived in Cuba since the 1920s, and reported neutrally on thirty-three Jamaicans, the first of some fourteen hundred *brigadistas* going to Cuba for training in construction and technology.[14] The *Gleaner*'s coverage of Manley's trip, while mentioning that Guyana's Burnham and Trinidad's Williams had visited Cuba earlier in the year, also included some threatening symbols: a clenched-fist salute by Manley and a 'bear-hug' between Manley and Castro.[15] After the Cuba trip, Manley assured Jamaicans that a difference between the two countries was that the private sector had a permanent role in Jamaica. At the Denbigh agricultural show on 4 August, he announced that the Cubans were going to help in constructing micro-dams in drought-plagued parts of Jamaica; a *Gleaner* editorial, while calling for more details, called the micro-dam idea 'commendable'.[16] As the year went on, anonymous anti-Cuban, anti-Communist articles began to be featured in the *Gleaner*'s opinion columns.[17]

The Cuban and African strands of Manley's foreign policy came together in the matter of Cuba's involvement in Angola at the request of the latter's Marxist government. Cuba's defeat of an invading South African army on 14 November 1975 prevented the overthrow of the MPLA (Popular Movement for the Liberation of Angola) government. Later, Manley praised Cuba's effort, which appealed to his geopolitical loyalties and to his sense of the heroic: 'You have to go back to the days of Alexander the Great to find a parallel where so small a country by feat of arms has affected so profoundly the balance of forces on a continent.'[18] Jamaica's subsequent support for Cuba's involvement in Angola had a price since, as Manley wrote, '[Henry] Kissinger's sense of order was outraged at the thought of a mere pawn behaving as if it were a queen or at least a rook or a knight.' When Kissinger later visited Jamaica on his honeymoon, he indicated to Manley that he was aware that a hundred million (US) dollar trade credit for Jamaica was under consideration in Washington and expressed the hope that Jamaica would at least remain neutral on the subject of the Cubans in Angola. When, after investigation of the facts, the Jamaican government decided to support Cuba, nothing more was heard of the hundred million dollar credit, although there was no evidence of 'the famous Kissinger theory of linkage' in the matter. Shortly after the news broke about Jamaica supporting Cuba in Angola, Manley wrote,

> a Kissinger confidante, James Reston of the *New York Times*, wrote . . . [an] an utterly inaccurate article about Jamaica. The article marked a turning point in Jamaica's image in the United States. Reston's wild charges about violence in Jamaica, the alleged presence of Cuban troops and Cuban secret agents, all added up to the impression of a Cuban takeover. This started off a chain reaction in the US press which never ceased until we were finally defeated in the elections of October 1980.

Battling with the media, domestic and foreign, became a major task for Manley and his supporters.

Battling with the Media

The myth of media objectivity persists despite the claims by politicians of all stripes in recent decades that the media is biased against them. The Manleys and the *Gleaner* had been antagonists for decades.During Manley's years as Prime Minister, problems occurred first in 1973, when unfavourable reports about the treatment of tourists in Jamaica began circulating. Ironically, it was a foreign journalist, Endre Marton—a 'distinguished writer on foreign affairs for Associated Press'—who first wrote that his personal experience was exactly the opposite of negative news accounts circulating in the United States.[19] Reflecting a common Third World concern, in July 1974 Manley accused foreign news services of 'brainwashing the region' and called for the establishment of a Caribbean news service.[20]

As Manley and the PNP government moved left in 1975, clashes with the domestic and foreign media increased. The *Gleaner* may have sought to establish a moral base for coming battles when it admitted that 'in Jamaica the *Gleaner* is the fourth estate in its influence on how the country operates.' In the light of that awesome declaration, Edward Ashenheim, an attorney representing the *Gleaner*, pledged the paper's impartiality in politics.[21] Many, if not most, PNP activists and Manley himself doubted this 'impartiality', a misgiving shared by more distant observers.[22] The Parliamentary Secretary in the Ministry of Education. Dr Upton Robotham, claimed that the *Gleaner* was guilty of 'shameful misrepresentation' of his 8 February 1975 speech, and he called the *Gleaner*'s account 'a most disgraceful display of irresponsible journalism.'[23] Gleaner headlines often seemed to focus in an unfair manner on the aspect of stories most damaging to the government. For example, one headline on the Gun Court read 'Full Court Div. Unconstitutional'. The first sentence of the Reuters story read: 'The Privy Council judicial committee said yesterday [5 November 1975] that Jamaica's 1974 Gun Court Act, after being shorn of its unconstitutional provisions, still remains a sensible legislative scheme for dealing with people charged with any firearm offence.' The Full Court Division referred to in the headline was one of three divisions established by the act and never met.[24]

There were other media, but none commanded the authority and respect accorded the *Gleaner* by the influential Jamaican middle class (working-class and poor Jamaicans relied more on the radio). The *Daily News*, published from 1973 to 1983, had a circulation of only about fifteen thousand compared to the *Gleaner*'s fifty thousand, and was regarded by those polled in 1978 by Carl Stone as the least honest and fair in its coverage of news.[25] The government's Agency for Public Information (API) began 1975 with a significant shake-up, when longtime Manley friend, novelist, and journalist John Hearne was removed as its chairman and named special adviser to the Prime Minister; Hearne would later become a vitriolic *Gleaner* columnist dedicated to bringing Manley down. API was of limited effectiveness, as 'many Jamaicans, who were or could have been, direct beneficiaries of the PNP's programs. . . . were unaware of what the government was doing.'[26] Reliably 'objective' radio and television would have benefitted Manley and the government, but an impartial JBC—established and maintained by Norman Manley in the '50s and perverted by the JLP in the '60s—was impossible to resurrect in the polarized '70s. In July 1975 the JBC board dismissed its general manager, Dwight Whylie, in part, Whylie said, because of political pressure. Manley had 'asked that the JBC adopt a more positive attitude toward the PNP's policy of Democratic Socialism, while retaining its role as an independent corporation.' Whylie thought this was a fundamental change of policy which required constitutional means. The JBC board denied Whylie's version, citing his deficiencies in budget procedures, modernizing plans, and staff confidence necessary to implement workers' participation at the station.[27] In November, the *Gleaner* criticized Manley for interfering in the JBC by forcing it to cancel the popular show 'Public

Eye' if John Maxwell could not be its host.'[28] The Press Association of Jamaica—which would itself later fall victim to polarization—joined the *Gleaner* in repudiating PNP attacks on the media.[29] In a speech to the PAJ in December Manley said there was a conflict between majority aspirations and 'establishment prejudices overwhelmingly displayed by the Press'. He rejected government intervention in the media, but was concerned that 'two particular elements of the Press had ceased to be concerned about law, illegality, criminality and possible subversion and were only concerned about finances.'[30]

'A Voice at the Workplace'

In October 1975, Manley's second book, *A Voice at the Workplace,* was launched in Jamaica. Manley gave credit to his mother Edna for inspiring it: 'More than any other human being in my life she has influenced the shape and direction of my life's work'.[31] Once again the *Gleaner* serialized the book, giving the reading public the opportunity to gain some insight as to Manley's activities in and thought about trade unionism and related matters. Manley himself regarded *A Voice at the Workplace* as a better book than *Politics of Change* since it was based more directly on profound personal experience and was less abstract. *A Voice at the Workplace* was also much more personal and autobiographical than *Politics of Change*. In addition to Manley's activities in the sugar and bauxite industries and in the JBC strike,[32] *Voice at the Workplace* presents Manley's views on the relationships of Jamaican trade unionism to law, to politics, and to broader concerns.

Manley's basic attitude regarding the law, expressed in an address to the Jamaica Bar Association on 14 November 1975, is that 'law must express the will of the people of a society.'[33] In *Voice at the Workplace* Manley described how the legal system—through such statutes as the Master and Servants Law and the Essential Services Law—favoured employers over employees and the élite few over the many. (An interesting reflection on the father-son relationship is Manley's tacit admission that the previous PNP government of his lawyer father Norman Manley had left these laws in place, even though by 1962 the elder Manley regarded Essential Services Law 'monstrously unjust'.)[34] Both laws were replaced with more progressive statutes early in Michael Manley's first term.

Manley faced a rising tide of labour unrest in his first term. In discussing unions and the law the former trade unionist turned Prime Minister made it clear that 'no group in organised society has the right to act without regard to the interest of the society' and that 'employers and unions are going to have to learn that the economy is not a private cake into which they are entitled to cut as they wish without regard to the general sharing of that cake between all members of society.'[35] It is not too surprising, thus, that Manley had some problems with trade unions, including the NWU, whose vision of society was more restricted and selfish than his own.

Manley's broad vision also encompassed the relationship between unions and politics. Manley asserted that in the '70s in Jamaica 'political considerations have little more than a marginal effect upon union action,' and that the unions had become 'more and more conventional in their response to worker needs.'[36] Thus political unionism, so notable a feature of the '50s, had receded in favour of economic unionism. Manley recognized several weaknesses in the union movement: competition leading to squandered resources; divisions among the NWU, the BITU and other unions which made it almost impossible to meet workers' educational needs; failure to organize large sectors of the work force; neglect of workers' housing 'upon which so much else in human development depends'; wage differentials; failure to address the special problems of women; the unions' lack of 'fundamental objectives' and unionism's tendency 'to infect the political system with its own lack of fundamental purpose.' Manley noted that 'the union movement is a product of the capitalist system' and worried that the union movement was helping perpetuate that system. The fundamental purposes Manley had in mind were a nationalist political challenge to the 'total framework of colonialist oppression' and the building of a society consciously organized on equality and social justice.[37] This was Manley at his best, proceeding from practical experience to articulate a vision which transcended partisan politics and narrow sectoral interests, despite his passionate attachment to the workers' cause.

Ideology

In 1975, ideological issues contributed to political polarization. A school-aged acquaintance of Manley's wrote him that 'the population' tended to think 'you are going communist.' 'I know from the vibes I get from younger people, that they just don't understand what you are doing. . . .'[38] In June, Leslie Ashenheim, chairman of the Insurance Company of Jamaica, claimed not to believe in either capitalism or socialism, but criticized the Manley government's rhetoric:

> The thing I regret about the Government most is not so much the things that they say and do, as their insistence in trying to impose this Socialist slogan. They could have done what they want to do—among them many good things—without trying to place the label of Socialism on us.

Ashenheim—a member of one of Jamaica's leading capitalist families and a director of the *Gleaner*—said that history demonstrated that communism grows out of socialism, a 'fact' which was driving many productive Jamaicans away. Ashenheim also said that egalitarianism 'was one of the most insidious things' being imposed on Jamaica.[39]

The opponents of the Manley government, notably the JLP (now led by Edward Seaga) and the *Gleaner*, pushed the line that socialism is merely a transitory state leading to communism. In September 1975, the idea that Manley was a prisoner of a 'communist' faction within the PNP and that there was an imminent threat by that faction to overthrow him was added to the opposition's propaganda. The PNP's National Executive Council denied all these allegations.[40] In his speech to that body on 28 September, Manley described what differentiated his party's and his government's democratic socialism from communism. The Jamaican model of democratic socialism aimed not at dictatorship, but at truly participatory democracy, not at more power for the State, but at more power for the people. Instead of the class warfare advocated by communism, PNP democratic socialism advocated an alliance of all classes. Whereas communism was against the bourgeoisie, 'to suggest that we are against the middle classes is total madness' since 'the middle classes are the vital repository of higher skills in a society.' Manley also stated that the PNP and the Government took no instructions from outside forces. Manley concluded that 'we are against Capitalism; but we are not Communist.'[41]

Notwithstanding such disclaimers, Seaga claimed—as he would for many years—that Manley and the PNP government were subtly paving the road to communism. In December he announced that 'nationalism' was the policy of the JLP. The PNP's position in the ideological struggle was helped in December by the declaration of the Caribbean Council of Churches that there was much overlap between Christianity and socialism, a crucial selling point in deeply religious Jamaica.[42] In November, in a lengthy address to the World Council of Churches meeting in Nairobi, Kenya, Manley identified himself with the Christian community and strongly attacked 'the moral squalor of the capitalist system.' He called on the WCC to oppose all forms of oppression and domination and support Third World liberation movements. He chose not to introduce the issue of 'socialism' into his address.[43]

Two additional matters that affected Manley and Jamaican politics in 1975 were issues of women's equality and violence.

Women's Equality.

Manley grew up in a family of remarkable women: his grandmother Margaret, his mother Edna, his redoubtable aunts Muriel and Vera, and others. Nonetheless, Manley inevitably absorbed the patriarchal and sexist values of his time and place, and indeed of Western civilization. With class and colour divisions, sexism was a traditional obstacle to equality in Jamaica. One example of this sexism, throughout the '70s, was cartoons in the *Gleaner* legitimizing sexual harassment, exploitation, and even rape.[44] University of the West Indies researcher and lecturer Eleanor Wint criticized the Jamaican mass media for debasing and degrading the character of black women.[45]

Manley did not give much attention to women's equality until required to do so by the responsibilities of office during a time of aroused feminist consciousness. In an interview in 1972, he was unable to specify how women might participate in his government beyond 'social welfare services' and 'citizen advice bureaus', extensions of the traditional housewife role.[46] Esme R. Benjamin, reviewing *The Politics of Change*, criticized Manley for its limited attention to women and for his statement that women have been the 'backbone of the family', when they should have been recognized as the 'backbone of the nation'.[47]

An address to the PNP in September 1974 revealed, however, that Manley, while not free of patriarchal values, was becoming sensitized to the issue of women's equality. Manley was discussing as part of the 'horror' of modern industrial organization, the denial of the individual's right to self-expression, when his remarks took this turn:

> . . . The man may be man-a-yard in his home, if he is lucky, and the woman may enjoy equal 'yardism', if she is lucky . . . (You notice that I quickly recouped myself. . . . I have my home to get back to, too [Laughter] and my home is the beginnings of equality in Jamaica!) I hope you see how I guarantee my peace of mind. We both guarantee Natasha's equality. We watch each other carefully over it.[48]

To be sure, Manley had earlier committed his government to strong support for the *de facto* and *de jure* equality of the 51.2 percent of Jamaicans who were women, had made sure women were paid equal wages on government projects, and had urged privileged women to help their less fortunate sisters.[49] But as his remarks illustrate, deep, old habits do not die overnight.

Beverley Anderson-Manley played a major role in sensitizing her husband to women's concerns. She gradually emerged as the leader of the PNP Women's Movement (PNPWM) and helped make it an assertive, controversial body. Her concerns with women's issues helped move her to the left of Manley. A *Gleaner* editorial referred to her as leading a claque in the Senate gallery applauding a leftist politician.[50] On 10 December 1974 she gave a keynote address at a UNESCO conference on 'Woman and Her Human Rights' in which she noted that one-third of Jamaica's working women were the sole breadwinners for their families. Working women were routinely paid less than men, and a 1968 study showed two-thirds of them earning less than ten (Jamaican) dollars a week. Beverley concluded that the 'need for radical social change exists' and that the government was making efforts to promote it. However, in the Cabinet, men outnumbered women sixteen to one; in the Parliament, there were fifty-one men and two women; of the Permanent Secretaries in government ministries, only one out of twenty-one was a woman. The gossip columnist 'Stella' provided this portrait of the UNESCO affair:

> The Prime Minister, who was 50 years old that day, decided to show that he's ready for the reversing of some traditional roles. Whereas his wife usually accompanied him to functions, this time he accompanied her and never said a word in the programme for she was the main speaker. But

> when she done speak her speak, he got up with the others to give her a little kiss and ting [thing].[51]

The UN had declared 1975 the International Year of the Woman, and in January, Manley reaffirmed that equality was a basic policy of his government.[52] Beverley Manley explained that in Jamaica women's liberation was not against men, but for equality. Its key concerns were employment, discriminatory laws, day care and child care centres, children's libraries and play facilities, help in family management, financial assistance, and community services.[53] At the United Nations World Conference on the International Year of the Woman in Mexico from 19 June to 2 July, Beverley explained that a new world economic order was needed to achieve real equality and she called for the practical training of women in industry, technology, agriculture, and commerce.[54]

The Manley administration enacted several reforms which benefited women: equal pay legislation and the establishment of a women's affairs bureau (1973); the Family Court (1974); and paid maternity leave (1979). In addition, poor women benefited from a wide range of anti-poverty programmes such as minimum pay legislation and community health aides. Manley appointed Carmen Parris as Jamaica's first woman ambassador, assigning her to France, where she performed admirably. Manley said her appointment was a demonstration of Jamaica's commitment to equality.[55] Manley also appointed Ena Allen as the first woman Puisne (Associate) Judge of the Jamaican Supreme Court. These gains for women were threatened, however, by that traditional men's reserve, violence.

Political Violence.

The increasing polarization of Jamaica in 1975 expressed itself in 'tribalized' political violence. The conflicts of the '60s were only a prelude to still greater clashes of the '70s. In 1986 the respected scholar, pollster, and columnist Carl Stone noted that the violence of the '70s went through several stages.

> . . . The first round was one of the JLP using armed thugs to destabilize the PNP's popularity prior to the 1976 election.
>
> In the second stage the PNP organized retaliation but mainly as defensive responses. In the third wave both were on the offensive and pushed the country on the verge of civil war.[56]

It was generally believed that both political parties had connections with armed ghetto gangs, and opposition leaders attempted to tie Manley directly to political violence and gang warfare. (In considering the following allegations, their sources in the political opposition and in the increasingly hostile *Gleaner* should be kept in mind; it is impossible for an outsider to obtain reliable, objective information on this subject.) In January 1975 Robert Lightbourne, an independent who had earlier quit the JLP, queried Manley in Parliament as to whether he or any other government ministers had gunmen in their employ. Manley said the question was impertinent, but answered 'no'.[57] In the same month Seaga called for a State of Emergency to deal with violence in Kingston; he alleged that much of it came from the Garrison gang made up of PNP supporters in the PNP housing scheme, Arnette Gardens (also known as 'Concrete Jungle'.)[58] In April 'Thomas Wright' (Cargill) claimed that Manley had 'special friends in the Garrison gang'.[59] Later the JLP's Pearnel Charles, a bitter foe of Manley's, alleged that the Garrison gang was involved in efforts to remove JLP supporters from the South-Western St Andrew constituency by force.[60] Manley dispelled some of the controversy surrounding the Garrison gang by discussing it in an interview with David D'Costa. Manley spoke of three violent gangs in the Corporate Area; the Garrison gang had 'emerged as a very worrying force in the "Concrete Jungle", constantly at war with the Rema gang and the Tivoli Gardens gang.' The *Gleaner* editorialized that 'since the Garrison Gang has long been alleged to be a P.N.P. strong-arm squad or private army, this is clearly an important repudiation of these ruffians.'[61]

Another controversy erupted around an attack on 14 March 1975 on a funeral procession for a prominent West Kingston PNP supporter, Winston 'Burry Boy' Blake. Manley and Beverley led the procession which was fired on from the JLP stronghold of Tivoli Gardens. Manley commented:

> The marchers in the procession were restrained by myself and others from retaliating. . . .
>
> It seems evident that the Residents of Tivoli Gardens are well armed with guns and are being master minded to carry out political violence at all times. . . . This was obviously a blatant act of political violence aimed at me, Members of the Cabinet, members of the PNP, sympathizers and members of the family of the deceased.

Seaga did not deny the shooting, but blamed the PNP for staging the funeral march through the area, and the police for allowing it. Manley replied that the procession was on public thoroughfares 120 yards from Tivoli Gardens itself.[62] Manley was subsequently criticized by the Jamaican Council of Churches for showing and encouraging disrespect for law since 'Burry Boy' Blake had been said to be 'a known criminal', tried more than once for murder, but acquitted thereof.[63]

In November, Manley accused the JLP of spreading fear and panic. Seaga, in turn, criticized the police commissioner for sending 'handpicked political warriors . . . on sporting missions into Tivoli Gardens shooting recklessly like drunken cowboys, discharging tear gas bombs like pirates sacking a city.' The police responded that Tivoli Gardens residents attacked routine police patrols.[64] When such events transpired against the backdrop of controversial foreign policies, media hostility to the government, labour unrest, ideological issues, and a radicalizing trend in the PNP Women's Movement, the stage was set for more violence in the crucial election year of 1976.

CHAPTER 16

Destabilization.Violence and Re-election
1976

One-two Manley
Three-four Beverley
Five-six Natasha, Natasha.
Seven-eight Shearer
Nine-ten Seaga, Seaga, Seaga

Jamaican Rope-skipping song[1]

Destabilization

In his memoir *Jamaica: Struggle in the Periphery*, Manley was understandably preoccupied with destabilization. The book begins and ends with mention of the Monroe Doctrine, the United States's unilateral declaration of its Western Hemisphere hegemony; three chapters are devoted to destabilization in 1976 and 1980 and to psychological warfare; and a long 'destabilization diary' documents 'unprecedented levels of violence, terrorism, organized public protest and a propaganda campaign locally and overseas' in 1976.[2] Manley saw one element of destabilization as the opposition JLP: 'Under its new leader [Edward Seaga] red-baiting was the order of the day'. Another element was the *Gleaner*, whose concern with Cuban communists dated back, Manley noted, to 1938, long before the Revolution of 1959. In 1975, Manley asserted, the *Gleaner*, had gone 'on the warpath' and had become 'indistinguishable from the opposition, its tactics similar to that of the notorious *El Mercurio* in Chile.' Another element was the press in the United States, including the *New York Times*, the *Wall Street Journal*, *Newsweek*, and other publications. Their articles describing (non-existent) Cuban troops on the island and Jamaica as the cat's paw of the USSR and Cuba were, to Manley, 'part of a plan and not an element in the spontaneous reaction of a free press.' Manley added that 'gunmen owing their allegiance to the opposition had mounted a reign of terror.' He acknowledged that 'of course some [PNP] elements hit back, some giving as good as they got. But it was very clear that we were not the aggressors.' To Manley Jamaica in 1976 was like Iran under Mossadegh and Chile under Allende, both undermined by the CIA.[3]

> They deny it to this day, but I prefer the judgments of the heads of the Jamaican security forces at the time. Police, army, and special branch concurred that the CIA was actively behind the events. My common sense left me with no option but to agree.[4]

Destabilization became a major partisan issue. In his 12 May 1976 budget speech, Manley appealed to the nation to help end destabilization. He referred to the activities of Jamaican migrants and the Mafia, both engaged in the drug trade, and to sophisticated, unexplained violence, industrial unrest, organized letter-writing to the press, a squeeze on the economy and an 'internationally orchestrated' campaign of anti-Jamaican articles following 'this country's support of Angola'. In a press conference on 13 May, Manley revealed that the security forces had been sure since 1975 that destabilization was underway, that it was skillfully orchestrated and almost militarily planned.[5]

The *Gleaner* (alleged by Manley and others to be part of the destabilization plan) devoted an editorial to the subject in June, which noted (1) that Manley told a constituency meeting 'that he could not say in a court of law that the CIA is in Jamaica, but "strange things have been happening" '; (2) that Dudley Thompson had said at an OAS meeting in Chile that there were CIA agents in Jamaica; and (3) that Manley had showed a film to the constituency meeting on the CIA's role in the downfall of Allende. The *Gleaner* said that Manley should present evidence of CIA involvement in Parliament and protest formally to the United States government, rather than conduct an inflammatory campaign of innuendoes.[6] In Chile, Secretary of State Kissinger denied United States interference to Thompson.[7] Vivian Blake, Minister of Marketing and Commerce and PNP moderate, noted Kissinger's denial, but added that 'the diplomats do not always know what the CIA is up to.'[8] United States Ambassador Sumner Gerard later denied allegations of destabilization as 'scurrilous and false'.[9] The controversy was kept alive by a visit to Jamaica in September by the CIA apostate, Philip Agee, who was invited by the Jamaica Council on Human Rights.[10] In November, *Gleaner* columnist Malcolm Sharp called 'destabilization' a 'Big Lie', since 'no shred of proof' had been offered to substantiate it.[11]

Over time, those who debunked the charge of destabilization in 1976 lost interest in it, but a diverse group of journalists and scholars have tended to agree with Manley's view. In 1977, two American journalists, Ernest Volkman and John Cummings, published an article entitled 'Murder as Usual' in *Penthouse* magazine. Volkman and Cummings alleged that, motivated by Jamaica's bauxite strategy and its ties to Cuba, Kissinger and the CIA, with President Gerald Ford's approval, engaged in a destabilization campaign against the Manley government. Part of the campaign (whether with the authorization of United States officials or not) was three attempts to assassinate Manley, the last on the night of the 15 December 1976 elections.[12]

On 10 December 1977, Theodore Sealy, a former editor of the *Gleaner*, spoke to the Press Association of Jamaica of 'the people in Washington—who lend us money sometimes and destabilize us at other times'.[13] In 1985, an experienced, respected Caribbean journalist,

Rickey Singh, wrote matter-of-factly of United States destabilization of the Manley government.[14]

Two careful scholarly accounts in the mid-'80s, while noting the difficulty of documenting destabilization, support Manley's view. Michael Kaufman noted that 'there is some evidence that a modest . . . [Chile-type scenario] was tried' in Jamaica in 1976. He noted that 'fabricated stories began to appear in the North American media,' and concluded that '. . . the balance of evidence from 1975 to 1980 indicates that destabilization greatly aggravated existing economic strains and social tensions.' Kaufman distinguished between spontaneous actions, such as individuals leaving the country or withdrawing funds, and a destabilization campaign organized by the JLP, the CIA, and multinational corporations: 'it is impossible to judge to what an extent an actual destabilization campaign played a role.' 'Yet the sheer mass of evidence suggests that there was such a campaign, even if we are not able to outline its parameters clearly.'[15]

Evelyn Huber Stephens and John Stephens referred to 'a series of unexplained violent events which could only be interpreted as part of a campaign of destabilization.' This series included a new kind of political violence, 'destructive acts without an apparent motive.' There were also attacks on policemen and firemen, representatives of state authority. The Stephenses noted the detrimental role of the United States media and the reduction of US aid in 1975–6. However, based on interviews with Carter Administration officials (who would have been disinclined to minimize skullduggery in the Ford Administration), the Stephenses believed the CIA was not involved, though their evidence was not conclusive. They speculated that although destabilization occurred, there was no single plan or group involved: rather two groups 'might well have been involved': (1) elements of the JLP, 'certain leaders and activists who had a long history of association with gunmen' (they named a JLP candidate and former JDF officer, whose case is discussed on p. 173), and (2) Cuban exile groups based in Miami, who claimed responsibility for bombings in Jamaica in 1974 and 1976.[16]

Violence, January to June

In January 1976 the JLP was 'clearly on the offensive', conducting disruptive protests during an IMF meeting in Kingston. Manley said some local capitalists had declared war on socialism.[17] Manley called on PNP supporters to form *unarmed* defense groups. This led to a storm of protest by the Jamaica Council of Churches and other groups.[18] The *Gleaner* reported Manley as saying that the JLP could not win power through elections 'so they have been hiring armed gunmen to shoot their way to power.' Manley reportedly referred to the 'thugs and murderers of the JLP' and to 'these people . . . trying to bring fascism to Jamaica'. 'I have made up my mind,' he was quoted in the hostile *Gleaner* as saying, 'If they come with the gun—then he who lives by the gun shall be buried by the gun.' The JLP denied involvement in violence, and the *Gleaner* called on Manley to produce evidence to back his allegations.[19]

Violence and crime escalated throughout the first half of 1976. In February, the government ordered a curfew and house-to-house searches, but such efforts were hampered by the inefficiency, corruption, and increasingly pro-JLP sentiments of the police.[20] The violence forced Manley into a defensive position that resulted in some unconvincing statements. With his mind on the American tourist, perhaps, Manley denied that Jamaica was a violent society: 'if you look at the history of our labour movement, for example, there's been none of the bloodshed that marred the early labour movement in America.' Also unconvincing was Manley's response to the following question:

> **Q.** It is also said that some violence may be caused by youth elements in your own party.
> **Manley:** Absolutely untrue. The youth elements of the PNP are very idealistic. The real complaint is that, like all young people, they're too idealistic, too theoretical. They see violence as a total betrayal of everything that socialism stands for. For them, it is a horror, the greatest disaster of Jamaican society today.[21]

After a 9 May meeting, a joint police/military command was formed to deal with increasing violence. Ten days later occurred the Orange Street Fire (or Massacre), in which fifty well-organized gunmen set fire to a block-long tenement, and prevented occupants from leaving and firemen from fighting the blaze. Ten or eleven people died. Five hundred were left homeless.[22]

Reverend C. S. Reid, an independent Senator who believed the Manley government had accomplished much, reflected on the rising tide of blood and fire.

> We may have been stratified by colour and cash in the past, but we were not at war with each other. The nature and intensity of our current divisions constitute a new situation.

Reid said he had received reports that aspirants to political power in both parties used criminals to gain power.[23]

In June, a machine-gun attack on a PNP club killed six and wounded fifty. Manley's 'destabilization diary' carried this entry:

> *June 5–6:* Weekend of violence, 2 killed, 5 wounded in the City's west end necessitating a 24-hour curfew.
> Discovery of a small hospital used by terrorists in Tivoli Gardens, the constituency of the leader of the opposition.
> Security forces confirm the existence of 2 prison cells used by terrorists in Rema, opposition stronghold.[24]

On 15 June, the Peruvian Ambassador, Fernando Rodríguez, was brutally stabbed to death; his murderers were never identified.

On 18 June, Herb Rose, a JLP executive member, resigned from the party, saying that it was 'training young men to commit violent crimes as part of its strategy to undermine confidence in the government'. Rose alleged that the JLP was giving ammunition to 'half starved and maltreated youngsters and encouraging them to take up a life of crime and violence in an effort to win the election,'[25] Rose had been an organizer for the PNP in 1971 and was hired by the JLP in June 1975 for a similar job. Asked to name names at a press conference, Rose refused, but did give names to the security forces. Pearnel Charles, the deputy leader of the JLP (detained under the ensuing State of Emergency), responded to the charges by stating that Rose had been frustrated because he wanted to be a JLP candidate, that he had been under investigation by the party and expelled from it on 17 June for 'mischief' and inability to work with other JLP officials, and that Rose was only 'a minor employee [of the JLP] who was for a period, invited to attend meetings of the Central Executive.'[26] *Gleaner* columnist Malcolm Sharp, consistently anti-PNP, noted that because of Rose's shifting allegiances he was not 'a bloom without thorns', but that 'if one half of what he alleges is true the conclusions are devastating.'[27] On the next day, 19 June, Governor General Florizel Glasspole declared a State of Emergency at the urging of the Cabinet.

The State of Emergency

Manley issued a brief statement explaining the circumstances which had led to the State of Emergency. He said that the violence had been carefully designed and staged. Under the State of Emergency, declared in accordance with the constitution, the security forces were authorized to arrest and detain 'all persons whose activities are likely to endanger the public safety'. The detainees were processed by a special tribunal and either released or detained longer. The State of Emergency would not affect the normal operations of schools, businesses, and transportation. Elections would proceed on schedule.[28]

The State of Emergency, which enjoyed broad public support, lasted for about one year. 593 people were detained, amongst whom were three JLP candidates (including the deputy party leader, Pearnel Charles), two JLP executive committee members, a prominent JLP West Kingston constituency organization member, one PNP candidate (Edwin Singh, caretaker for South Clarendon), and a prominent PNP member of the South Saint Andrew constituency. The press toured the detention facility and issued favourable reports on conditions there, including visits of family and lawyers to the detainees. Ray Miles (JLP) who was released on the same day he was detained, described the comfortable conditions of his detention.[29]

In the House of Representatives on 22 June, Manley explained at length how important it was that the country understand the government's action. He carefully noted that the JLP had called a State of Emergency in October 1966, invoking 'identical constitutional powers'. The Prime Minister summarized what had led to the State of Emergency.

> January, 1975 [*sic*, for 1976], marked a new and dangerous trend. Tragically, the events which have taken place since then have added a new dimension to violence: the dimension of urban terrorism, confrontation with the Security Forces and other agents of the State, and widespread arson. This terrorism has been organised and is designed for the double political purpose of embarrassing the economy of Jamaica and undermining the confidence of the Community in the democratically elected Government.

Manley noted that following the JLP's defeat in by-elections in 1975, there was 'a serious attempt to discredit the existing electoral system on which our democratic institutions rest,' the failure of the JLP to contest a subsequent by-election, and a 'concentrated attempt to spread a wave of hysteria throughout the country based on the oft repeated allegation that the Government was Communist.' 'At the same time, the local press, particularly the *Daily Gleaner*, reproduced a spate of articles emanating from foreign sources, the vast majority of which were unfavourable to the elected Government and the state of the economy.' Manley said that security forces intelligence reports indicated a new wave of violence was planned for late June and July. The objectives of the State of Emergency were to move against gunmen and 'the new breed of terrorist', to smash the link between politics and violence, to create breathing space to allow a 'return to the path of commonsense', to create the security necessary for the economy to function, and to allow Jamaicans to enjoy peace in their homes and streets again. Manley explained the provisions of the State of Emergency. He again expressed the government's determination that the State of Emergency should interfere with politics and elections as little as possible, and announced that the JLP had accepted an invitation to discuss the matter with the government. Manley appealed to 'the great majority of decent people in both the Jamaica Labour Party and the People's National Party to turn away from violence.'[30]

Inevitably, Seaga and the *Gleaner* alleged and implied that the State of Emergency was used for partisan purposes. On 26 June, the *Gleaner* editorialized that twelve important JLP persons had been detained, compared to one PNP caretaker and nine PNP 'minnows'. 'It seems unlikely that the political violence could have been so one-sided,' the paper concluded, acknowledging that all detention decisions were taken by police officials on recommendation of the Criminal Investigation Division and the Special Branch. Although Seaga had called for a State of Emergency in 1975, he frequently raised doubts about its application in 1976.

On 29 June, Prime Minister Manley delivered to the House of Representatives a measured statement whose contents were sensational. Manley responded, in part, to the fact that 'one daily newspaper and the Leader of the Opposition have thought it fit to suggest that the Government is acting in a partisan manner.'[31] He informed the House that the security forces had acquired 'late last week' (after the declaration of the State of Emergency) a briefcase belonging to a JLP candidate who had been a JDF officer.[32] The briefcase contained documents in that candidate's handwriting. Some pages were headed 'Anti-Communist League' and some were headed 'Werewolf', and they contained statements such as:

> Michael Manley and his Government are dedicated Communists and we intend to destroy them at all costs. . . . Werewolf is now willing to take up arms against the Communist regime and purge them from our shores. . . . The High Command will coordinate both the conservative and militant wings and be responsible for liaison with any other forces, local or foreign, who sympathize with the cause.

Another document read:

> ST. ANN'S AREA
> 22 Trained men
> 100 ideological indoctrinated [*sic*]
> 300 Supporters
> SUPPLIES
> 200 rifles
> 100 Sub machine guns
> 2 Barrels of gunpowder
> 50,000 Anti-government pamphlets.

In addition, Manley announced that the security forces had found two tapes 'of the secret transmissions of our Security Forces when engaged in joint Police-Military operations', one of them made on May 14, five days before the Orange Street Fire.[33] Since the security forces themselves made no such tapes, Manley reasoned, 'the recording of these conversations can thus only be a deliberate and plainly subversive act involving a form of espionage on that aspect of the work of the security forces that is so vital in the struggle to maintain law and order. . . .' The tapes were found in the home of a JLP leader. Although the JLP candidate and JLP leader were detained for some time, neither was ever tried. As Manley explained years later:

> Neither the document nor the tape was by itself enough to make a successful case in court. Equally, neither document nor tape has ever been explained.[34]

Seaga denied JLP party involvement in the apparently subversive activities.

The State of Emergency was supported by such groups as the Jamaica Agricultural Society, the Junior Doctors' Association, the Chamber of Commerce, and even the *Star*, the *Gleaner*'s afternoon tabloid.[35] It was ruled legal by Chief Justice Kenneth Smith.[36] Inevitably, it had its critics. Morris Cargill wrote with extreme exaggeration that Jamaica 'has recently decided to have what amounts to a totalitarian government,' that the country was under 'martial law' which would probably continue until the opposition was 'disorganized and intimidated.'[37] More significantly Senior Puisne [Associate] Judge Parnell censured the excesses of some of the security forces during the State of Emergency in staging their 'own private war among themselves.' Reverend C. Evans Bailey, chairman of the Methodist Church in Jamaica, criticized the length of the State of Emergency and the 'apparent reluctance' of the security forces to establish the detainees' guilt, although he praised the physical conditions at the detention centre he saw.[38] Nonetheless Carl Stone reported in November that his polls of 'the masses' revealed overwhelming support for the State of Emergency, qualified by concern about police abuses, especially against youth. He noted that the State of Emergency had given the PNP 'or at least its top leadership' a clean image.[39] By restraining violence the State of Emergency helped rather than hindered the campaign and elections that followed.[40] The State of Emergency reduced, but unfortunately did not end political violence, which continued during the campaign and elections in the autumn.

Political Violence July to December

In July, *Gleaner* columnist Malcolm Sharp wondered 'how can any sane man have confidence in a country when its political leaders are widely suspected of engendering violence?' In September, Aaron Matalon, member of a big business family associated with the PNP, stated that 'some of the senior members of both Political Parties must stand condemned' of promoting violence.[41] A Stone poll of why voters switched parties revealed that nine percent of those who voted for the JLP in 1972 switched to the PNP by 1976 because of the JLP's association with violence, while none of the voters who switched from the PNP to the JLP did so because of the association of the PNP with violence.[42] Nonetheless 'Thomas Wright' (Cargill) kept attacking the PNP as a violent party, and reported several assaults by 'PNP thugs' on JLP political meetings. He claimed in November that the PNP had started a new round of violence and thus 'at last it might be clear which party is and has for years been, the real party of violence.'[43]

Manley frequently had to address the problem. On 7 September he broadcast an appeal to the nation calling on both sides for peace. *Gleaner* columnist Sharp praised him for 'some very commendable sentiments' and a measured, calm tone; yet Sharp thought Manley's facts sounded partisan, and were not likely to appeal to the JLP. Sharp also criticized the JLP for a statement with the phrase 'swimming in blood' after an incident at York Town.[44]

Meanwhile, the Manley government studied the roots of violence in unemployment and political victimization. A memorandum sent to Manley in November recognized that although women bore the burden of raising children (and thus deserved increased resources), violence was almost entirely a male reserve, linked in urban centres to 'political allegiances fuelled by poverty, unemployment, and desperate need'. 'We have seen men kill one another over the allocation of jobs, contracts, houses, and other public benefits, with and without reference to assumed political affiliations,' the memorandum continued. The only way to solve 'this disastrous cannibalism of the poor by the poor' was the total and abrupt separation of the allocation of public jobs and benefits from political affiliation. This would require an 'absolutely independent and impartial body' to administer jobs and benefits programmes, with a minority of members representing the two major political parties. The memorandum recommended that government work be distributed according to age, gender, and place of residence.[45] The problem of such proposals was getting the JLP and the PNP's own cadres to cooperate; the PNP was able to make little progress in this area, except in housing. Perhaps for this reason, Manley, at UWI in November, emphasized 'heavy manners' to deal with the problem of 'killers . . . hired by a part of the political system to try to smash up the confidence of the society in the rest of the political system.'[46]

Stone's polls indicated that although both major parties' hard cores blamed each other for violence, most people had no firm views on who was to blame except 'crooked politicians' and 'criminal gunmen'. Stone noted, however, that the fear of violence could affect the turnout for elections, which his polls suggested would hurt the PNP.[47] As noted above, there were three attempts on Manley's life in 1976: on 14 July, 23 September, and 15 December.[48] In addition PNP leftists D. K. Duncan and Hugh Small were shot at in December. Reggae superstar and PNP sympathizer Bob Marley was also wounded by gunmen who broke into his residence in that month.

Victory: The 1976 Election

By law, general elections were due within five years of the PNP victory on 29 February 1972, but the PNP decided to hold them before the start of the winter tourist season because of the violence. The atmosphere in which the elections would be held was described by Malcolm Sharp. People in both parties sincerely regarded the other side as treacherous and held exaggerated convictions that their own party was good and the other evil. This was:

> politics shaping for a final confrontation between Good and Evil. Between a progressive and democratic PNP and a reactionary and repressive JLP. Between a JLP which stands as a last hope for Freedom and a PNP that has been subverted into a Communist instrument.[49]

Perhaps devoutly held views of 'good' and 'evil' on both sides were more significant than the formalities of the PNP's 'democratic socialism' and the JLP's 'nationalism' and capitalism upon which the election nominally turned.

Manley's personality was also crucial. Sharp tried to turn Manley's charisma into a liability when, borrowing a line from the Prime Minister's Independence Day speech, he argued that 'Joshua becomes Job as he assures us that "all progress rests upon struggle finally." '[50] More positively, Donald Morgan, a Jamaican sociologist critical of the PNP's policies, noted that 'in all fairness to Prime Minister Manley, I must say that he has a keen sense of what Jamaicans need.'[51]

Manley's appeal was displayed in his speech to the thirty-eighth PNP annual conference on 19 September 1976.[52] The National Arena proving too small, the speech was moved outdoors to the National Stadium. Speaking as 'Comrade Leader' Manley characterized Jamaica during the first ten years of 'independence' under the JLP:

> Women would work their hearts out beside men and draw less pay. When children were born, most of them were called illegitimate, and had no real rights in law. The land was writhing under the lash of victimization. Corruption was everywhere in the nation. And there was no Jamaican to be found anywhere who could have pride to say I am a Jamaican.

Thousands then heard the Prime Minister of Jamaica analyze capitalism and imperialism:

> The system of capitalism is based on an economic arrangement in which the surplus of wealth that is made out of production is held by the private hands that own the productive machines. And only those hands decide what to do with that surplus. . . .
>
> And when we talk of imperialism we are talking of international capitalism and the way it works. And if we look at our country, when you talk about how a few get richer and richer and the many get poorer and poorer. . . . it happens because that is the only way the system can work. . . .

Manley denounced the strategies of the 'clique' fighting against his government: its capture of the JLP; propaganda; 'gunmanship and violence'; and economic sabotage. With his voice rising, drawing energy from the crowd, Manley thundered:

> Understand, comrades, if the price of their expanding . . . business is to turn back the progress and to betray the people, and betray the masses, that is a price we will never pay. We will never pay that price. *We will never pay that price!* WE WILL NEVER PAY THAT PRICE!!

The Prime Minister summarized the accomplishments of his first administration and concluded by emphasizing his concern for people and invoking the spirit of Jamaica's National Heroes (except Alexander Bustamante).

> . . . In the name of every child in every hovel anywhere in Jamaica or in the world, in the name of every black man working out his days in the indignity of his servitude in Southern Africa, in the name of every mother whose heart is in agony as she gazes upon her baby and asks, 'What is there for you, my little one?' In the name of every Jamaican who cares for justice, in the name of every youth who has died, in the name of Sam Sharpe, in the name of Paul Bogle, in the name of George William Gordon, in the name of Marcus Garvey . . . (*applause*), in the name of Norman Manley . . . (*applause*) in the name of every victim that ever lived, in the name of every martyr that ever died, we will not fail. We shall overcome. (*thunderous applause*)[53]

This was the politician as Afro-Caribbean man-of-words, as teacher, as artist, as poet, as chorus-master, as prophet, and the JLP had no match for him.

The JLP's Edward Seaga overcame a lack of charisma by stringent organization of his West Kingston constituency, control of the JLP, a reputation for financial wizardry. Seaga was the first major Jamaican political leader without a trade union base, and tension existed between his wing of the JLP and the trade union wing loyal to Shearer. Seaga often criticized the PNP government for overspending, while Shearer, responsive to his trade union base, proposed expanding social programmes. Seaga's early leadership of the JLP was marked by losses in by-elections in 1975 and the Herb Rose revelations of June 1976. In the autumn of 1976, the JLP was rocked further by the expulsion from the party of Ian Ramsey, a forty-six year old attorney. Ramsey claimed he was dismissed to get rid of the sole independent voice in the JLP. He had criticized Seaga's decisions not to contest the by-elections and to let Rose get close to the party leadership, noting that because of the Rose affair 'the JLP as a party may have been gravely compromised.' Ramsey rejected Seaga's 'growing dictatorship of one man' in the party.[54]

The JLP election manifesto explained its ideology of 'nationalism' (including 'local control of the economy'); criticized PNP policies and results; promised a 'people programme' to revive the economy, promote agriculture, reduce unemployment, and provide 'a full education for all' (while avoiding over-taxation); and extolled the JLP's record from 1962 to 1972.[55] Seaga often emphasized the threats of communism and Cuba allegedly posed by the PNP's policies and practices. *Gleaner* columnist 'Petras' argued that the JLP's rhetoric on Cuba and communism had crossed the line between 'legitimate propaganda and distortions that undermine the public interest.'[56] October polls by Carl Stone revealed that most of 'the masses' did not buy Seaga's line that the PNP was leading Jamaica to communism. Most regarded Cuban technical assistance as valuable, despite suspicions about the Cubans. Stone also found overwhelming, non-partisan popular support for the government's social and economic programmes which were directed at 'the masses', despite Seaga's attacks on some programmes.[57]

In response to Seaga's attacks, Manley emphasized that 'Socialism is love.' In 150 meetings before the election he began his speeches, 'Comrades, brothers, sisters: The word is love.'[58] He reached out to working-class members of the JLP, long noted for their deep attachment to its founder, by praising Bustamante's 'contribution to the working class movement in Jamaica'.[59] On 21 November, at a rally in Sam Sharpe Square in Montego Bay, before a crowd Manley estimated at 120,000, about five percent of Jamaica's national population, elections were set for 15 December. At UWI the next day, Manley articulated a deeper vision of 'democracy' than the widely-accepted Western model—'a society is only truly democratic when people themselves are able to affect decisions in all the important aspects of their lives'—and listed the steps his government had taken to widen democracy: the vote for eighteen-year-olds, greater autonomy of local government, community councils which in the future would supervise the price of goods, rent tribunals, democratization of the school system.[60]

Beverley Manley added her considerable intellectual and communications skills with a television address on 3 December, in which she spoke as 'a wife and a mother' to concerns 'posed by some of my sisters'. She argued that change is inevitable, that independence in 1962 did not mean true freedom, that 'money and "management" did not work' from 1962 to 1972, that owing to its history and culture the possibility of communism in Jamaica was remote, and that building a new society would entail 'sacrifice, hard work and disappointments'. She concluded with a plea for viewers to vote for the PNP not only for themselves but 'for those waiting to be born'.[61]

Two days before the election Manley broadcast a description of the conditions under which his government had attempted serious changes: the terrible, convulsive world economic crisis; stagflation; the oil crisis; and an opposition which 'clearly set itself the task of undermining the Jamaican society with lies, malicious propaganda and violence' and which bore 'all the familiar hallmarks of fascism'. He listed among the numerous accomplishments of his government its attempts to control violence. He urged that there was a 'place for everyone in democratic socialism', which required only a few basic commitments: opposition to foreign domination; 'pride and principle'; and a commitment to social justice as 'the core of everything we do' and the great link between Christianity and democracy. Stating that there was 'no easy road ahead', Manley called for austerity and self-reliance: 'this country has now got to learn to build its economy around simpler things like feeding ourselves, clothing ourselves, housing ourselves, building the basic schools . . . and the day care centres.'[62] After viewing one of Manley's campaign rallies, the critical observer Malcolm Sharp noted:

> There's just no doubt about it: this man is the most effective crowd-speaker we've ever had. He outdoes his father and Bustamante. He plays his crowds like a conductor and they respond to him like violins.[63]

On 15 December, the PNP won 56.8 percent of the vote and forty-seven of sixty seats in an expanded House of Representatives. A record 85.2 percent of the 870,972 eligible voters cast ballots. The PNP increased its share of votes from the very poor (to 71.6 percent) and the poor working class (to 60.6 percent), but saw its support decline among the better-off working class (to 64.5 percent), the lower middle class (63.1 percent), the middle middle class (to 40.2 percent), and the upper middle class (27 percent).[64] In a fundamental realignment, the PNP was becoming a mass-based party, while the upper working and middle classes shifted to the JLP. Commenting on the election, the defeated Seaga said, 'I think the PNP scored a very clear and decisive, victory' and 'the people . . . had made a clear choice.'[65] Months later, the none-too-happy *Gleaner* added that 'the majority of voters have endorsed the PNP and its programme of democratic socialism.'[66]

CHAPTER 17

The Descent Begins 1977

Manley began his second term facing mixed prospects. The 1976 election was a strong popular mandate for his administration, but political support waned surprisingly soon after the election. The Rema Incident in February showed that political 'tribalism' always simmered just beneath the surface, ready to erupt with disastrous consequences. The PNP had temporarily overcome serious internal tensions for the purposes of the election, but had not really healed such problems as ideology and the rifts between its right and left wings. The economy had deteriorated critically, which would eventually require controversial negotiations with the International Monetary Fund. More positively, the advent of the Carter administration provided, for a while, a more liberal and tolerant atmosphere in Washington. Jamaica's relationship with Cuba was momentarily less controversial, and Fidel Castro made a successful visit to Jamaica in October. The year would end with Jamaica failing an IMF test, with painful consequences for the future.

The Rema Incident

In 1977 violent 'tribalism' continued to rend Jamaica's social fabric and undo Manley's hopes of serious reform. The Rema Incident in February hurt the Manley government. Rema was (and is) one of several government housing projects in urban Kingston. As Manley later said, these projects—Rema, Tivoli Gardens, Arnett Gardens (Concrete Jungle), McIntire Lands (the latter in Manley's East Central Kingston constituency), and others—became 'political fortresses' in the bitter political struggles of Kingston's poor people.[1] Rema was a JLP-dominated area in PNP Minister of Housing Tony Spaulding's Southern St Andrew constituency. Spaulding faced a challenge from the JLP's Pearnel Charles, who, Manley said, wished to 'do a Seaga' in the constituency. For reasons disputed by the JLP and the PNP, the residents of Rema refused to pay their rent. In the flush of electoral victory, Spaulding moved to evict the non-paying tenants. In the violence that followed one man was killed, several were wounded. PNP loyalists began to throw the possessions of JLP people out of apartment windows, and gang violence ensued: it was, Manley recalled, 'an awful scene'.[2]

The Rema Incident was widely perceived as a grab for political territory involving the replacement of a thousand JLP supporters with a thousand PNP supporters. Groups such as the Jamaica Council of

Churches, the Jamaica Teachers' Association, and the Jamaica Council on Human Rights called for an impartial inquiry. Unfortunately, the report of the inquiry by a retired Supreme Court judge was delayed seventeen months, during which the view of the incident as blatant PNP victimization festered, damaging the government. When the report did come out it traced the cause of the incident to long-standing political rivalries, aggravated by the 'overwhelming defeat of the JLP, thereby giving rise to the unquenchable thirst on the PNP side for further fields to conquer', and on the JLP side, to resentment and the desire to embarrass the government by non-payment of rent.[3]

When Manley visited Rema to discuss residents' grievances, several complained that he held meetings in a schoolroom rather than participating in a more open exchange, and that he talked to older people, not youths.[4] With the Rema conflict in mind, Manley discussed the problem of violence and the division of 'the masses' in a speech at PNP headquarters on 18 February:

> . . . It is already a sad and difficult problem that there is so much division and bitterness and strife at the mass level. Sometimes one has to defend oneself against the consequences of this bitterness and strife. But the first objective of the Socialist must be to unite the masses and not to divide them.[5]

In early March Manley wrote to an English friend that he was trying a personal initiative to bring peace to Rema, Concrete Jungle and Tivoli Gardens.[6] This led to no lasting result. The Rema Incident got Manley's second term off to a tragic start.

The Intra-Party Struggle

In January Manley wrote to a critical supporter of the PNP about the difficulty of understanding the inner workings of a party.

> The truth is that the internal dynamics of a political movement can only be fully understood by those who live within the particular environment. . . . These dynamics . . . take on a life and reality of their own which exists quite apart from the nation of which it is a part.[7]

Manley led a party whose social composition—middle-class leadership and bureaucracy, and lower-class base—inevitably led to intra-party contradictions and tensions. This has frustrated earlier attempts to define 'Democratic Socialism', which led to ideological confusion until at least 1979. The competing camps consisted of a right wing (sometimes called 'moderates') of mostly older members who predominated among MPs and Parish Councillors and a left wing of mostly younger party activists and intellectuals with stronger ties to 'the masses' and an interest in more radical, rapid change than the right wing. Manley occupied a central, conciliating, and negotiating position, intellectually sympathetic to the leftists, but tied by long-term personal and political interests to the party right. Each faction feared that the other would 'capture' and manipulate Manley.

The selection of party candidates for the parish elections of March 1977 (which the PNP swept) indicated the depth of intra-party strife. In February a telegram to party headquarters from a local PNP Youth Organization (PNPYO) group read:

> IRREGULARITY IN SELECTION OF CANDIDATES FROM— —DIVISION CAUSING DISUNITY AMONG SUPPORTERS. EXTREMELY IMPORTANT. INVESTIGATE IMMEDIATELY.

Other charges included undemocratic procedures, murder threats, bribes, violence, alleged plots to overthrow Manley and others, attacks by 'fascist elements', and moves to undermine 'democratic socialism'. There were also claims of 'internal tribal conflict' and conflict between youth and 'older members of the party'.[8] The intra-party conflicts bore a striking resemblance to inter-party conflicts, and there is little that the PNP accused the JLP of, that party factions did not also accuse each other of committing.

At the 29 July PNP National Executive Council retreat, the attention Manley gave to the intra-party situation implied that it was almost more critical than the ongoing conflict with the JLP or the economic crisis. He noted that to many PNP leaders democratic socialism was only a palliative and that its deeper implications of a fundamental restructuring of the nation had eluded 'many of us'. He said the basis of party unity must be those committed to the struggle for socialism and against imperialism. He urged the need for constant and consistent study, instead of decrying those who studied as 'arrogant'. There was a need for structured discussions to arrive at decisions 'to which all members must subscribe.' 'We are destroying ourselves,' Manley warned: 'The only solution is an uncompromising redefinition of our ideological direction and intensive political education.'[9]

The PNP's internal struggle reached crisis level in September with the confrontation of the left and the right over election of party vice presidents. Manley suspended the elections and took over personal supervision of the party secretariat. Leftist D. K. Duncan resigned as PNP General Secretary and Minister of Mobilization. Manley received letters such as the following:

> Mr. Manley, honourable Prime Minister, are you really aware of how low down in the gutters some of the more influential and *powerful* members of the party have dragged its name . . . ? Because of their selfishness, greed and dishonesty the party deserves to lose the next general election whenever it should come around.[10]

Moreover, the party had failed to consolidate its recent victories in the

JLP stronghold of St Thomas. Its MPs were inattentive to constituents, the PNP Parish Councillors quarrelled with each other, and there was minimal commitment to the government's programmes.[11]

A bright spot was the PNP's accreditation committee for candidates for government posts, which advised ministers of government. It was part of the party's commitment to participatory democracy. Its report of 16 September stressed equal concern with four criteria: commitment to the party's philosophy, policy, and programmes; loyalty; ability; competence. Its deliberations—forty-two meetings were held between 13 January and 15 September—were highly confidential since they discussed candidates' unpleasant features. Most ministers cooperated with the accreditation committee, but some tried to minimize or bypass it. There was a high degree of agreement between the committee and the Cabinet.[12] On 15 September the committee wrote to Manley disagreeing with his choice for chairman of the Radio Jamaica board and suggested another choice; Manley apparently accepted. He also crossed out one name proposed for the board, substituted another, and suggested the need for a representative of the Church.[13] The committee, unfortunately, had neither mandate nor resources to address the larger internal problems of the PNP.

Economic Crisis and the IMF

Underlying the political picture was a severe economic crisis as a result of Jamaica's chronic underdevelopment; oil prices; world inflation; the cost of Jamaica's large imports of food; the decline of tourism, bauxite, and bananas; the flight of capital; and a tightening money market. Gross Domestic Product had declined 3.9 percent in 1974, 0.4 percent in 1975, and 6.3 percent in troubled 1976; it would decline another 2.4 percent in 1977, before rising a miniscule 0.3 percent in 1978, and then falling 1.4 percent in 1979 and 5.3 percent in 1980.[14] As a result of the crisis, secret negotiations were opened with the IMF in late 1976, before the elections. Only Manley and a handful of top officials, knew of these negotiations. Edward Seaga, however, found out about them and the IMF's demand for a forty percent devaluation of the Jamaican currency.[15] Manley later wrote that 'obviously, this could not have been done in the middle of an election campaign.'[16]

The IMF was established in 1944 at the Bretton Woods, New Hampshire conference to rebuild and stabilize the world economy after the depression and World War II. The developed capitalist countries in general and the United States specifically have enjoyed a veto power over its important decisions, owing to their contributions to the fund from which loans are drawn. Countries experiencing severe economic difficulty (most Third World countries in the '70s and '80s) apply to the IMF for credit which is granted in slices called 'tranches'. The first tranche is automatic, but beyond that tranches must be negotiated between IMF officials and the borrower. The IMF claims that its ideology is neutral and that it provides technical advice to member countries. The many critics of the IMF deny its neutrality, pointing to a bias toward free-market ideology that tends to reinforce economic neocolonialism. Critics also argue that the IMF's 'advice', strict adherence to which is often a condition of loans, requires a surrender of national sovereignty and is designed to remedy the problems of developed industrial economies, not Third World ones. The report of the Independent Commission on International Development Issues, the so-called 'Brandt Report', noted in 1979 that

> The Fund's insistence on drastic measures, often within . . . only one year, has tended to impose unnecessary and unacceptable political burdens on the poorest [nations], on occasion leading to 'IMF riots' and even the downfall of governments.[17]

Jamaica between 1977 and 1980 was a case study of that process.

The magnitude of the PNP's 1976 electoral victory temporarily encouraged Manley and the left to consider a non-IMF solution to the economic crisis. In a speech on 5 January 1977, Manley argued that the IMF would impose harsh terms that would mean cutting government programmes.

> This government, on behalf of our people, will not accept anybody anywhere in the world telling us what to do in our country. We are the masters in our house and in our house there shall be no other master but ourselves. Above all, we are not for sale.[18]

On 19 January, in a speech to parliament carried on live nationwide radio and television, Manley again rejected applying to the IMF on the grounds that grave issues of policy and 'the sacrifice of important elements of a nation's sovereignty' attended such a move. He called for a programme of economic reconstruction based on 'democratic socialism', national self-reliance and the mobilization of the people.[19] As a personal example of sacrifice, Manley agreed to cut his salary from J$22,000 to J$16,000 a year.[20]

The leftist, non-IMF alternative, the Emergency Production Plan (EPP), became a major battle between the PNP left and right. On the left were several UWI economists, notably Norman Girvan, who became Chair of the National Planning Agency. On the right was the PNP financial management leadership, including Deputy Prime Minister and Finance Minister, David Coore and Bank of Jamaica head, Arthur Brown. The left and right clashed over ideology, the IMF (opposed by the left), the bureaucracy (which the left said needed to be overhauled to implement its economic plans), and even style, as many leftists dressed in jeans and 'tams' (knitted caps) and thus offended the more traditional rightists. The preparation of the EPP was a tremendous effort. It proposed radical changes not only in the economy, but in mobilization, decision-making, and political power.

The EPP rested on three elements: self-reliance on Jamaican natural and human resources, diversification of foreign economic relations including stronger relations with the Eastern bloc, and changing the economic structure of the country with an emphasis on agriculture. The two stickiest issues were financing the plan in the absence of IMF support and the loss of jobs in the manufacturing sector (an estimated 25,000 jobs, one-third of the manufacturing work force).[21] Many of the manufacturing workers were PNP supporters, and the loss of jobs had serious political implications for the party and government.

According to Manley, the EPP was debated in the Cabinet (where the balance lay with the right and centrists) in late March. Manley wrote that the Cabinet after an 'agonizing session' 'decided that there was nothing for it but to resume the negotiations with the IMF.'[22] Other careful accounts argued that Manley made the decision himself and sold it to the Cabinet, so that the decision to go to the IMF was not actually made by the leadership per se, much less debated or discussed downward throughout the party, to the rank-and-file level. The party left shared this view, which for them explained not only their own subsequent alienation but that of the rank-and-file and 'the masses' as well. An interesting sidelight is that Cuban leaders, Carlos Rafael Rodríguez immediately, and Castro himself in October, were brought in to sell the IMF approach to the Jamaican left. The revolutionary Cubans played a moderating role urging accommodation with the international capitalist order.

A party moderate offered this characterization of Manley's leadership style in this connection:

> Michael's trade union background gives an interesting insight. He was a supreme performer. He had no equal in his facility with figures or debate. He was also tough. Now, the system requires that negotiations be ratified by the workers. Michael's ratification would be obtained by explaining—or selling—his negotiation. Consensus was not by debate but rather by salesmanship. He took this system to the political arena[;] . . . he was the quintessential advocate.[23]

In deciding to turn to the IMF, Manley was guided by several considerations: what he judged to be the EPP's unrealistic estimates of foreign exchange, the resultant drop in imports, the manufacture of which would cost many thousands of politically-sensitive jobs; the possibility of influencing the more moderate administration in Washington; and his influence with Britain and Canada. In fact, Carter in Washington, James Callaghan in Britain, and Trudeau in Canada all put pressure on the IMF to extend moderate terms to Jamaica, which it did, nursing meanwhile resentment for Manley's political pressure.[24]

The compromise People's Production Plan, announced on 22 April, retained the rhetoric of self-reliance, but took the heart out of it by turning to the IMF. No one could foresee the severe suffering and political collapse of the government that would result from accepting the money and tutelage of the IMF and no one can know if the EPP would have worked. Emphasizing the loss of jobs under the EPP, Manley later reacted angrily to the idea, advanced by leftist critics, that turning to the IMF involved breaking his 'we are not for sale' pledge.[25] The pledge represented Manley's sincere conviction, but that it was based on an insufficient recognition of the economic realities. In siding with the right wing of the party in the Spring of 1977, Manley adhered to the pattern established in 1952 and 1964–5, although the party as a whole had clearly moved left in the 1970s under his influence.

By 1977, with one exception, the Manley government's socio-economic reforms were over, and most of the remainder of Manley's term would be spent wrestling with the IMF and the economy and dealing with political problems. The exception was the State Trading Corporation (STC), introduced in November 1977. The STC was a government-owned holding company with a monopoly of importation of items consumed in sufficiently large quantities to warrant bulk purchase. It lowered costs, saved foreign exchange, and provided a cheaper 'basic needs basket' to consumers. Like each extension of state power into new areas, the STC worried the private sector, but under managers like O. K. Melhado, who served both his friend Manley and the succeeding Seaga government, the STC was successful. The People's Production Plan, however was not. In December Jamaica failed a quarterly IMF test by a small amount. The IMF used that to force a very harsh economic package on Jamaica that eventually helped undo the Manley Government.

Women and Workers

The victims of the economic crisis were mostly women, workers, and the poor. One of the most important influences on the PNP and Jamaican left was the pressure exerted by progressive women, including, notably, Beverley Manley. Their cause had been stimulated by the UN International Year of the Woman in 1975, and they had added a new dimension to prior critiques of Jamaican inequality which focused on race and class (although some leftist women preferred to denounce capitalism and imperialism rather than sexism). On 17 April 1977, Ralph Brown, PNP Minister of Local Government and MP for West Central Kingston, said that four out of five PNP workers were women.[26] That gave a significant thrust to such critiques as that of Bea Lim. She noted that the Jamaican Constitution was silent on the subject of sexism, although it banned discrimination on account of race, colour, creed, or political opinion. She said that 'discrimination against women is a way of life in Jamaica,' and by no means limited to conservatives:

> . . . It is often the very 'brothers' who rail the hardest against 'imperialism' and other evils (foreign ones of course) who are the most heavy-handed imperialists on their own home fronts.

Lim noted that only two percent of employed women were unionized, 'a shocking statistic in a union-oriented country like Jamaica'. She provided details of laws which 'blatantly' discriminated against women, and noted that there was no woman in the Manley Cabinet.[27]

Also in April, the seventy-member Committee of Women for Progress (CWP), described as a democratic, bipartisan and anti-imperialist organisation, was founded. It denied being feminist, insisting that women's problems are not caused by men, but by imperialism. Whereas others worked for women through projects and community work, the CWP intended to emphasize agitation and monitoring the overall progress of woman.[28] A further sign of the times was the swearing in of Jamaica's first woman soldiers, fifty-four in number.[29]

Manley's government had already done more for women than any other administration in modern Jamaican history, but he recognized the need to do more. In his budget speech in May Manley addressed the subject of women as an aspect of mobilization. He said mobilization itself was 'the true key to survival and later to progress in the Third World'. He noted that his government had created a Women's Bureau and 'an apparatus of advice' in the office of the Prime Minister. These were designed to redress 'the most historic exploitation in all human civilization: the exploitation of women by society'.[30] Beverley Manley, a forceful advocate of women's causes, gave Manley high marks for his consistent support of this aspect of Jamaica's multitude of problems, but obviously much more remained to be done.

Poverty was the most important and pervasive reality of all. The suffering of ordinary Jamaicans was partly revealed by a meeting on 23 October 1977 between Manley and government workers who, although they were more fortunate than some of their unemployed brothers and sisters, nonetheless suffered greatly under prevailing economic conditions. The workers were not overawed by their Prime Minister. Their sympathy was with him, but tested by their hard lot. Manley listened carefully and at various points made polite inquiries. Post Office workers complained that they had only two wage classes, while other agencies had up to nine, and that most were at the lowest level. One worker with thirty-three years' service received $J49.20 per week. Five or six workers remembered pledges made to them by Manley and his father in 1966; Manley had said the Post Office employees would be first on his list whether Norman Manley wanted it or not, but now 'it is as if we have been forgotten entirely.'

The workers made eloquent and specific appeals. A worker in the public cleansing department said:

> . . . We have a lot of casual workers which were 700 and odd. And it cut down to ten. And that is very bad. Can't get nothing to eat, that simply mean to say that those people have to go out and thief and whereas you don't want that. You see, that system supposed to be removed.

Another described the indignities suffered by sanitation workers in terms that affected Manley powerfully:

> I personal know that none of them is treated like a human being. They are treated like a swine. . . . You have a good pen that you put your swine into, chiefly the English hogs and you don't have a good pen to put Public Cleansing Workers. . . . (*applause*.)

Concerning one of Manley's favourite programmes, a woman worker at the Jamaica Hotel School in Casa Monte observed: 'Worker Participation, Board of Directors say no.' Someone said getting medical care at Kingston Public Hospital was very time consuming and couldn't there be a special service for government workers. Manley responded that he wanted to think about it, but was concerned that a special service would undermine the principles of equality; the workers applauded. Toward the end a worker addressed his Prime Minister. 'I am you and you I am.'[31]

Later, Manley revealed a hard truth: 'every dollar spent on a Government worker's wage is a dollar withdrawn from the obligation of schools for the children, health centres for the sick and jobs for the unemployed.'[32]

Relations with the United States Government

While dealing with political problems, economic crisis, and human problems, the Manley government enjoyed a warming of relations with the new Democratic administration in Washington. One of the most progressive people in the Carter Administration was Andrew Young. A *Gleaner* editorial commented that one purpose of a trip by Young to Africa was 'to spread the word that America no longer was a racist regime opposed to the aspirations of black Africans.' Another source of support for the Manley government was the US Congressional Black Caucus. Its Chairman, Congressman Perrin Mitchell, reaffirmed the caucus's friendship for the Jamaican people and support for the government's just policies. Of bauxite, Mitchell said it was only to be expected that countries would want to regain control of their natural resources from foreign hands.[33] Manley wrote to a personal friend in March: 'I watch Carter with fascinated interest. I hope he lives up to some of the early promise.'[34]

A further sign of improving Jamaica-United States relations was the one-day stopover of Rosalynn Carter, the President's wife, at the end of May, as part of her Latin American tour on behalf of her husband. Her State Department briefing book described Manley as 'a leader "as magnetic as any that exists," a fiery orator, suave and sophisticated.'[35] She and Manley spent seven hours in substantive talks, which focused on the world economic order: Manley advanced his view that the system 'condemns the third world to endemic poverty.' Carter told a news conference that she was 'very much impressed' by Manley: 'he really hurts for people who lack food and clothing and housing.' Mrs Carter ended her tour with a walk through a poor section of Kingston. Manley found her to be a 'first class' diplomat.[36] In her memoir, Mrs Carter recorded that Manley 'defended Cuba's policies, in spite of the facts,'[37] indicating the gap that existed between her government and Manley's.

In August, Andrew Young visited Jamaica and said that 'the United States respects his efforts to steer Jamaica along political and economic paths that Manley calls "a third way" between capitalism and communism.'[38] Although the change in Washington was welcome, the PNP reacted cautiously. An internal PNP memorandum referred to Carter as 'a man of some moral fibre', but reckoned that his presidency was not a change in the economic direction of United States imperialism. It went on to say that the PNP was at an early stage of the liberation struggle.[39]

Cuba

Manley's Cuba policy and Cuba itself remained controversial in Jamaica, and not just with Manley's enemies. Reverend C. S. Reid, a former Senator and critical supporter of the Manley government, was among those Caribbean churchmen who as early as 1970 urged the Jamaican government to recognize Cuba. However, he criticized the 'style' of Manley's Cuba policy: 'We seem bent on becoming the Chief Apologist of the Cuban Revolution in the Western Hemisphere.'[40] Nor were Manley critics necessarily blind to the benefits of Jamaica's relationship with Cuba. *Gleaner* columnist David d'Costa remarked, somewhat sarcastically, that Jamaica had received some excellent condensed milk from Cuba at low prices and, more seriously, that the excellent José Martí School at Twickenham Park was built at minimal cost to Jamaica. D'Costa also noted the hypocrisy of the human rights policy of Carter, which overlooked Iran's record: 'The Shah runs a terror state of such despicable dimensions as to make Cuba seem a positive model of human-rights rectitude.'[41] Indeed, contrasts and contradictions seemed to mark Jamaican attitudes toward Cuba: a drawing originating in the Jamaican exile community in Toronto showed Manley in combat fatigues with a Castro-style cigar in his mouth, but when a Cuban classical dance company—which would perforce appeal mostly to the middle classes—visited Jamaica, it was sold out well in advance.[42]

From 16 to 21 October 1977, Jamaicans got to experience Fidel Castro up close, as he returned Manley's visit of 1975. In private, Castro endorsed the IMF path, attempting to reassure the Jamaican left. He also advised the more ardent leftists *against* following the Cuban path, because of its very heavy costs; this advice shocked and dismayed some on the left.[43] On 17 October, Manley and Castro addressed a crowd of a hundred thousand in Sam Sharpe Square in Montego Bay. Castro recalled their meeting during the trip to Algiers in 1973, and said that from that time he admired Manley's 'sincerity, honesty, strength, love of the people, and revolutionary principles.' Fidel said Manley had made a tremendous impact at Algiers, and had made the world aware of Jamaica.[44] Later Manley wrote to his son Joseph, who had decided to attend university in Cuba:

> Fidel's visit was an absolute triumph from every point of view. His own performance was a masterpiece of mature diplomacy and has done tremendous good from the point of view of the process.[45]

Manley expanded on his private thoughts on Castro's visit in a long letter to an English friend, which provided this unusual view of Castro:

> He had a tremendous two and a half hours meeting with thirty-five church leaders and only ducked one question in the entire period—though it took a keen observer to notice that he had ducked! A noted Jesuit asked him to explain why a certain text book for ten year olds asserted that Jesus Christ was an important figure in literature and the evolution of ideas and philosophy but had never actually existed. His reply circled the issue in ever increasing concentric circles until it had safely transferred the argument to safer terrain. Halfway through the exercise I very nearly succumbed to an attack of hysterics from which I was saved by my admiration of the skill with which the exercise was carried out. Apart from that the whole exercise was frank, friendly, blunt and enormously effective. He confessed that it was the first time he had a meeting of that sort in more than seventeen years. He was very apprehensive beforehand and seemed extremely exhausted afterwards. But then he is like that. He throws himself into what he does with almost ferocious concentration and 'gift of himself.' I think he found Jamaica interesting, and all agree that it was the most significant and important visit by a foreigner in Jamaican history.[46]

Joseph Manley's attendance at the University of Havana added a personal dimension to Manley's interest in Cuba. Joseph's letters from Cuba to his father conveyed two broad themes. As a child of the Jamaica consumer society he was constantly writing home for such things as sports and stereo equipment unavailable in spartan Cuba; on the other hand, as he came to understand Cuba from years of residence and study, he appreciated the Revolution more and more. In September 1977 Joseph wrote that 'the University is fantastic,' that he was doing very well, that he understood everything in Spanish, and that his guard-duty and voluntary work was about to begin.[47] On the last day of Castro's visit to Jamaica, Joseph wrote that 'my really hard year will be my 3rd, with cybernetics, computer econometrics and political economics.' Of current coursework, 'the most interesting subject is Marxist-Leninist philosophy. We really go into that one in depth, as you can imagine.' He thought Fidel was brilliant in his televised Jamaican press conference.[48] In December, Joseph wrote of Beverley Manley's trip to Cuba:

> . . . Bev's trip is a marvelous and—words fail! Superb. Unequivocal success. She gave a fantastic speech at the mass rally in Santiago last night—very well received. The speech got front page coverage in the papers under the headlines. 'Combative Act. . . . Very militant.'

Joseph congratulated Manley on the success of the bauxite venture at that point: 'For a "bungling socialist idiot/Messiah" you're doing rather nicely!' He added that playing tennis in Cuba on dirt, or muddy, courts was a mess.[49]

The End of Year

In a broadcast to the nation on 15 December, Manley characterized 1977 as 'one of the toughest years we have ever had to get through.' Crime was a serious problem, and special attention had been given to 'the awful problem of rape'. Responses to the crime problem included the increased presence of the security forces on the streets, the reorganization of the Criminal Investigations Department with help from the Royal Canadian Mounted Police, and plans for legislation providing mandatory life sentences for gang rape or rape involving use of a weapon. Manley noted that unemployment was especially hard on youth and women. He also said that the 'opposition Press has intensified its campaign of negativism, and now even hate.' The Prime Minister said in the 1960s that Jamaica had managed to live beyond its means because of favourable world conditions and heavy expansion of the bauxite industry, circumstances that no longer prevailed.[50]

A week earlier, Manley had written to a friend in Canada who was confused about events in Jamaica. Manley gave his friend an optimistic view of the economy, stressing positive developments in foreign exchange budgeting, tourism, bauxite, domestic agriculture, the visible trade balance, and non-traditional exports. 'We are beginning to live less luxuriously which is only fifty years overdue. That is not too bad as things go in Jamaica.' Manley stressed that the real problems were crime and unemployment: 'Crime particularly is my most heart-rending worry.' His optimism returned in discussing foreign relations, as he discussed 'good' relations with the Soviet bloc, 'excellent' relations with the Carter and Trudeau administrations, and the strengthening of 'our position of leadership' in the Third World. With respect to the latter Manley described two important, but as it turned out, short-lived victories. The first was the decision of Mexico to go ahead with the Javamex alumina complex 'in spite of attempts of the multinationals to smash the venture.' The 'greater triumph' was getting the International Bauxite Association to set up a common minimum price for sales to North America: 'This represents another triumph for Jamaican foreign policy over the multinational corporations who have prospered up to now by playing us all off against each other in the Caribbean, and the Caribbean as a whole against Australia.'[51]

On the whole, foreign affairs went much better for Manley in 1977 than domestic matters. A hinge between the two was the IMF. In December Jamaica failed its IMF domestic assets test by the 'tiny margin' of 2.6 percent, despite the fact that 'in most cases a small margin of failure is allowed.' Notwithstanding liberal pressure from Washington, Ottawa and London, the IMF resolved to take a hard line: Manley had no doubt that the IMF had decided to punish Jamaica.[52]

CHAPTER 18

'Evil... Stalks the Land' 1978

On 6 February 1978 Governor-General Florizel Glasspole addressed the members of the New Testament Church of God. 'Evil of every description,' he said, 'stalks the land. Like other countries we are beset by vices of every sort.' Earlier, in January, the gossip columnist 'Stella', pictured the conditions that led to one of the evils, urban violence, especially in West Kingston:

> You [youth] man just stan round pon sidewalk, not moving, like when fly settle pon ripe mango. Yout dawta park wid baby pon hip. Madda siddown a gate, father satta by the corner. No wonder kaskas [malicious gossip] and hate grow.

Shortly after rioting had occurred on 17 April Carl Stone added a third view:

> Almost daily. . . . grown men and women confront you with tears in their eyes and withered pride, impeaching in the most sorrowful tones for a 'smalls fi help out a yard for the pickney them no 'ave food.' Children wander the street like stray dogs with nothing to eat.

Stone thought neither the government nor the opposition had effective programmes to deal with such problems, and that Jamaica was heading toward an even deeper chasm of social conflict.[1]

The Economy and the IMF

Following the failure of the IMF test in December 1977 Manley went on the air in January to explain what further devaluation would mean. His ambivalence about the IMF was revealed in a statement that agreement with it was 'almost an indispensable part' of the government's production plan. Manley explained that devaluation would facilitate exports but by making imports more expensive would increase the cost of living six percent a year (overall inflation reached 34.6 percent in 1978). Subsidies on gasoline could not be maintained. The government would try to minimize the impact on the poor, but the proposed increases affecting them reveal the low levels of minimum wages, pensions, and public assistance.[2]

Manley said the subsequent IMF negotiations were 'one of the most ghastly experiences of my life'. Part of the trauma was caused by the necessity to make changes in the government's economic team, resulting in the resignation of Finance Minister and Deputy Prime Minister David Coore, Manley's 'closest friend for all of 42 years', since they were ten-year-olds at Jamaica College. To Manley the IMF agreement of May 1978 was one of the most savage packages ever imposed on any client government by the IMF.' It included an immediate fifteen percent devaluation, followed by further mini-devaluations every two months, resulting in a decline of the Jamaican dollar from slightly better than parity with the United States dollar to a rate of J$1.76 to US$1.00. Restrictions on state spending would have a severe impact on social programmes. The most massive tax package in Jamaica's history was introduced. Wage increases were limited to fifteen percent in the face of estimated inflation of forty percent. Toward the end of the negotiations, 'the most conservative and pragmatic members of the Cabinet were in revolt, appalled at the wholesale slaughter that was being insisted upon.' The private sector and unions were in shock.[3]

A *Gleaner* headline screamed 'IMF SHOCKER!.' In a two-hour live telecast on 11 May Manley said there was no alternative to the IMF package. He admitted that some people were concerned that 'fundamental questions of sovereignty' arise when dealing with the IMF; he found it 'a very difficult route [. . .] to take,' although he thought it the right one. The Marxist Workers Liberation League, close to some left elements in the PNP, announced that it would campaign against the IMF agreement. The accord put the burden on the poor; the non-IMF alternative would mean hardship for all, but 'mainly on the rich and in the end the people, not the IMF, would be in charge of Jamaica.' One economic analyst called for Manley to step aside. A Leandro cartoon turned even Manley's beloved cricket against him: the Prime Minister was pictured surrounded with dropped balls including one labelled 'IMF 1977', while a ball labelled 'IMF 1978' sailed his way and a spectator yelled, 'Do Mike, don't drop this one.'[4]

Carl Stone defended the IMF decision. He noted that 'extraordinary ignorance and misinformation pervades the society on the nature of our economy.' This made people vulnerable to mindless propaganda from the left about the non-IMF alternative. The 'alternative', as it was known, would have a severe impact on manufacturing, commerce, and the public sector; only peasants would be relatively unaffected. One hundred thousand would lose their jobs. Moscow was not willing to support the alternative, despite efforts of the PNP. The fact that the people and the politicians still looked for outside salvation—be it from the United States, the Soviet Union, or the IMF—showed that their minds were still colonized: 'colonized minds cannot decolonize anything,' Stone concluded:

> IMF or no IMF the task facing us is that of decolonizing and restructuring the Jamaican economy to render it more self-reliant. Abusing the PNP and the Prime Minister for errors is escapism. . . . Outside help can only buy

> us time to get on with that task that in the final analysis we will have to painfully do ourselves.[5]

Unfortunately for the Jamaican people, the PNP government and Manley, the IMF agreement not only extracted a heavy sacrifice, but it also failed to help the Jamaican people turn the economy around.[6]

Green Bay and the 'Peace Treaty'

The economic crisis was compounded by a resurgence of violence following the lifting of the State of Emergency in mid–1977. The violence produced a tragic and controversial event known as the Green Bay Massacre. The basic facts are as follows. On 5 January 1978 members of the army's Military Intelligence Unit killed five young men at the Green Bay firing range. They had lured them to the range with the promise of guns. The Jamaica Council of Human Rights called for a public inquiry. On 22 May an inquest jury of five men and three women issued its unanimous verdict: that unspecified 'person or persons are criminally responsible' for the killings.[7] Criminal proceedings were started in July. After many delays, one JDF member was freed on 30 June 1981 for lack of evidence. In February 1982, the remaining seven defendants were freed after the government prosecutor entered a plea of *Nolle Prosequi*.[8]

Controversy developed around three points: whether the five young men who were killed were vicious criminals or JLP supporters, or both; the closely related question of the army's motivation; and Manley's relationship to the Massacre.

Manley maintains that the five victims were leaders of 'the largest and most vicious criminal gang in Jamaica', who coincidentally were based in Southside, a JLP stronghold in his constituency.[9] He vigorously disputes others' identification of the victims as 'JLP gang members', 'labourites', or 'JLP supporters who were allegedly gunmen.'[10]

Whether the gunmen were common criminals or political activists, or both, bears on the question of motivation. Manley maintains that the army's motive was to proceed against criminals.

> What is absolutely certain is that the Green Bay operation was not politically motivated against five members of the JLP. There is no question that it was directed against this criminal gang because they were wreaking havoc in the Kingston area.[11]

Those who state that the victims were 'JLP gang members', 'labourites', or 'JLP supporters' imply, without directly stating it, a partisan motive behind the killings. For some of the ghetto persons Carl Stone interviewed at the time, the killings 'symbolized the brutality of the middle class run system against them the poor people'. (Stone noted that his interviewees were unconcerned, however, when gunmen killed 'businessmen, motorists, policemen, white people,' and others.)[12]

In a long account, David d'Costa—a *Gleaner* journalist considered by Manley to be part of the cabal attempting to bring his government down—suggested an institutional-bureaucratic motivation for the massacre. D'Costa noted that police violence had long been a 'problem', but that the army had been well-regarded for its discipline. The 1976 State of Emergency had confronted the army with ghetto gunmen, whom the soldiers regarded as 'animals' and 'enemies'. Soldiers were killed in gang ambushes. This occurred against the background of rivalry between the police and the army over such things as budgets and preeminence in internal security. Internal security was normally the province of the police Special Branch, but the army's Military Intelligence Unit competed with it in that area. Manley, as Prime Minister and Defence Minister, naturally, under the conditions of Jamaican political life, took an intense interest in internal security. D'Costa argued that the Green Bay killings were carried out by the Military Intelligence Unit to convince Manley of its internal security skills. He said that the murdered JLP 'gunmen' were carefully selected from Manley's constituency (D'Costa was vague on the grounds for selection, be they anti-crime or political), but the operation was 'carried through with a degree of naiveté and incompetence which virtually guaranteed failure.' D'Costa characterized the army's defense at the inquest as 'a farrago of lies and fairy tales'. He was severely critical of the roles of army Chief of Staff Major General Rudolph Green, who claimed to know nothing of the crime, and Minister of Security Dudley Thompson, who in the *Gleaner* columnist's view, tended to legitimize the killings by his outspoken views on the suppression of crime and violence.[13]

The third, less substantial, controversy, involved Manley's connection to the massacre. The JLP urged Manley to resign as Prime Minister on the grounds that he could not escape responsibility for the conduct of the military,[14] but made no case for Manley's direct involvement. D'Costa, a hostile journalist, in his long account, failed to state that Manley had any direct or indirect role in the planning of the operation or the killings. Manley stated in Parliament that he knew nothing about the killings until after they happened.[15] He first heard about the operation after the fact, during a Cabinet meeting; he also said that 'there is no question that the army . . . fellows were wrong to set up this operation.'[16]

Within the scope of this work, it is impossible to explore the Green Bay Massacre further,[17] except to note two consequences. The event tended to tarnish the image of the Manley government. It also contributed to rival gangs agreeing to a short-lived 'peace treaty'.

While the Green Bay controversy simmered through the first half of 1978, efforts were made to effect a treaty between the youth gangs that had warred since the mid-'60s. The peace effort began on 10 January, five days after the Green Bay Massacre, under the leadership of JLP supporter Claudius Massop of West Kingston and PNP supporter

Aston 'Bucky' Thompson of Central Kingston. It was later extended to Arnett Gardens and Rema (Wilton Gardens). John Hearne reported that when he observed the peace process the sentiments he heard most were hostility to politicians of both parties and pleas for 'work without discrimination' and 'community development'. Neville Toyloy reported that the youths

> have made it plain . . . that they will not be turning [guns] against one another for politicians.
>
> They are tired of being used. The key, as they see it, is unity. For it is they who have suffered and died.

Manley and Seaga subsequently met with Aston Thompson and Massop for three hours and discussed relations with the security forces, unemployment and the distribution of jobs, and discrimination against ghetto dwellers seeking jobs. Manley later reported that the PNP and JLP leadership had discussed the peace movement and established a Peace Committee representing the two parties, churches, trade unions, the Private Sector Organization of Jamaica (PSOJ), service clubs, the Friendly Societies Council, the Jamaica Bar Council, and the Jamaica Council for Human Rights. The *Gleaner* established a community Peace Fund, which Manley welcomed with a J$500 contribution. Not to be outdone, Seaga pledged J$5,000 to be deducted in instalments from his parliamentary salary.[18]

In the Senate, Minister of Security Dudley Thompson identified crime as Jamaica's number one problem and pledged that 'mad dogs in the society will be destroyed.' Later in February he reiterated his 'mad dogs' statement and added that he believed that 'the degree, nature and tempo of crime today is a hangover of the imperialist efforts that have tried to destabilize Jamaica.' The Mayor of Montego Bay, Cecil Robinson, reinforced the latter theme by asserting that the tourist trade in his city was being destablized by 'known criminals from Kingston'. Thompson continued his hard line by stating that the security forces would not be deterred by criticism.

> I have no apologies for those who shout about human rights and police brutality. We have a definite challenge to authority in the form of sheer brutality and terrorism.

Later Thompson denied that poverty was the root of crime; instead he found its main cause to be indiscipline and 'disrespect for authority'. Thompson's stern view seemed to have some community support: a Stone Poll rejected the idea that the police used too much violence. On 17 April, rioting and looting broke out in Kingston, in which three looters were killed and two policemen were wounded. The *Gleaner* praised Manley, Thompson, Seaga and the security forces for their handling of the riot.[19]

The climax of the Peace Movement came at a 'One Love' concert before twenty-five thousand on 22 April. The concert featured reggae superstars Peter Tosh and Bob Marley, and was attended by Manley and Seaga. Tosh chastised both leaders in a speech 'coloured with indecent language' and called on them to legalize ganja, protect natural resources, and safeguard equal rights and justice for the whole nation. Marley invited Manley and Seaga to the stage. Seaga went first, joined by Manley 'a few minutes later'. On stage with Marley, the two bitter rivals 'locked hands and danced a short jig,' after which they walked in separate directions to their seats.[20]

But since 'tribalism' was so deeply rooted, the Peace Movement was already beginning to crumble. Stone reported that its leaders were being seen as 'puppets manipulated by party figures'.[21] According to Kaufman, 'the designs of the JLP, the weight of patronism and tribalism and the growing hardships of the poor had more clout than any peace.'[22]

Hearne and the 'Gleaner' vs Manley

One of the personal costs of Manley's politics was damaged friendships, especially with John Hearne, who had been a friend of Manley's since Jamaica College. Hearne, it will be recalled, had been removed in 1975 as head of the government's Agency for Public Information. Hearne was undergoing a political metamorphosis. In 1977, he referred to himself as an 'unrepentant, convinced Democratic Socialist' who believed that capitalism 'is a system that must end if our true *human* history is to begin.'[23] By 1984, he announced himself 'a conservative free-marketeer in ideology'.[24] Manley and Hearne began to part ways when Hearne refused to accept Manley's version of the Rema affair of 1977.[25] In 1978, Hearne became a leading, (and in at least one case, libellous) critic of his erstwhile good friend Michael Manley.

Early in 1978, Hearne surveyed the Jamaican media scene, noting the demise of *Public Opinion*, long identified with the PNP and the Manleys, 'after forty years of brave and honest service'. The *Daily News*, after four years of existence, was ailing. Jamaica was about to become a one-paper society once more (the *Gleaner* owned the afternoon tabloid, *The Star*).[26] *Gleaner* editor Oliver Clarke presented a rather dismal survey to the Inter-American Press Association meeting in Cancun, Mexico on 27 February: all radio and television owned by the government (this was untrue); the *Daily News* in receivership; and the *Gleaner* facing its second large reduction of staff, after a large operating loss in 1977, a year in which the government cut its advertising by sixty-eight percent.[27] The inaccurate implication could be drawn from Hearne's and Clarke's comments that 'freedom of the press' was threatened by the Manley government.

One major fact disputes the implication: the notable freedom, and even seditious license, enjoyed by the *Gleaner*'s news staff, editorial writers, and columnists. Hearne himself is a good example. Having generally defended Manley in 1977, he attacked him in 1978. In January Hearne wrote that the Manley government had little regard for the truth, emphasizing one word calculated to anger Manley, whose commitment to justice in South Africa was legendary: 'Has there ever been a Government, in our history, that has practised such a policy of APARTHEID between its utterances and the Truth?'[28] In February, Hearne, still a member of the PNP (although two years in arrears of dues), wrote that he was disturbed by the growing corruption, authoritarianism, and expediency of the PNP, by its bypassing of Parliament, and by its use of civil servants to do party work.[29] Shortly thereafter, Hearne wrote that 'we are not likely to hear the truth about anything from this government,' and added a personal attack:

> Emotionally and morally, our Government is now operating at about a four-year-old's level.
>
> Our Prime Minister, for example, is quite convinced that his excursion to the Socialist International and Hungary is of more importance than staying at home to do the boring, difficult job of negotiating with the International Monetary Fund.

Hearne called Manley's trip to Hungary—which resulted in a J$10 million line of credit for medical equipment and other agreements—an 'educational holiday'.[30]

In late May Hearne published a column entitled 'Snoopy-Go Home', in which he likened Manley to the cartoon beagle. Hearne said Snoopy was his favourite character; 'the only trouble with Snoopy is that he will not—cannot—believe he is but a dog.' The *Gleaner* columnist wrote that Manley 'can be one of the most lovable and entertaining of creatures to have around—so long as there is no real work to be done.' He accused Manley of being obsessed with 'imaginary enemies', the CIA, the IMF, and capitalists; implied that he was responsible for the Green Bay Massacre; and said that he was worse than King Leopold of Belgium.[31]

In the 26 August 1978 *Gleaner* John Hearne accused his erstwhile close friend Manley of 'approving and personally ordering torture' and 'using "official murder" '. The latter charge presumably referred to the Green Bay killings, while the former referred to the allegations of a former JDF lieutenant colonel, Leslie Lloyd.

Manley issued a detailed denial. In 1976, relations had deteriorated between Lloyd and Brigadier (later Major General) Green, JDF Chief of Staff. On 17 December 1976, four submachine guns in Lloyd's custody disappeared from Up Park Camp. The alleged torture, first charged by Lloyd in March 1977, was said to have occurred during a JDF interrogation in January of the fourteen soldiers alleged to be responsible for stealing the guns.[32] The Defence Board asked Green to investigate and he reported that some force had been used in the interrogation. A JDF doctor who saw all of the soldiers alleged to have been tortured, found no evidence of torture, although one had an abrasion on his back 'consistent with blows from a wire'. On 27 April, the Defence Board invited Lloyd to retire (on the basis of 'various complaints', but not because of his torture allegation). Lloyd retired and was given a non-JDF government job. On 20 March 1978, Lloyd wrote to Manley stating that he wished to be reinstated; the thrust of his comments concerned the allegedly unfair procedure against him, rather than the alleged torture *per se*. Manley refused to reinstate Lloyd. He categorically denied that he would ever approve or use torture or murder.[33]

Manley regarded the Hearne column as part of its author's and the *Gleaner*'s campaign to discredit him. He brought a libel suit against Hearne and the paper, which he won in 1985, when the *Gleaner* was forced to issue an apology, which stated, in part:

> The accusations made against the credit and reputation of Mr. Manley are withdrawn and the Gleaner and Mr. Hearne wish to make it clear that the said accusations, and in particular, the use of the word[s] torture or acquiescence in the use of torture, were unfounded.[34]

The result, years after the fact, was a moral victory for Manley which did not undo the political damage nor personal hurt done at the time. Meanwhile, in the late '70s, Hearne and the other *Gleaner* columnists continued their incessant attack on Manley and the PNP, complaining all the while about the threat to 'freedom of the press'.

Foreign Affairs

As in 1977, and with the important qualification of the IMF negotiations and agreement, Manley faced a somewhat more favourable situation with respect to foreign affairs than in domestic matters. In January, the *New York Times* published his opinion piece calling for the United States to play a leading role in new, more just relations between the rich and poor nations.[35] The *Gleaner* found nothing new in Manley's article, but congratulated him on it and endorsed its ideas.

> . . . It is well that Mr. Manley should have restated our anxiety for respect for our sovereignty and our right to real self-determination. For alas! in the international world, small countries which live in the geographic shadow of large powers find it almost impossible to live with the dignity of their independence.[36]

The key to the sovereignty of poor, small Third World countries lay in the proposed NIEO, of which Manley had become a foremost global spokesman. In one of his kinder columns, Hearne noted that the NIEO might be the most central and 'constantly developed' of Manley's concepts since 1969: 'When Michael Manley dies, the initials NIEO will probably be found engrained on his heart.'[37] Carl Stone was more critical:

> History will record the vaunted foreign policy of the Manley government as so much hot air, symbolism and posturing if after so much involvement on the international scene we do not find ourselves in the group of favoured nations with sufficient credibility to negotiate for investment money like Algeria, Morocco, or Tanzania.[38]

Manley defended his foreign policy in the annual budget speech in June 1978; Jamaica did not have the power to change the world or to do anything, 'but Jamaica together with 107 other nations and more than 2 billion people [has] the capacity to change it little by little.'[39]

Relations with the United States were still good in early 1978. The Cuban connection continued to be a problem for Manley, as even Cuban gestures of goodwill could provoke controversy in the heated, often hysterical Jamaican political context. Such an event occurred in February when Castro returned J$273,000 found in a package attached to a parachute on the Isle of Youth. It was unclear why the money had fallen on Cuba, but by international law Cuba could have kept it. Instead Fidel returned the money to Manley with a suggestion to invest it in an arts and trade centre in West Kingston.[40] A *Gleaner* editorial, 'Fidel the Santa Claus', raised doubts about the origin of the money, but continued:

> . . . The honesty and friendship of Dr. Castro must be commended, despite his indicating how he would like the money spent. For [a] quarter million Jamaican dollars could so easily have been kept for the Cuban Embassy's expenses in Jamaica. But a sovereign government of integrity like Cuba's would not behave in that manner and this is praise-worthy.[41]

While commending Cuba, *Gleaner* news stories carried criticism from both PNP and JLP supporters criticizing Manley's failure to consult with the central advisory committee of the ghetto Peace Movement on how best to spend the money. Later it was noted that the Jamaican government had spent J$307,614 on Castro's 1977 visit, much more, for example, than on the visit of President Samora Maciel of Mozambique.[42]

A welcome and richly-deserved foreign affairs and personal triumph for Manley was the gold medal awarded him on 11 October 1978 by the UN Special Committee Against Apartheid. The Jamaican Prime Minister was one of seven world leaders (four posthumously) awarded gold medals for their 'significant contributions to the anti-apartheid movement and the international struggle against apartheid in cooperation with the United Nations and with the South African liberation movements.'[43] In his keynote speech on the occasion, Manley noted the contributions of two other Jamaicans, Marcus Garvey and Norman Manley, in the struggle against South African racism. In addition to his co-honorees, he invoked a long line of liberation heroes: Mahatma Gandhi, Martin Luther King, Toussaint L'Ouverture, Simon Bolivar, José Martí, Julius Nyerere, Fidel Castro and Ho Chi Minh. Manley argued that it was impossible to understand oppression in South Africa without understanding imperialism, 'the subjugation of three-quarters of the globe by a technologically triumphant minority'. He called for a peaceful but 'total mobilisation of the world community' against apartheid. His speech was received with tumultuous applause.[44] It was one of Manley's finest hours.

A Socialist and Political Art?

A strength characteristic of Manley is a sense for the wholeness of things, for 'the round' as Michael Manley puts it, which has led him as politician to take an interest in virtually all human activity. This broad vision has caused problems: those with a limited concept of politics have accused Manley of dabbling in marginal areas to the neglect of 'real politics'; some have seen Manley as the embodiment of an ever-encroaching state which recognized no limits. But assuming that all things *are* interconnected, assuming there is a 'politics' for all human activity, and assuming that national development is a total process, then Manley has been on the right track. With parents like Norman and Edna Manley, politics and art to Manley are interconnected realms.

Manley made this clear in a lecture on the role of the artist in a developing country on 1 February 1978, amidst all the troubles of that time. Manley had concluded that patriotism could not be created through the Jamaican political process, because 'the system is designed to make it impossible.' It institutionalized differences and quarrels and depicted authority as inevitably tyrannical. Not only was it impossible to create patriotism through Jamaica's political system, it was impossible to mobilize the people behind great national goals: 'I have tried for the past six years and I have failed.' Manley challenged artists to do what the politicians could not: inspire patriotism and be 'the prophet[s] of the future to which we aspire'. Since the political process had an inherent counterproductive element, it needed 'powerful, purposeful and continuous support from the artists who must, themselves, be patriotic and politically aware.' Artists, Manley knew, were special, 'the race horses in our society', but Jamaica could not afford 'the luxury of a race horse who is not also willing to pull the dray.'[45]

The speech elicited a thoughtful private critique from a UWI historian 'committed to what you and Jamaica stand for, and are trying for, no matter what the odds. . . .' He disagreed with Manley's notion that the Jamaican artist could create patriotism where the politician could not. 'Our politicians may not all have done very well in helping to create a sense of local patriotism; but in our situation politics is the first and main thing we've got.' Artists had not always supported the politicians in England; when the state seemed to fail, the artist became critical and individualistic. There was no point in challenging artists to help create patriotism; 'all you will get is catch-word journalism.' It was up to the political leader to project and create 'a climate of patriotism to which the artist will respond.'[46] In a brief reply, Manley said, 'I see what you mean but am not sure that I am fully convinced.'[47]

Some of Edna Manley's art, particularly that of the 1930s, had the prophecy and social consciousness that Manley regarded as essential in Third World art. But his mother's art was more typically individualistic, even mystical and, until after her husband's death in 1969, focused on him and their relationship. Although she did her political duties as an inevitable necessity, bearing the title 'Mother of the Nation' with quiet dignity, she tended to be suspicious of ideologies, including socialism.[48]

Manley's daughter, the poet Rachel, inherited Edna Manley's tendencies. Manley's views on the role of the artist in a developing country thus encountered resistance from his daughter. A review of Rachel's second book of poems, *Poems 2*, noted their intensely personal nature. The reviewer said one poem, 'Predestined Calvinist', symbolized London as 'a devouring mother with whom [Rachel] has an ambivalent relationship', through which the poet 'projects strongly a sense of guilt at remaining attached to her by an umbilical cord which she cannot bring herself to cut.'[49] Manley wrote to Rachel praising *Poems 2*. He noted that 'all poetry has an interior and an exterior aspect' and could express the interplay between the poet's inner world and 'the broad social experience'. 'Your work could develop in terms of a wider awareness of the exterior experience,' a point he carefully advanced not as criticism, but as invitation.[50] Rachel's long reply wrestled with her father's view of the politically-committed artist without accepting it and came close to nullifying the artistic function:

> We do not create at all. The artists are the craftsmen of the form of a truth that is there. Mankind is the artist and the inventor. We are the scribes, the recorders.'[51]

In May Rachel sent Manley a poem with these final lines:

> it is I, who knows the price,
> sobbing in prayer
> for a still born generation

The intensely personal poem was called 'The Ancestor'.[52]

Glimpses of Private Life

Manley's private life provided occasional respite from the storms of public existence, but also reflected the broader social drama. Manley's rambling country home, Nyumbani, grew like Topsy on a piney, cool mountain adjoining his parent's retreat, Nomdmi. A grower of prize-winning roses, Manley spent at least one day each weekend working with Beverley in the garden at Nyumbani, 'with Natasha as a constant source of interest and joy'.[53] High officials of Guyana and Belize sent

Manley orchids to grow. Combining business with pleasure, Manley sold roses, making a profit of almost J$550 in the months of July, September and October 1978.[54]

Private correspondence often brought satisfaction difficult to obtain in public life. Joseph wrote from Havana thanking his father for making him solvent: '. . . What are fathers for if not to spoil sons. There you have the history of the human race.' Joseph was reading *Voice at the Workplace* again, finding it 'very clean and analytical, especially the last chapter'.[55] In late May, in the midst of controversies over the IMF and the Green Bay Massacre and the breakdown of peace in the Kingston ghetto, Manley wrote to Joseph that 'we are all dying for July to come as we miss you and long to see you.' He was making sure their boat was in good shape. 'When you come, ice cream and guava cheese would be wildly appreciated.'[56] A letter from the PNP finance officer acknowledging Manley's donation of J$1,230 from sales of *The Search for Solutions* remarked:

> The Party Secretariat is constantly amazed at your dedication to the Party and the struggle, as is expressed by your often ascetic approach to our problems and that [*sic*] of the people.[57]

Manley routinely devoted all proceeds from his writings to the PNP.

Manley's private life was also touched by tragic aspects of the Jamaican experience. One of these was emigration. A close relative wrote from London that while he had not abandoned Jamaica, 'for my personal development. . . . I found it best to get out for a while. I found Jamaican "Society" and people were increasingly assuming roles I must fulfil and dictating what direction my life must take.'[58] A friend who had gone to Florida, motivated by finances and the desire for professional education for his children, described the human side of emigration:

> Only the experience of this physical & mental 'transplant' can be understood by the actual people that have done it—for whatever reason.
>
> It is not pleasant; it is very heart rending, very, very emotional.[59]

Crime was another general problem which affected Manley directly. Not only were those close to him victimized, his own house was broken into nine times in two years by 'the simple expedient of removing a burglar bar.'[60] The evils which Governor-General Glasspole had depicted as stalking the land left no Jamaicans untouched, not even the Prime Minister and recipient of a UN gold medal.

CHAPTER 19

Deepening Crisis (1979)

Michael Manley, his party, government, and nation continued to wrestle with crisis in 1979. The economy declined further, as Jamaica struggled with a highly unfavourable international and domestic situation, culminating in another crucial failure of an IMF test in December. In politics, the PNP struggled with declining support and internal disaffection, while it finally defined its ideology in the *Principles and Objectives of the People's National Party* and attempted to begin implementing the ideology via political education. The year witnessed numerous momentous international events, among them: the deepening of revolution in Iran and the initial success of leftist revolutions in Grenada and Nicaragua, the Soviet invasion of Afghanistan, and renewed hostility by the United States against Cuba as the thaw of the early Carter years ended under right-wing pressure. Manley played an extremely active international role in 1979, including a visit to the USSR in April and a thundering address to the summit of the Non-Aligned Movement in Havana in September. The PNP Women's Movement, led by Beverley Manley, was prominent and controversial in agitating for a maternity leave law and in supporting political education and solidarity with Cuba and the USSR. Manley's personal life reflected many of these broader public developments.

The Economy

The new year began with street protests over the rising cost of petrol. In a PNP National Executive Committee meeting in the aftermath of the demonstrations Manley reckoned that everything depended on getting the economy moving, but that there was no alternative to the IMF. The Cabinet unanimously agreed. The Marxist Worker's Liberation League figures on an alternative to the IMF strategy did not add up. The only way out of the IMF trap was to increase production and exports. It was noted that in such a crucial area as the allocation of scarce items, the party had no policy based on socialist principles applicable to large private distributors, despite the serious implications for both the State Trading Corporation (STC) and small distributors.[1]

Economic recovery, not to mention national development, depended on cooperation from diverse segments of society. On 25 April 1979, in a broadcast to the nation, Manley proposed a new social contract to 'the people', which he defined as 'farmers, businessmen, managers, fathers, mothers, children, soldiers, policemen, churchmen, social workers, unemployed, the poor, everyone'. An attempt to overcome the chronic and worsening economic divisions in Jamaica, the social contract specified the responsibilities of government, employers and workers. Government would limit recurrent expenditure as much as possible, not increase overall taxes, and seek to ensure necessary foreign exchange. Employers were expected to expand production, provide job security for workers, and keep price increases to ten percent. Workers were expected to keep wage increase demands to ten percent. Together employers and workers were to negotiate incentives for increased productivity to justify wage increases beyond ten percent. At the close of the address Manley noted a hard fact of Jamaican economic life: the inverse relationship between the costs of production (including wages and profits) and the competitiveness of exports.[2] The proposed social contract failed to turn the economy around.

A major reason for this, one that must not be forgotten in assessing the Manley government's management of the economy, was that Jamaica was a prime example of the ravaging of many Third World economies by OPEC's stiff increases in the price of oil, which tripled again in 1979. In his 'social contract' speech, Manley observed that without oil 'every light would turn out, every bus and car stop and every factory shut down instantly.' As he later observed, part of the way in which the Puerto Rican economic model 'flatter[ed] to deceive' was to encourage dependency on the low-cost oil of the '50s and '60s. As a result, Jamaica had, for example, substituted oil for sugar cane refuse as fuel for sugar factories, sacrificing jobs and intensifying oil dependence.[3] By 1979, ninety-nine percent of Jamaica's energy for industry, electricity and transport came from imported oil; Venezuela, an OPEC member, was the main supplier. In 1972, Jamaica had imported 16.1 million barrels of oil at a cost of J$44 million; in 1979, it imported the same total of 16.1 million barrels for $J331 million.[4] The UN's energy programme reported that Jamaica's energy options were very limited and inflexible. Conservation, which the government stressed, would have to play an important role, but depended on 'total political commitment and [a] non-partisan approach', conditions difficult to achieve on the island.[5]

Also beyond governmental control was the availability and cost of money borrowed abroad, and the political strings on it. Before the Board of Directors of the Inter-American Development Bank (IDB) in Montego Bay, Manley reviewed Jamaica's economic problems and reform policies. He said he was pleased that the IDB continued to be concerned about the quality of life, but mentioned that the measures adopted to cope with Jamaica's economic crisis 'will seriously hamper our efforts to create a more equitable distribution of the nation's resources.' He asked the bank to review Jamaica's sound economic policies and help restore investor confidence. Jamaica faced a common dilemma in that a country working for a more just society 'may' find private investment drying up.[6]

The terms of trade also worked independently of Jamaican needs and desires. In his Independence Day news release Manley had to admit that UN trade talks (UNCTAD V) and negotiations between the African-Caribbean-Pacific (ACP) nations and the European Economic Community 'have been difficult and frustrating'.[7] In a long address to the Third World Foundation in London on 29 October 1979 Manley pointed out that, whereas in 1965 20.84 tons of Jamaican sugar had been required to purchase a tractor, in 1979 57.87 tons were required to purchase a more powerful tractor which did essentially the same job. According to figures from the International Sugar Organization, the export value of manufactured goods had moved from an index of 100 to 245 during the years 1970 to 1979, while sugar had declined from 3.68 cents a pound to 3.30 a pound. Given such facts, the NIEO was needed to remedy trade structures that hurt poor, Third World nations. In addition, new international financial organizations were needed. Technology was expensive and involved an unequal exchange with the First World. Reform in information systems was also needed since 'control of technology and economic power is supported by a comprehensive systems of propaganda' that protected the status quo and gave it mythic attributes. In his closing words, Manley referred to the elimination of poverty as the supreme challenge, a challenge that required collective self-reliance and internal social justice the Third World, the transfer of resources from the First World, and the design of new economic structures by the whole world.[8]

In December, Jamaica failed the IMF international reserves test by US$130 million and fell short of budget cuts by J$100 million. The government blamed causes beyond its control, which a confidential IMF report partly agreed with.[9] In response to the deepening economic crisis Manley shook up his Cabinet, reducing the number of ministers from twenty to thirteen. He explained to the nation that the crisis consisted of four main elements: the failure of the IMF test; impending increases in the prices of oil, sugar and food; financing the government budget, of which seventy-one percent went on wages, salaries and emoluments; and a continuing lag in production in both the public and private spheres. To end the crisis, bauxite, tourism and manufacturing had to improve, but the 'great national effort has to come in agriculture.' Manley himself assumed the agriculture portfolio, while continuing to personally supervise culture, information and defence.[10] As Manley well knew, solving the economic quandary required unified political purpose which the Jamaican system was ill-suited to provide.

Politics

The political year began with massive demonstrations on 8 January 1979 over the price of petrol. The protests were organized by a new front group of the JLP, the National Patriotic Movement. There had been a steady rise in dissatisfaction with the PNP since early 1977, and the police refused to quell the protests. One analyst notes that although planned and coordinated by the JLP, 'the petrol-price protests struck a responsive chord among wider layers of the population.'[11]

At the PNP National Executive Council meeting on 28 January, Manley commented that the protests 'reminded us about the pressures facing our people, allowed our people to see the character of the JLP leadership, [and] revealed the poor state of organization, communication, morale and political understanding in the party.' P.J. Patterson noted that the PNP's organization was slipping; in 1978 it had 1,600 recognized groups; by January of 1979 there were only 560 active groups, with 340 under warning and 770 in abeyance. Manley noted the need for political education, despite the fact that some in the party thought it a waste of time. Such education ought to include the history and origins of the current crisis, the role of imperialism, why prices increased, and reminders of PNP successes; co-operatives, the STC, the advocacy of the NIEO, anti-apartheid activities, 'worker democracy' and nationalization.[12]

A basis for political education was finally established with the publication of *The Principles and Objectives of the People's National Party* in February. Although the PNP had rededicated itself to 'democratic socialism' in 1974, attempts to define the party's ideology were divisive and it took over four years to produce the *Principles and Objectives*. Some observers have seen the delay in ideological definition as fatal, since it encouraged confusion in party ranks, and when ideological definition finally did come the party had been weakened and demoralized by the economic crisis. Manley saw much truth in this view.

'The book', as *Principles and Objectives* is known in the party, is sixty-seven pages long and divided into six chapters covering the party; its ideological, economic, social and international policies; and a glossary of terms. The PNP is said to strive for land for the small farmer, national liberation, racial dignity, the international solidarity of the oppressed, the rights of working people (Bustamante is mentioned as a combatant in that struggle), national democracy and socialism. The party seeks to combine internal democracy with external discipline, to have leaders set good examples, and to encourage criticism and discussion. The chapter on ideological policy declares socialism as the PNP's objective, describes socialist society and enumerates principles of democratic socialism.[13] The crucial glossary gives party definitions of twenty terms such as 'capitalism', 'communism', 'scientific socialism', 'social democracy', and 'democratic socialism'.[14]

Principles and Objectives reflects Manley's commitment to politics based on ideology and political education, but party rightists or 'moderates' have not generally shared this dedication. 'The book' was approved after three days' debate by two thousand delegates at the PNP annual conference in September. Two party policies were particularly divisive: that in the alliance of classes comprising the party, the interest of the working class must predominate, and that in the mixed economy, the public sector must lead. 'Imperialism' was also a divisive issue. Manley later noted that 'to the extent that the document represented an advance in seriousness and no retreat from commitment, it would have been noted by the oligarchy and filed as further evidence of the importance of removing us at any cost. The CIA would doubtless have fed similar advice to the US establishment.'[15]

Meanwhile, problems were brewing in Manley's own constituency in East Central Kingston. A supporter warned him in January of opposition people spreading gifts and money. She pleaded that 'for us to lose would mean death to me and many others.'[16] The United Africans' Youth Group complained that the 'Member of Parliament, Councillor, Area Leader, and Section Leaders are totally non-functional. As a result . . . there are no organizations within the community. No political representatives are being seen.' The minutes noted that in the McIntyre housing scheme houses were leaking, lighting and public communications systems were out and residents had been frustrated by broken promises of repairs. The youth group had to stop many demonstrations by resentful McIntyre inhabitants, the community was being 'infested by non-residents', and policemen (named in the group's minutes) harassed the residents. Opposition speeches in the area went unanswered.[17]

Notes for a speech to be delivered at the statue of Norman Washington Manley in Kingston on 22 March 1979 reveal Manley's understanding of the crisis his government faced. He likened the road to democratic socialism to the struggle of a family: the dream was of home, the children, holidays; the obstacles: sickness, financial worry, 'child that let you down.' He reviewed the government's social and economic reforms and assertive foreign policy. History and imperialism had made Jamaica vulnerable to the economic crisis. The struggle facing democratic socialism had three elements: the opposition; the economy and the IMF; and the role of the workers and the people generally. The opposition, including the *Gleaner*, was exploiting the crisis by lies and confrontations to divide and confuse the nation. Charges of corruption should be met by arguing the record of the administration. Charges of communism should be countered by explaining PNP policy, the real role of community councils and the purposes of Manley's trip to the USSR (discussed below). To deal with the economy Manley would have to explain the role of foreign exchange and the resultant need to export in order to earn it, and the need to get a new IMF agreement without the devastating 'crawling peg' devaluations. But the nation itself must deal with remaining problems in which Jamaican owners, managers and workers all had a patriotic role to play. Abroad it was necessary to struggle against the world economy built by imperialism, featuring the exploitation of nation by nation. Workers and the people could advance the struggle as home guards and price inspectors and through Community Enterprise Organizations (CEO's), cooperatives, Project Land Lease, Pioneer Farms, and worker participation in factories, utilities, bauxite, sugar and tourism. Throughout,

Manley stressed the importance of struggle, both overseas and at home, in order to create more just and equal systems.[18]

Opposition to the government intensified throughout the year as the overall crisis deepened. In Washington and Miami, Edward Seaga accused the government of violating human rights and abandoning the democratic process. Manley, his government and independent observers regarded these accusations as irresponsible. In late October, Minister of Justice and Attorney General Carl Rattray made a national broadcast defining the legitimate role of the opposition and criticising Seaga's behaviour. 'In a parliamentary democracy such as ours', Rattray said, 'when the Opposition begins to talk of the "underthrow or overthrow" of the government and to lay the basis for inciting civil disruption and even foreign military intervention, it is time for serious public analysis of an Opposition in a parliamentary democracy. . . .' Rattray decried the opposition's 'sensational allegations' against the government, made 'without one shred of evidence presented to the Constitutional authorities'. He went on to criticize attacks on the security forces by Seaga and other JLP leaders: 'if these attacks result in an undermining of the Security Forces . . . then national security is gravely threatened. . . . The consequences could be chaotic and result in anarchy and the illegal overthrow of the lawful government'. Rattray reaffirmed the constitutional rights of the opposition and called on the JLP to follow prior examples of a 'strenuous, alert and patriotic' opposition to help in the building of the nation.[19] His measured words, however, had little effect on the JLP.

Foreign Affairs

Manley's activist foreign policy led to difficulties in Jamaica's foreign relations that added to his government's economic and political problems.[20] Relations with the United States deteriorated as the superpower felt itself threatened by events in Grenada, Nicaragua, Cuba, Iran and Afghanistan. Against this background, Manley travelled to the Soviet Union, delivered a defiant speech at the conference of non-aligned nations in Havana, and defended the controversial Cuban ambassador in Jamaica. Pursuit of Jamaica's national sovereignty and independence collided with the Carter Administration's defense of perceived US interests, as Carter succumbed to conservative pressure.

In his speech at an official dinner given by the government of the USSR at the Grand Kremlin Palace in April, Manley expressed warm greetings to Leonid I. Brezhnev, 'a true friend of all national liberation movements'. Manley recalled Jamaican author Claude McKay's positive visit to the USSR in 1923 (overlooking McKay's latter disillusionment with communism) and Marcus Garvey's eulogy of Lenin. The Jamaican Prime Minister observed that the symbols of independence 'have often served the purposes of imperialism by masking its true nature,' and tried to convince his audience that the NIEO represented 'a new and critical phrase in the liberation process'.[21]

On his return from Moscow, Manley informed Jamaicans of the trip's results. The USSR agreed to purchase fifty thousand tons of alumina a year, beginning in 1980, and signed a long-term contract to purchase another 250 thousand tons a year, beginning in 1984, to replace the ill-fated JAVAMEX venture. Jamaica-USSR trade was to be based on the principles of mutual advantage and the fostering of 'economic growth and development'. A long-term fishing development plan and a study for expansion of cement production were also being considered. Manley cautioned that the deals with the USSR should not be regarded as the salvation of Jamaica. He wrote the word 'key' next to the following passage: 'Our salvation . . . requires hard work, discipline and patriotic commitment to the national struggle.'[22]

Privately, Manley was less restrained. To one correspondent he wrote that 'the visit to the Soviet Union was . . . both successful and enjoyable—we had to work like hell!' To another he wrote that 'the visit to the Soviet Union was an inspiring success.'[23] The cost of such success, of course, was the increasing discomfiture of the Jamaican opposition and the United States.

Manley thrived in his role as Third World leader, which may have served as respite from the more difficult job of solving Jamaica's domestic problems. In late July and August he made 'a wild trip', beginning with the Commonwealth Heads of Government meeting in Zambia. From there he went to Mozambique for a short official visit. From Mozambique he jetted to Norway for a bauxite conference. He went from Europe to New York by concorde airliner for a meeting with bankers.[24] From the bankers he flew to Havana for a meeting with Castro; after an all-night, eight-hour talk he got the Cuban leader to accept the provisional settlement of the Zimbabwe (formerly Southern Rhodesia) liberation war.[25]

Manley returned to Havana for the Sixth Summit of the Non-Aligned Nations. His speech on 4 September 1979 praised Tito, Nasser, Nehru, Sukarno, Garvey, Nkrumah, Lenin and Castro and paid tribute to Cuba for its 'clear uncompromising and yet always humane leadership.' Manley referred to the controversial Forbes Burnham of Guyana, widely accused of electoral fraud and dictatorship, as 'an astute friend of mine.' The Jamaican leader drew special attention to the significance of the western Hemisphere for the struggle against imperialism, using words to describe the Cuban leader that he had just applied to Lenin:

> No area of the world has had a more extended exposure to, experience of, nor proximity to imperialism than Latin America and the Caribbean. We have seen forces of progress extinguished in Guatemala, snuffed out like a candle in the Dominican Republic, undermined and finally overwhelmed in Chile; and yet, I dare to assert that, despite these tragic reversals, the forces committed to the struggle against imperialism are stronger today than ever before. We believe that this is so because our hemisphere has had a Movement and a Man: a catalyst and a rock; and the Movement is the Cuban Revolution, and the Man is Fidel Castro.[26]

Noting that four of the seven new NAM members were 'Latin American' states—Bolívia, Grenada, Nicaragua, and Surinam—Manley saw 'a common thread of radicalism which reflects the need for change in our hemisphere.' This would require 'a fundamental and long overdue shift in United States policy.' He then turned to economic issues, renewing his call for the NIEO and calling for Third World self-reliance. Manley reckoned that 'an infinitely complex structure of inter-locking, corporate power . . . largely runs and controls the bulk of the world's economic power' and was 'largely beyond either the will or the capacity to control of the political leadership of the developed world.' He had no illusions about the long, hard road ahead, but concluded on the optimistic note that 'our Movement is irreversible because our cause is just.'[27] One sympathetic critic described the speech as 'not particularly appropriate to the situation in Jamaica.'[28]

The Havana speech was followed by a controversy centering on the Cuban ambassador in Kingston, Ulises Estrada, who took up his post on 25 July 1979.[29] Seaga's consistent anti-communism and anti-Cuba propaganda and the *Gleaner*'s adoption of those themes was of obvious concern to Estrada, and he met with the *Gleaner's* editorial board on 10 August for a discussion of the issues; no public outcry followed that meeting. The controversy began when Alfonso Hodge, a member of the Central Committee of the Cuban Communist Party (PCC); attacked Seaga and the *Gleaner* at the twentieth congress of Guyana's opposition People's Progressive Party. Hodge referred to Seaga's 'fascist style' and called the *Gleaner* 'ultra-reactionary'. Seaga and the *Gleaner* protested to Estrada, accusing Hodge and Cuba of interfering in Jamaica's internal affairs. Estrada met privately with Seaga and again with the *Gleaner*'s directors, before a press conference on 18 September in which he denied the accusation of Cuban meddling. He said that Cuba sought friendship, but if war was declared the Cuban revolution would accept the challenge and 'as Fidel says, when Cubans fight we fight seriously.' That statement was taken by the *Gleaner* in a front page editorial as 'issuing threats against Jamaicans in the Jamaica Labour Party and this newspaper.' Although Estrada later explained that he had in mind ideological and not physical fighting, his words were needlessly provocative, demonstrating that Cuban diplomacy can be as maladroit as that of other countries.[30]

Manley defended Estrada and Cuba. Cuba's work in Jamaica, he said, was its best defense. The main issue was not that, but the campaign of 'lies, propaganda, misrepresentation, distortion and manipulation' by the *Gleaner* and its 'stooges'. One lie was that there were five thousand Cubans in Jamaica; the truth was that there were 449, 400 of whom were building a sports college near Spanish Town and a secondary school near Vernamfield. Manley noted that attacks on Estrada started even before he took office, that Cuba supported Jamaica's government (whichever party was in power) because it represented the wishes of the people, and that Estrada's comments about war and fighting referred to propaganda attacks, not military ones. He regarded the *Gleaner*'s account of Estrada's 18 September press conference as 'one of the most vicious, dishonest and wicked pieces of calculated, lying, newspaper representation that has ever been made in the history of Jamaica.' The *Gleaner* delayed Estrada's second statement that no threat to the JLP or the *Gleaner* was intended, for thirty-six hours, and downplayed a third statement by the Cuban ambassador in which he apologized for any misunderstandings. The real objective of the JLP and the *Gleaner*, said Manley, was 'power at any cost, by any means, whether fair or foul . . .'[31]

Events in revolutionary Grenada connected with those in reformist Jamaica. On 13 March 1979 a popular insurrection had overthrown the government led by the eccentric authoritarian, Eric Gairy. Grenada's People's Revolutionary Government, led by Maurice Bishop, asked for Jamaican support in radio and television, banking, tourism, security and the bulk purchasing of imports, and the Manley government provided what support it could.[32] In a speech on 18 September, Grenadian leader Maurice Bishop referred to the role of the press in destabilizing progressive regimes. He cited a specific case in which the *Gleaner*, 'a well known destabilizing influence in Jamaica', misquoted him during a press conference on 2 September at the Kingston airport where Bishop stopped on his way to the summit of the NAM.[33] In December Manley wrote to Bishop of a meeting he had with United States government officials—Philip Habib, Andy Young, and Robert Pastor—in which he spoke strongly about the 'stupidity' of the hostile United States attitude toward the Grenadian revolution. 'Obviously', Manley advised Bishop, 'the "hawks" are gunning for you as indeed, they have been gunning for us, particularly since Havana.' Manley suggested elections in Grenada, as the United States and others were obsessive about them.[34] Manley may have hoped still to sway more moderate Americans. On Christmas Eve he wrote to President Carter offering his support in the Iranian hostage crisis, including support for sanctions against Iran, despite the reservations on them.[35]

Beverley Manley and the Women's Movement

As Jamaica's crisis deepened in 1979, one of the forces developing on the PNP left was the Women's Movement (PNPWM), led by Beverley Manley. Born in 1941, Beverley Manley was the daughter of a hard-working mother and a father who was a station master on the Jamaican railway. Seventeen years younger than Manley, she was almost of a different generation, one less imbued in British colonialism. She began her university education as a mature adult in the '70s, and at the University of the West Indies Mona campus responded to radical intellectual currents. Eventually she would write a master's thesis critical of the PNP in its early years, and she would draw broad parallels between the early organization and the PNP of the '70s, of which she was an active leader.

Michael Manley supported the PNPWM and women's causes, although his advocacy was tempered by his role as moderator of competing factions and his intellectual rather than visceral understanding of women's oppression. Speaking to the thirty-fifth anniversary meeting of the Jamaica Federation of women (JFW) on 30 May 1979, he referred to women as 'the rock-like foundation of every single organization and movement that attempts to serve the people.' 'The shop, the market, the school, the church, the political meeting, the JFW, 4-H or Friendly Society Meeting—all are meeting grounds usually dominated by women,' and thus women usually knew the problems of a community. Despite PNP programmes such as the Women's Bureau, the Status of Children Act, the national minimum wage, and equal pay for equal work, eighty percent of women made less than J$30 per week. Working mothers faced many problems with the health and education of their children, and vocational training was inadequate.[36]

Under such conditions it was not surprising that some PNP women would become a radical force in the party. Beverley Manley, Maxine Henry, and the PNPWM generally helped lead the developing struggle in party and government against the IMF path. Beverley Manley pressured the government to act quickly on one of the few new programmes of 1979, a proposal for paid maternity leave for working mothers. Her husband agreed with the idea in principle, although he was not sure as to when and how to implement it.[37] The maternity leave law was passed on 31 December 1979. It provided paid maternity leave for working women eighteen years or older, married or single, and protected women from dismissal, demotion, or loss of benefits because of pregnancy. Twelve weeks of leave were provided, with eight at full pay. The law protected groups traditionally subjected to extreme exploitation, such as domestic helpers, casual or seasonal agricultural workers, and barmaids and waitresses. Penalties were established for employers who refused to comply.[38]

Frequent requests for help kept Beverley Manley in touch with the hard lives of 'ordinary' women. A typical example of a letter addressed to Beverley reads as follows:

> Dear Madam.
>
> Howdido. I respectfully asked, this letter will be confirm. I am a 24 years old woman having 4 children, oldest one is 7 years old and the youngest is 1 month, it like a bad luck as woman with man, when I had the first on[e] the Baby Father went away and leave me. I talk to a next one to See if I could get Such a Feeding and clothing for the child[.] it so happen that I got the Second one and until it reaches Four. Dear Mrs Manley, I am do asking you if you could send something in my behalf to help me or I am asking you to put me on a pension Roll So that I can help the children. I don't intend to look any more man, please help.[39]

Beverley would answer such letters, referring the sender to the appro-

With Pierre Trudeau at Jamaica House.

With Helmut Schmidt and Willy Brandt: The Runaway Bay Planning Meeting in Hamburg, 1978.

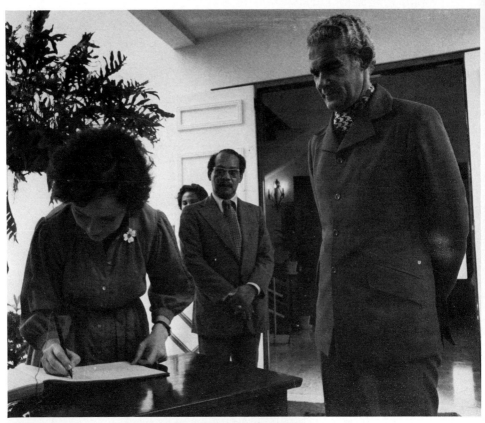

With Rosallyn Carter at Jamaica House, 1978.

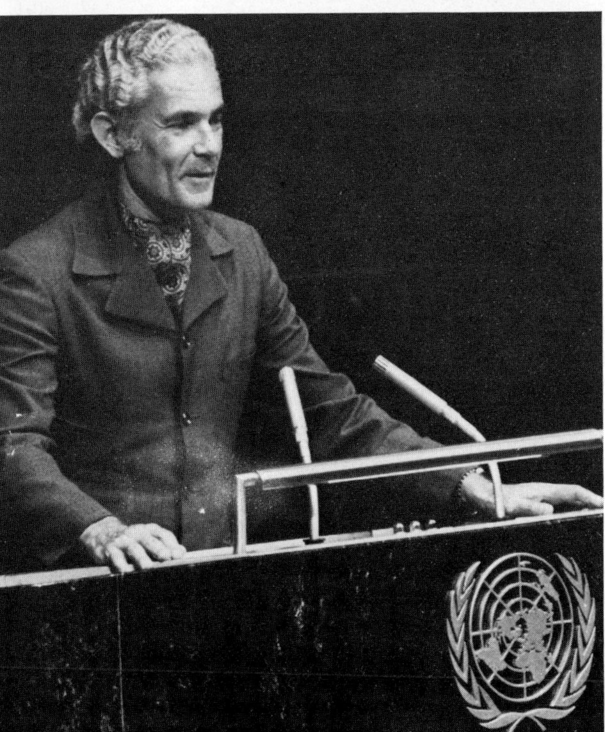

At the United Nations in New York, October 1978, Manley collects his Special Award for Distinguished Service against Apartheid. Other recipients included Pandit Nehru, Kwame Nkrumah, Olaf Palme and Paul Robeson. Kurt Waldheim, as Secretary-General, presented the awards.

Meeting the Soviet Delegation in Moscow, 1979.

Manley, Perez, Nordli, Obsanju, Schmidt and Fraser at the Runaway Bay Conference, Jamaica December, 1978.

priate person or agency for help. Often she sent small parcels of supplies.

Beverley Manley's duties included an active role in the controversial foreign policy of the PNP. She was part of the Jamaican delegation to Moscow in April, and served as a patron of the Jamaica-Soviet Friendship Association. She delivered a speech to the association at its Kingston inaugural on 6 May. The Prime Minister's wife described her impression of the Moscow trip:

> Contrary to the anti-Soviet propaganda that has almost inundated this society in recent times, we Jamaicans on that visit caught a glimpse of a happy, confident people in the Soviet Union—a people so dedicated to world peace that the topic was a dominant feature of every discussion in which I was privileged to participate.

Beverley Manley went on to mention the struggles of oppressed peoples 'in Southern Africa, in Uruguay, in Vietnam, in Nicaragua, in Puerto Rico' and to list the inspiring victories 'in the name of liberation' in the Soviet Union, in Cuba, in Vietnam, in Angola, in Uganda, in Mozambique, in Ethiopia and in Grenada. She stated that Jamaicans had long received a distorted view of socialist society, 'particularly the socialist society of the Soviet Union, the first and most advanced'. She denied that there were 'two imperialisms' in the world, since the essence of imperialism was economic exploitation: 'economic relations between Socialist Countries and Third World countries are based primarily on the political discussion and determination of bi-lateral agreements with the stated objective of just exchange between the participants.'[40] The PNPWM and Beverley Manley were also strong supporters of Cuba, and publicized Cuba's contributions to Jamaica in education, public health, construction, agriculture, fishing, sports and culture.[41]

Because of her outspoken positions on the Soviet Union, Cuba, and imperialism—and because she was an assertive woman in a male-dominated society—Beverley Manley displeased liberals and conservatives, including PNP 'moderates' or rightists. She was assailed by the *Gleaner* and its afternoon tabloid the *Star* for her forthright anti-imperialism. Beverley in turn attacked the 'fascist section' of the JLP 'headed by Seaga' which she claimed had organized the January demonstrations without support from 'Shearer and his large following'. She claimed that 'women are a main target of Seaga.' Since women as mothers and homemakers experienced the economic crisis with special sharpness, the PNPWM needed to educate women and organize them against 'Seaga's fascist politics'. This included 'work with decent JLP supporters'.[42] Although Michael Manley supported the PNPWM, Beverley because of her gender, her relative youth, her university experience and her work with poor women, was moving to a more radical position that threatened the always tenuous ties between PNP right and left. At times the ideological differences between Michael and Beverley, Michael being the more moderate of the two, placed an ideological strain on the marriage. In any case, two strong personalities at times found marriage a rather confining convention.

In other ways, Michael Manley's personal life continued to reflect developments in Jamaica at large. In a letter to his daughter Rachel in May, Manley referred to the 'hellish pressure' on him.[43] On another occasion Manley had to return an expensive awning given to him by a businessman who supported him and later felt betrayed by him.[44] As Manley encouraged Jamaicans to produce more, he himself netted J$780.75 from the sale of Christmas trees in one month and J$1,785.77 from lumber in another.[45] The Prime Minister's passion for sports seldom flagged. He wrote a friend that he had seen John McEnroe play tennis: 'I am still a [Björn] Borg man, though.' To another friend he wrote in December that 'we have to see the Benitez/Leonard fight.'[46] A bigger and more fateful fight faced Manley himself in the coming year of 1980.

CHAPTER 20

'Hovering at the Edge of . . . Hell' 1980

. . . Say the owl's
A lonely sentinel
Hovering at the edge
Of what we now call hell

Rachel Manley[1]

Michael Manley's government and the experiment in 'democratic socialist' change in Jamaica came to an end in 1980 'in a hail of bullets and a river of blood'.[2] The culmination of years of tribalized, violent politics and economic crisis, 1980 would demonstrate decisively—as 1973 had in Chile—the obstacles faced by a government attempting a democratic path to socialism in the hemisphere the United States jealously regards as its own sphere of interest.

For Manley, as for Jamaica, it was a hellish year, although one not without its pleasures. In April he received a certificate of appreciation from the Jamaica Exporters' Association and in August was awarded the Institute of Jamaica's centenary medal for his 'sustained and outstanding contribution to Cultural Development in Jamaica in the field of History and Social Science'. Manley did not listen to much of his beloved classical music while he was in office, but gardening for pleasure and profit seems to have been an important relief from the constant pressures of politics and government. The Jamaica Horticultural Society awarded him prizes for his flowers, most notably the Mary Lock Cup. He ordered 360 bareroot roses from a firm in Medford, Oregon at a cost of J$1,763.80. In September he asked two florists to cooperate in an increase in the price of roses to J$5 a dozen. None-too-secure personal finances were helped by the $J6,000 Beverley Manley made on a lecture tour in the United States, which went towards paying a $J29,000 home loan the couple had. Their son David, Manley's fifth child and the couple's second, was born prematurely in September, and Manley donated $J1,000 to the university hospital in gratitude for their skilful care of Beverley and the baby. Despite such good fortune, however, the frustrations of the outside world were never far away. In August Manley wrote privately that he despaired of changes in education in Jamaica's complex pluralistic situation.

It is all so slow and lost in frustration. What with the political system, the

> J. T. A. and God knows what else, I sometimes feel that we are in the middle of a great chocolate mousse that has neither bottom nor sides.

The uncharacteristic pessimism betrayed a fatigue bordering on exhaustion which Manley recognized only years later after he had been voted out of power.[3] That defeat came as the result of long-standing trends described in preceding chapters—economic crisis, violence and destabilization, and determined political opposition from the *Gleaner* and the Seaga-led JLP—the final months of which are recorded below.

The Break with the IMF

After the IMF agreement of May 1978, real wages had fallen by thirty-five percent in 1978 and ten percent in 1979.[4] Although Jamaica had received the IMF's 'seal of approval' for the 1978 agreement, in September 1979 commercial banks deferred consideration of a US$650 million proposal to refinance Jamaica's external debt. Moreover, of the US$429 million in IMF assistance approved in May 1978, only US$172 million was actually disbursed. As mentioned above, in December 1979 and January 1980 Jamaica's foreign exchange performance fell short of IMF standards by US$150 million. The shortfall was mostly due to factors beyond Jamaica's control.[5]

On 3 February 1980 Manley announced that the government was unable to reach a new agreement with the IMF. He stated that there had been deficiencies in expenditure control and management. Business and consumer subsidies would have to be cut and public utilities rates increased. The IMF was insisting on a US$150 million cut in government expenditure, but the government thought cuts of more than US$100 million were impossible. The US$50 million difference would mean the layoff of 11,000 government workers, a social and administrative disaster, or the removal of all subsidies for food, the Jamaica Omnibus Service, fertilizers, and the literacy programme, and the reinstitution of full tuition at all post-secondary schools. The government planned to call for elections to decide Jamaica's economic path once the Electoral Commission had a new elections system in place. Courageously, despite his constitutional right as Prime Minister to call for elections 'without prior notice', Manley stated that it was not the time for 'clever tactics', but for 'serious national debate'. Whatever the vote might show, 'it is vital that everybody accept the results of this election and unite around a course of action chosen by the majority of the people.'[6]

On 8 March the PNP's Economic Affairs Commission presented its critique of the IMF approach and its alternative. Its report pulled no punches. Polls showed that the PNP could lose thirty-six of its forty-seven seats in the forthcoming national elections.[7] Since the 1976 elections the government had lost the people's confidence: 'Worse than that, the Government has managed to erode the self-confidence of the people and has placed the continued development of the Socialist process in severe jeopardy.' The report said that 'steps should have been taken long ago in the direction of tightening up the management of the State,' and because of the delay such steps might be counter-productive at such a late date under IMF pressure. Following IMF programmes had failed to attract support from the private sector, was incompatible with PNP goals and objectives, heaped economic hardship on the people, and 'instead of gaining greater economic independence as promised, the people now have to go to Washington to have their budget approved.' It was impossible to achieve real development by following the IMF: 'Without development there is no future. If there is no future, who needs the IMF?'[8]

The PNP National Executive Council met on 22 March, voting unanimously to break with the IMF and 103 to 45 to do so before the election; the decisive factor in the vote was 'profound and . . . engrained suspicion of the IMF.' Manley offered to resign since he had led the party and government into the discredited IMF agreements, a proposal which the council rejected. Manley later wrote that in spurning the IMF 'the PNP had recovered its soul—at least as far as the rank and file were concerned.'[9] On 25 March Manley announced the decision in Parliament.

Five days later Manley explained the break with the IMF to the nation in a broadcast announcing the government's non-IMF programme. 'Operation Reconstruction'. He said that dealing with the IMF had required 'sudden and savage adjustments of the economy in return for the promise of foreign exchange that the country needed so badly to survive.' The government had worked hard to meet its commitments to the IMF, 'even though these have meant levels of hardship and suffering for the people beyond those which we thought absolutely necessary.' Manley explained some harsh realities: that Jamaica could not continue to borrow heavily to pay for oil and to pay the high interest rates on money already borrowed. The government's new priorities would be firstly to produce local foods; second, to get export industries the foreign exchange they needed; third, to keep tourism in good shape; fourth, to start new programmes in bauxite and alumina; fifth, to closely supervise 'prices and other problems which hurt people'; and sixth, to prevent the leakage of foreign exchange. Manley appealed to Jamaicans to end political violence which could only make economic recovery 'a thousand times more difficult.'[10]

A memorandum from university economist George L. Beckford to PNP General Secretary D. K. Duncan shows what those most committed to the non-IMF economic strategy thought that it was necessary to make it work. Beckford wrote that breaking with the IMF gave the government an opportunity finally to implement the democratic socialist alternative mandated by the electorate in 1976. To do so 'capitalism/imperialism' would have to be put on the defensive. 'The people, and the people alone' could ensure that democratic socialism succeeded. This required being truthful with them, admitting that the IMF path had been a mistake. National unity would have to be emphasized since political tribalism was inimical to democratic socialism. Immediate political education was required. Members of Parliament—who themselves would have to be politically educated and committed—should be in the field to educate the people.[11]

The PNP Women's Movement pushed hard for the non-IMF path. At the PNPWM's seventh annual conference in April, its President, Beverley Manley stressed the importance of getting food to the people and of monitoring food distribution. The women's movement also announced a ten-point programme to deal with food shortages and consumers' rights, and publicized a list of seven actions by supermarkets which aggravated the situation.[12] The PNPWM suggested that the government begin distributing food, give price inspectors more power, promote consumers' rights education, and appoint women who shopped to the Prices Commission Board.[13]

In early August Manley reported that no economic disaster had occurred after the IMF break, thanks to help received from friendly countries, the 'sympathetic understanding of our overseas creditors', and 'mainly through the efforts of the Jamaican people'.[14] Then, in mid-August, hurricane Allen passed just north of the island, doing at least US$150 million damage, destroying bridges and roads, damaging fifteen thousand houses, and wiping out seventy-five percent of banana export cultivation and breadfruit and other tree crops.[15] As if nature's awesome power weren't enough, rural pipelines in St James, Trelawny and St Mary were sabotaged, apparently by striking parish council water workers represented by the NWU, the BITU and JALGO.[16] And despite two years of intensive government study of alternative energy sources, the energy import bill in September was still 'astronomical'.[17] Added to all the economic difficulties was chronic political violence and a new effort to destabilize the government.

Political Violence and Destabilization

After Manley announced in February that elections would be held as soon as new electoral machinery was in place (they were held on 30 October), political violence escalated. In 1980, 889 murders were reported in Jamaica, an increase of 538 over 1979; much of the increase was due to political violence.[18] Rumours of attacks and actual attacks on leading politicians of both parties increased. Partisans of both parties argued that the other party was primarily responsible for the violence. It was commonly believed that leading figures in both parties were connected to violence.[19] Sometimes political violence occurred without direction from above. At other times it was hard to distinguish from criminal violence and personal feuds.

The JLP argued that since it was ahead in the polls, it did not need violence to guarantee its election. Manley dismissed that argument by explaining that the more violence spread fear, the harder it was for the PNP to carry its side of political debate. It was the PNP which had lost political support, the PNP which was victimized by rumours and propaganda.[20] Scholar Michael Kaufman agreed:

> The JLP had distributed heavy arms to its gangs. It had consistently mobilized its supporters in sometimes violent actions against the government and its supporters. PNP supporters dubbed the Leader of the Opposition CIAga.[21]

One contemporary account linked Edward Seaga to gun-running and the CIA.[22] The *Gleaner* published an article by PNP General Secretary D. K. Duncan in which Seaga's tendency toward violence was illustrated by quotations attributed to him: 'it will be blood for blood, fire for fire and thunder for thunder' and '. . . on that day I will dip mi finger in PNP blood to mark mi x.'[23] Shortly before the election Manley alluded to a statement by the Joint Monitoring Committee of the Jamaican Council of Churches that confirmed that 'the majority of reports of political violence relate to acts being perpetrated against candidates and supporters of the People's National Party.'[24] On the other hand, a poll carried out by Carl Stone in June 1980 (but released only in August), revealed that more people thought the PNP was behind the violence than thought the JLP was.[25]

Manley admitted one major incident of PNP violence against the JLP, the Gold Street Massacre in his East Central Kingston constituency in April. Twenty to thirty men dressed in military fatigues attacked a JLP fund-raising dance, killing four and wounding eleven. Manley called the event 'ghastly', said that it was properly denounced by churches and the private sector, and declared that it was the only massacre carried out by the PNP.[26] Shortly after the Gold Street tragedy, a special issue of the PNP newspaper included a full-page appeal by Manley to end political violence. It stated that 'it is wrong for poor man to kill poor man and for ghetto youth to snuff out the life of another ghetto youth in the name of politics.'[27] A week later, the party organ included an appeal from the PNP Youth Organization for an end to the violence and listed two attacks by the JLP on PNP youths.[28] Frequent appeals for peace by Manley, Seaga, Governor-General Glasspole and other leaders sadly had little impact.

Significant parts of the security forces turned against the Manley government. On 22 June members of the army arrested three JDF officers, twenty-three soldiers and three civilians who were charged with plotting a coup. The aborted coup's leader was H. Charles Johnson, the head of a small right-wing group called the Jamaica United Front Party. The plot included plans to take over two tanks, capture Manley, and force him to resign. The JLP at first denounced the abortive coup, but Seaga, in Washington, proclaimed it a 'comic opera'. The highly respected commander of the JDF, Brigadier Robert Neish, made a rare television appearance to state that the coup attempt had been quite real, a statement supported by Police Commissioner William Bowles; both took the wind out of Seaga's 'comic opera' claim.[29]

Behind the aborted coup was the general politicization of the army and police against the Manley government.[30] Among JDF officers anti-communism played a significant role in this, and the general economic and political malaise of the society alienated some rank-and-file soldiers and police officers. Others were estranged by what they considered the prior politicization of the security forces by the Manley administration.[31] The turning of the army and police against the government, may have been (as Manley believed) part of a plan of destabilization by the JLP and the Unites States.

On 4 July—the United States' Independence Day—the house of the United States embassy's first secretary in Kingston, N. Richard Kinsman, was fired upon and allegedly grenaded. Neither Kinsman nor his maid were awakened by the firing, and in the morning his first call was to the *Gleaner*, not the police. The 4 July 'attack' was two days after a news conference in Jamaica, in which the editors of the *Covert Action Information Bulletin (CAIB*) divulged the names of fifteen CIA employees in the Kingston embassy; Kinsman had been identified by the editors as CIA chief of station a year earlier. The editors of *CAIB* regarded the 'attack' on Kinsman's house as 'a phoney' pretext for United States Congressional action against *CAIB*.[32]

Manley received a typed, two-page letter dated 5 July from an American who, with his wife, overheard a conversation on 3 July between two other Americans, one of them drunk, in a lounge of a hotel near Rose Hall great house. One of the Americans was heard saying that they were 'finally going to get those turncoat bastards [Louis] Wolf and [Philip] Agee,' ex-CIA agents who collaborated on the *CAIB*. They also reportedly said that 'Manley will sure be in more trouble after tomorrow.' On the next day, of course, the 'attack' on Kinsman's house occurred.[33]

The kind of trouble Manley might be in as a result of the possibly phoney 'attack' on Kinsman's house soon became evident. A *Gleaner* story dated 7 July quoted Seaga as follows:

> The terrorist attack . . . is the latest in a series of anti-American attacks led by government institutions, the ruling People's National Party, and the Communist Workers' Party, all working in orchestration to spread hate and violence as part of their strategy of effecting a 'military solution.'

The *Gleaner* claimed that the government proposed to add professors of Black Studies 'from various universities' and 'selected members of the Black Caucus in [the U.S.] Congress' to the effort.[34] Such claims would have been comic were it not for the deadly seriousness of the political struggle in Jamaica.

Throughout the summer and early autumn, there were rumours of assassination plots and actual attacks on political leaders. In July a rumoured plot on the life of Seaga was reported to Manley. He immediately informed the United States Ambassador, Loren Lawrence, requesting that all possible protection be provided to his bitter political foe. Two Jamaicans armed with high-powered rifles were reportedly arrested in Miami.[35] A few weeks later, Manley's friendly adversary Hugh Shearer was slightly wounded by a missile thrown at his motorcade; Manley expressed 'shock and outrage', feelings deepened by his great affection for Shearer.[36] The government later received word of a threat on the life of JLP Senator Bruce Golding; Ambassador Lawrence was again advised and the cooperation of United States police requested.[37] Manley and his touring party were fired on in early October.[38]

The most serious incident involving a political leader was the murder of PNP candidate Roy McGann and his bodyguard, police corporal Errol White on 13 October 1980, the day before official nominations, less than three weeks before the general election.[39] McGann was an MP, Parliamentary Secretary in the Ministry of National Security, and candidate for East Rural Saint Andrew. His murder was the first time in Jamaican history that a candidate had been killed. The crime was committed during a late night confrontation between PNP supporters and a much larger group of JLP supporters. McGann and corporal White were shot after police had arrived on the scene; McGann was heard over the PNP's private radio network shouting 'the Police are firing at us' while corporal White desperately tried to identify himself by yelling 'Police, Police!'[40] Those shooting were not deterred.

Manley's first response was a short, general appeal for peace. He demanded that all PNP candidates conduct their campaigns peacefully and appealed to PNP workers and supporters to 'assume a non-violent role . . . in the face of whatever provocation.'[41] On 14 October, Manley vowed that 'we will not bow to terrorism from any quarter' and that nominations and elections would proceed as scheduled. He asked PNP supporters 'to ensure that their activities are not marred by feelings of revenge or reprisal.' Acknowledging that it was not a time to cast stones, Manley noted that 'it is only the blind who cannot see the pattern that has been emerging in the last two to three months.' The only way to ensure that Mcgann had not died in vain was 'if every well thinking Jamaican takes an oath in the sight of God to stamp out this evil thing which is threatening to destroy everything worthwhile of which our country is capable.'[42]

On 22 October, Manley addressed the nation on the 'terrible developments' taking place in Jamaica. Political terrorism had been unleashed against women, old people, school children, candidates, and 'even our little defenseless children and babies.' (Edna Manley created a chilling sculpture, 'Ghetto Mother', in 1980 which captured the horror of the violence unleashed against women and children.) The Prime Minister announced that on the very evening of his address, after Roy McGann's funeral, five people had been killed in McGann's constituency and the driver of another PNP candidate, Jack Stephenson, had also been murdered. Tension and fear was greatly increased by 'unfounded and unsupported statements by the Leader of the Opposition [Seaga] and other JLP spokesmen that the People's National Party plans to incite riots and violence' as a pretext to cancel the general election. Manley vowed that the election *would* take place on 30 October, as it did. He closed his address by noting that 'what is happening is nothing short of madness.'[43] Although another State of Emergency was indicated, Manley felt that he could not get the consensus required for it.[44] It is also questionable that the security forces would have enforced it.

The most troubling aspect of the McGann-White killings, other than the death of two good men, was some ominous involvement of the police and army in this active opposition to the government and the PNP. A further event occurred about one week before the elections when soldiers twice fired their guns during a PNP rally in Spanish Town. The first episode was caused by some youths throwing rocks who could have been handled by less drastic means. Manley witnessed the second episode of gunfire and said that there was absolutely no reason for it, but it effectively broke up the meeting.[45]

The Gleaner

Throughout 1980, as Jamaica's economic and political crisis worsened and elections loomed, the *Gleaner* hammered away on the Manley government, attacking not only its policies but its very legitimacy. The *Gleaner's* role was magnified by its enormous power inside Jamaica—Andrew Kopkind referred to it as the most important surviving institution from the old colonial era—and by its international acceptance: the *New York Times*, the *Washington Post*, the *Miami Herald, Newsweek* and the London *Daily Telegraph* accepted its reports at face value. In a lengthy and perceptive article Kopkind noted that 'the *Gleaner* is no mere mouthpiece for Seaga and his JLP. It also formulates strategies and orchestrates their execution. . . .' Kopkind quoted Kingston lawyer and journalist Ronald Thwaites as saying that 'when you come down to it, the *Gleaner* is the opposition.' Personalizing the *Gleaner's* campaign, ex-Manley friend and *Gleaner* columnist John Hearne reported to Kopkind that

> this has been a ferocious campaign. It would be idle to pretend that there has not been a systematic attack on the government by the *Gleaner*. For myself, my one intention is to get this man Manley out of office, by any fair means at hand.[46]

Hearne's comment was typical of the *Gleaner's* approach in its use of truth and the appearance of candour (the *Gleaner's* ferocious anti-Manley campaign) to cloak the use of falsehood (that the means used were fair).

Fred Landis, who had written a doctoral dissertation on the media in Chile and denounced the role of *El Mercurio* in the overthrow of Salvador Allende, performed a content analysis of the *Gleaner* for the Press Association of Jamaica in April and May of 1980. Anticipating the abortive coup of June and recognizing the disaffection of the security forces, Landis believed that the *Gleaner's* attacks were directed primarily at military officers. 'The *Gleaner* reflects Chile all over again. You only have to study the standardized techniques in one country and you can recognize it anywhere. . . .' Landis accused Jamaica's largest newspaper of carrying out psychological warfare against the Manley government.[47]

The *Gleaner* used a combination of techniques. Its editorials often presented a posture of impartiality, while the business of savage critiques were left to columnists such as Hearne and Wilmot Perkins. Supposedly objective news stories were presented in a one-sided way. Manley claimed that after former JLP chairman Hector Wynter joined the newspaper's editorial staff in 1975, he was never asked his views on stories to be printed about government or PNP policies and programmes.[48] As Landis had noted, page composition and the interspersing of innocent items with those designed to spread fear, was a common technique.

It is difficult to evaluate the specific impact of the *Gleaner's* partisan propaganda. The hard core of either party was unlikely to be much affected by it, although the general tendency would have been to delegitimize the government and legitimize the opposition. Despite *Gleaner* distortions, people's direct experience of Jamaica's deepening economic and political crisis would have lessened support for the Manley government. An important bloc of voters was independent of either party, suspicious of both, and could provide the swing votes to change a government. It would be natural for such independent-minded voters to look to Carl Stone for a more objective analysis of the situation in mid–1980. Stone's views on Manley and Seaga may have influenced events leading up to the election.

Stone on Manley and Seaga

Stone had worked for the PNP government in the mid–70s, had done or would do polling for both parties, and was at different times regarded by each as biased against it. His polls were accurate, he emphasized empirical data, and his position in the political spectrum was somewhere between Manley and Seaga. The general good sense of his columns provided a welcome relief from the mostly partisan fare stemming from both parties. It is necessary to bear in mind, of course, that even the relatively objective Stone could not be indifferent to and unaffected by the crisis afflicting Jamaica in 1980, which coloured his assessment of Jamaica's two major leaders. In 1980 he leaned toward Seaga, but the characterization of Stone as 'a rabid JLP supporter'[49] was exaggerated and ill-informed.

Stone began his comparison characterizing Manley's and Seaga's images.[50] Manley's positive image was as 'the champion of the cause of poor people and Third world dignity and self respect.' Seaga's positive image was that of a 'financial wizard'. Manley had a negative image of 'an incompetent and vain ego tripper in search of personal glory but far from equal to the task of governing the country.' Seaga's negative image was that of 'a shadowy and suspicious figure who is likely to do evil things if entrusted with power.' Stone noted that these images were the result of propaganda and that there were few unbiased sources for an accurate picture.

Noting that Manley was the more complex of the two, Stone analyzed him at length. To Stone, Manley was a 'natural politician who gets turned on by crowds.' He had a natural flair for political drama, like that of Bustamante, and knew how to upstage opponents 'by carefully staged events and dramatic moments'. Manley was 'neither an original thinker nor a dogmatic person'. He took advice from many sources. Stone denied that Manley was the idealist he was often made out to be, but instead a 'tough political infighter who never gives up and has a keen sense of always doing what is necessary to survive politically.' Stone claimed that like most major Jamaican politicians Manley did not attribute much to the intelligence level of the general population. This caused him 'a big credibility gap' because he explained Jamaica's problems in ways the man in the street did not find credible. Manley was 'a bright man who grasps things quickly and is able to put over any point of view with eloquence and conviction.' Stone made clear that Manley was a socialist, not a communist. Contrary to a common opinion, Manley's views had been fairly consistent since he became PNP leader in 1969.

According to Stone, Manley had four major weakness as Prime Minister. First, his concept of change focused on political changes 'seeking to bring Jamaica closer in line with Third World socialist trends.' Thus, Manley and the PNP lacked a workable economic strategy. Second, Manley was incapable of taking unpopular decisions or taking responsibility for them. Third, Manley was a 'talker rather than a doer', immersed in speeches, 'projecting the right images', and propaganda; he lacked a strong, competent number two man as administrator. Finally, according to Stone, Manley had made the fatal error of 'beginning to believe his own propaganda that the people were so much in love with him that he did not have to do much to earn it beyond sporting the right ideology and the right rhetoric.'

Underlying the supposed weaknesses, Stone detected that 'Manley had a problem living in the shadow of his father's achievements.' He was heir to Norman Manley's accomplishments and wouldn't have become party leader without the family connection. At the same time he had tried to dismantle the PNP of his father and remold it in his own image, thus shattering the internally-divided party. 'What happened inside the PNP in terms of polarization, internal strife and division and confusion about goals is a microcosm of similar trends in the wider society. Manley's leadership has played no small part in these parallel developments.'

Stone admitted that it was difficult to compare Manley and Seaga since Manley had been in power for years and Seaga had not. Conditions in the '70s were different than those of the '60s, when Seaga had been government minister. Moreover, the two shared some qualities. Both believed in a mixed economy (although they differed on the relative weight of the private and public sectors). Both were 'overly sensitive to criticism'. The two leaders tended to be excessively preoccupied with Kingston and would forget the city was not Jamaica. 'Both men have quick tempers that often produce exaggerated reactions to political events, but Seaga appears to have matured far more than Manley in this regard.'

Otherwise, Stone drew a contrasting and sharply positive picture of Seaga. Stone found Seaga more of an administrator than a natural politician, a doer, not a talker. Stone praised Seaga's Tivoli Gardens showcase. Seaga was not as good a listener as Manley, but was likely to get twice as much done as him. 'Where Manley excited deep emotional loyalties, Seaga generates respect and awe because he always presents facts and figures rather than rely on rhetoric and ideology.' Seaga had a tough ego and the willingness to make unpopular decisions. Unlike Manley, Stone wrote, Seaga understood the economic alternatives facing the country.

Stone's final comparison came close to encapsulating the choice facing the electorate. 'Manley represents the ideology of economic independence, Third World Solidarity, and anti-imperialism. Seaga represents the prospect of pragmatic government seeking to re-build a shattered economy and restore international connections that have bread and butter value rather than cosmetic appeal.'[51]

The Campaign and Elections of 1980

Manley outlined the election issues to PNP campaign workers in July. Who ruled Jamaica, the people through their leaders or the old bosses? Would the new government be one concerned with the people or only with profits and economic growth? Would workers labour under conditions that encouraged their participation in decision-making and their dignity or their subservient obedience? Would Jamaica enjoy real independence and sovereignty or sponsorship by some foreign 'Big Shot'? Would the struggle for justice and opportunity for poor people and poor countries continue or would it be replaced by 'curry-favouring' with the rich and powerful? Who would win, the poor and their allies or the rich? Manley urged the campaign workers to be patient but firm when they talked to voters: 'a lot of the struggle is in your hands now.'[52]

C. S. Reid, a moderate observer, presented the issues differently. One issue was that of political leadership, which Reid saw as concentrated in a few hands; Reid was critical of 'politicians' in general, without mentioning names. A second issue involved Jamaicans' confusion about their identity. Were they Jamaicans or 'Africans', and if the latter, exactly which 'Africans'? 'Is our home Jamaica? Ethiopia? Miami?' A third issue was a crisis of purpose in the country: 'Can we really govern ourselves?' Reid claimed Manley had a chance to define national goals, but lost it in 1974 when 'democratic socialism' was resurrected as PNP ideology.[53]

The campaign, of course, was conducted against the background of horrifying violence and economic deterioration. As in elections in the United States, issues tended to recede before people's fears and hopes (however ill-founded) and fixation on the candidates' personalities. Seaga fanned the fear by promoting the patent nonsense that the Cubans were 'virtually directing all Government and controlling the Minister of Security.'[54] In response Manley tried to exploit divisions in the JLP. He argued that there were two Labour Parties, one representing the Bustamante tradition and carried on by Shearer, and the other, led by Seaga 'which was introducing a new kind of politics into the island.' The implication was that Seaga's new politics was outside Jamaican tradition.[55]

The views of scholars reflected the partisanship of the election. Michael Kaufman asserted that the JLP was 'on the warpath' between 1977 and 1980. Its leaders consistently circulated false and misleading stories. Seaga spoke frequently in the United States against the Manley government, alleging that it had embraced the Cuban model. In 1979, he circulated a document linking every government leader and PNP leader to the Soviet KGB or Cuban DGI. Meanwhile, 'the PNP went out of its way to respect the political rights of a party that had no respect for either the political system in particular or human life in general. People were gunned down every day, but moderation dominated the PNP response.'[56]

Closer to the events, more influenced by them, Carl Stone had a different view. Writing in early July, Stone referred to the 'vicious and vitriolic campaign of hate with which the PNP introduced its Corporate Area candidates.' He accused unspecified PNP leaders of 'fascist-mongering and hate mongering . . . against the JLP leader.' He maintained that the hard-core left had taken over the PNP, and blamed the party's 'message of hate' for fanning the forces of violence. 'Michael Manley', Stone wrote, 'has it in his power to shift the new PNP on to a more positive footing, but I doubt if he has the moral strength to do so before it is too late.'[57]

Later Stone wrote that the PNP and Manley had cultivated a myth of vast differences between the PNP and the JLP. The budgets drawn up by the two parties showed similar patterns of expenditure on health, education, and agriculture. Many PNP 'socialist' policies continued JLP programmes. The real differences in the parties were in their foreign policies and approaches to politics. The PNP's 'Seaga-baiting' was a tactic, Stone said, to distract the public from its miserable performance.[58] The JLP did its share of Manley-baiting, too, as reflected in graffiti around the island: IMF = Is Manley's Fault; Joshua = Judas; 'Manlie'.[59] The emphasis on personalities obscured issues, but in the end much broader issues were probably in the minds of the voters.

Despite fear and violence, seventy-seven percent of eligible voters turned out for the October elections. The people swung massively to the JLP, giving it fifty-nine percent of the vote and fifty-one out of sixty seats in the House of Representatives. The JLP increased its share of voters among all social classes, although the class realignment of 1976 was not reversed.[60]

Had Manley not led the PNP and the government in the '70s it is unlikely that such serious attempts at social and economic transformation would have occurred, and that such extreme opposition would have formed. At the same time, it is also possible that the dangerous alienation of black youth of the late '60s would have increased still further, leading perhaps to even more explosive events that those which actually occurred.

CHAPTER 21

The Enigmatic Wheel of History 1981-3

> So does the enigmatic wheel of history turn, casting down and lifting up. Some rise by talent and application. Some fall by twists of fortune that seem outrageous. All are subject to blind fate.
>
> *Michael Manley*[1]

The defeat of 1980, which led to eight years of rule by Edward Seaga and the JLP, left Manley deeply scarred. He felt he had been held up as an object of ridicule, that he had been hated by an entire society he had tried to love.[2] But surviving political defeat and personal misfortune also strengthened him, preparing him to lead Jamaica once again. As he and the PNP absorbed the shock of defeat and the lessons of the '70s, Manley wrote three books—*Jamaica: Struggle in the Periphery* (1982), *Up the Down Escalator* (1987) and *A History of West Indies Cricket* (1989)—and dozens of articles, pursued an active and lucrative lecturing schedule, was an influential Vice-President of the Socialist International, and with accelerating vigour from 1982 onwards, led the political opposition to the increasingly unpopular Seaga JLP government. The first step was analyzing the defeat of 1980.

Analyzing Defeat

Following the 1980 elections many PNP supporters were stunned 'almost into a state of stupor', and 'clouds of gloom thickened with the wanton destruction of the homes and livelihood of many of the Party's supporters.'[3] Manley felt the defeat in an intensely personal way and analyzed it in his memoir *Jamaica: Struggle in the Periphery*, completed ten months after the election and published in 1982. Manley titled the chapters dealing with the 1980 elections and his assessment of the PNP defeat 'Destabilisation Triumphs' and 'Psychological Warfare and Its Effects', respectively. He argued that 'the 1980 vote sprang from an essentially negative impulse. The majority of people were frightened, insecure, confused, unsure of the future, and reaching for safety.' Basic to everything he wrote was the economic situation with negative growth and increasing unemployment. Manley acknowledged problems of management as 'we overestimated the capacity of the bureaucracy to run the wide range of new institutions.' Violence was 'the greatest single threat to our political fortunes', and the role of the security forces was a part of that problem: 'In the end, they were more against than for us.' The 'communist bogey' was strengthened by the close relations of the PNP left with the Marxist-Leninist Worker's Party of Jamaica (WPJ). Despite efforts to establish distance between the PNP and the WPJ, the relationship 'did us real harm.' Manley discussed the US Army's manual on psychological warfare and concluded that, in addition to everything else, his government was destabilized by the efforts of the US government, the JLP, and the *Gleaner*.[4]

A PNP committee listed six major causes of the defeat—the economic situation; crime, violence and the role of the security forces; the communist scare; public perception of the PNP and its leadership; disunity within the PNP; and party leadership, MPs, workers and canvassers out of touch with the people—and added that it was meaningless to try and rank them in importance. The PNP Women's Movement and its Youth Organization listed similar factors, adding election day fraud, the 'ineffectiveness of the PNP organisation in many constituencies', and the 'actions of Imperialism'. The appraisal committee also solicited the opinions of fifty-two non-PNP individuals from various classes and occupations; the reasons for the PNP's defeat cited by these 'outsiders' were shortages (33 responses), unemployment and the communist scare (26 each), mismanagement of the economy (21), poor leadership (14), violence (11), and sabotage (5). The committee also credited the JLP with a 'devastatingly successful' opposition strategy against which the party and government never developed an effective, coherent defense, let alone counterattack. PNP disunity was a major problem which led to contradictory, confused policies and actions regarding the IMF, the private sector and capitalism, 'US Imperialism', transnational corporations and foreign investment, economic sabotage, the security forces, the *Gleaner*, the state bureaucracy, the inefficiency of the state sector, and relations with communists. Party disunity thus 'crippled the ability of the PNP Government to govern effectively, manage the economy, enforce internal discipline, maintain credibility, retain the confidence and morale of its own followers, and develop an effective propaganda and mobilizational campaign to deal with the JLP.'[5]

Violence preoccupied the PNP and its supporters long after the election. In 1981, a party report stated that violence by elements of the security forces and the JLP 'threw our organisation into disarray.' Journalist John Maxwell wrote in the PNP's newspaper *Rising Sun* that the 'M–16 carbine and its relatives' were the biggest factor in the PNP's defeat. *Rising Sun* described attacks on PNP officials and sympathizers. A week before the elections PNP candidate Donald Buchanan was fired on. The two police security officers assigned to protect him fired back. The police themselves were charged with 'shooting with intent' for defending Buchanan, but acquitted in court in 1982. In another case, D. K. Duncan was charged, but eventually acquitted, of alleged gun violence. In a third case, the inquest into the killing of Roy McGann was postponed at least four times since early 1982.[6]

Some laid the defeat on Manley's shoulders, an inevitable approach in the personalistic context of Jamaican politics and an inviting one given the awesome difficulty of unravelling complex economic, social, political and international forces. Winston Van Horne, a Jamaican professor resident in Canada, argued that Manley promised more than he could deliver, that he failed to dissociate himself from 'supporters who believed that politics is war in another form,' that he failed to stop migration and miscalculated the cost of the brain drain, that he 'underestimated the impact of public control on his social programs.' Van Horne also acknowledged broader, less personal factors. Most Jamaicans could not understand the difference between socialism and communism, and anti-communism was deeply entrenched in Jamaica's culture through its churches. The Socialist world could not provide adequate support for Jamaica's experiment. Van Horne praised Manley for his diagnosis of Jamaica's problems and for making greater social justice the fulcrum of his plans. But Manley's prescriptions were 'often unsound' and his actions 'injudicious'. Van Horne shared a common, though undeserved view of Manley as one who lived too much in his head and too little in the world around him. Van Horne noted that bad luck also played a role: the oil shocks of '73 and '79, world recession, floods in '79, and hurricane Allen in 1980.[7]

People in the streets had their own views. Under a large tree in the area of Kingston's Victoria Park known as the People's Parliament, the following debate occurred:

> 'Manley jus' for de people', a PNP supporter stated.
>
> 'Den how he lose de election?' shot back a JLP Seaga backer.
>
> 'Somebody tol' you he lost it. If all Jamaica voted for Manley, he would have lost it. The CIA, all a dem no wan' him.'
>
> A loud chorus of 'nos' broke out at this.
>
> 'Yeah, mon,' the PNP supporter argued, 'Seaga's a American. Spyaga!'
>
> Exasperated, the Seaga advocate demanded, 'What you mean? Him a Jamaican. Him represent de poorest constituency in dis here country.'[8]

A less partisan view noted that Manley's government lost support from all major groups. The PNP's share of manual wage labourer's votes dropped from seventy-two percent in 1976 to forty-eight percent in 1980; its portion of the farm labour vote dropped from fifty-six percent to forty-two percent in the same period; support from the unemployed and unskilled fell from sixty percent to forty percent; from white collar wage labour, from fifty-seven percent to thirty-seven percent; from small peasants, from forty-five percent to thirty-five percent; and from business persons, managers, and high-income professions from twenty to fourteen percent.[9] Several observers agreed with Manley's impression that the 1980 vote expressed sentiment against his government, rather than for the untested alternative proposed by Edward Seaga and the JLP. Seaga thus faced the difficult prospect of trying to consolidate the victory of 1980 with economic and social programmes that would win the positive endorsement of the

Jamaican people in time for the next general elections, although as it turned out, destiny in the form of the Grenada invasion postponed that reckoning for several years.

'Seaganomics', 1980–3

Seaga inherited a severely troubled economy. Jamaica's Gross National Product had declined twenty percent between 1972 and 1980. Unemployment stood at twenty-seven percent. An estimated eighteen thousand professionals and managers left the island during the Manley years. The three leading foreign exchange earners—bauxite, sugar, and tourism—were all in trouble.[10] Jamaica's economic problems affected the relatively privileged Manleys directly. Edna Manley recorded in her diary that Douglas was unemployed for months after the 1980 election and that 'our whole family is having a tough time financially.'[11] The problems of the poor were of course infinitely more excruciating.

Prime Minister Seaga pursued a two-fold strategy of improving relations with business and political sectors in the United States while attempting to maintain his commitment to state capitalism and regulation of the economy. Eventually the contradictions between the twin thrusts of Seaga's policy would lead to tense relations with the United States, but in the early '80s Seaga was depicted as the 'great right hope' of pro-Western business and political interests disturbed over pro-Cuban, anti-United States currents in the Caribbean. The Reagan Administration and the IMF extended much help to Seaga. Many investors believed that the IMF spared Seaga a forced devaluation of the Jamaican dollar in early 1981 to give him a domestic political cushion. Seaga also succeeded in lining up major loan commitments early in his first term, which prompted an unnamed Jamaican business leader to note the ironies of political economy. 'Before October [1980], every time Michael borrowed a dollar, everybody said: "Oh my God" . . . Now Eddie is borrowing everywhere, and everybody says: "What a great guy!" '[12]

Under Seaga's management, the Jamaican economy's performance in 1981 was mixed. Gross Domestic Product grew by four percent in 1981, and consumer price increases were cut from 28.7 percent in 1980 to 12.7 percent (and possibly lower) in 1981. On the negative side, Jamaica's trade deficit increased, as imports skyrocketed and exports stagnated. Seaga's borrowing also increased Jamaica's foreign debt from US$1 billion to US$1.3 billion in one year.[13] The economy continued to be very vulnerable to outside pressures. In his New Year's message at the end of 1981 Manley highlighted declining bauxite production—a developing nemesis for Seaga and the nation—and the widening trade gap.[14]

The economy stagnated in 1982, as agriculture declined by nearly eight percent, mining suffered nearly a thirty percent drop, the trade gap widened further, and foreign debt leaped to above US$1.8 billion.[15] Like Manley, Seaga was troubled by shortages of foreign exchange, lack of managerial talent (which remained abroad in large numbers), a sluggish private sector, and unreliable utilities. There were other similarities. Seaga continued a government agency initiated by Manley to stimulate exports, the Jamaican National Export Corporation (an agency more characteristic of state capitalism than the 'free-market' variety) and he began to have problems with the IMF. The Seaga government blamed its difficulties on the world recession, but its modified 'free-market' policies were to blame as well. Phasing out price controls was popular with the IMF but hurt the poor. Anthony Williams, the President of the Jamaica Manufacturer's Association (JMA) said in a speech to that body that 'deregulation has so far meant considerable losses for most of us.' Moreover, the Seaga government actually enlarged the scope of state economic activity by purchasing the Esso refinery in 1982, a move criticized by the private sector. Seaga's 'free-market' policies were catastrophic for small farmers, a sector that profited under Manley. Domestic food production fell nineteen percent in 1982, and Jamaica's food import bill rose to US$200 million. In response the government reimposed restrictions on the imports of food.[16]

With the shock of 1980 receding and Seaga in difficulty Manley became more active in economic matters and party journalism. Taking his first public union stance in ten years, he led an NWU bargaining team which gained a record contract for four thousand bauxite workers in February 1982. Hugh Shearer, Deputy Prime Minister in the Seaga government, chaired joint negotiations for several unions, so that two former prime ministers acted on behalf of labour.[17] In June Manley resurrected his old 'Root of the Matter' column in *Rising Sun* criticizing the government's policy toward sugar workers. In subsequent columns he criticized the JLP government's massive borrowing, 'dangerous totalitarian tendencies', and anti-labour policies.[18]

In June 1982 Manley credited Seaga with mobilizing a tremendous amount of loans for Jamaica, easing the foreign exchange crisis, and stabilizing the IMF situation. He denied that stabilization was recovery; however, since 'there has been no significant increase in production. . . .' Manley approved of the Reagan's Caribbean Basin Initiative (CBI) to the extent that it increased economic aid to the region and promoted one-way free trade with the United States. But Manley disapproved of the military aspects of CBI, its failure to funnel aid through existing regional institutions, its emphasis on bilateralism rather than regionalism, its neglect of smaller territories' infrastructural needs, and its exclusion of Cuba, Grenada, and Nicaragua.[19]

In 1983, the third year of Seaga's leadership, the economy grew by 2.3 percent, but consumer prices increased by 12.5 percent; exports declined while imports remained high resulting in a trade gap of US$438.5 million; and Jamaica's foreign debt increased to nearly US$2.4 billion.[20] The government and the IMF conflicted over the amount of government spending in view of anticipated revenues.[21] In 1984 Carl

Stone published production figures which showed a massive decline in the mining (bauxite) sector, a decline in domestic agriculture, and stagnation in exports. The strongest growth occurred in services rather than production. Stone commented that 'to use the word recovery to describe what has been happening is to confuse the bandaging of sores with the removal of impurities that cause them.'[22]

Manley agreed. He criticized Seaga as 'the most reckless and immature maker of promises that Jamaica has ever seen' who played fast and loose with the facts. He recalled that 'the JLP spent years [in the '70s] doing their best to destroy the Jamaican economy with a combination of violence and propaganda.' Now it was pursuing a neocolonialist economic model that 'has failed in Puerto Rico, is a near disaster in Brazil and is sliding to chaos in Chile.' The PNP drafted a major paper on economic strategy and Manley urged party members to study it in 'a calm, serious and objective way.'[23]

Social Conditions

Women were especially victimized by 'Seaganomics'. On International Women's Day, 8 March 1983, the Committee of Women for Progress launched the Women's Resource and Outreach Centre to help women cope with the hard times of the early '80s. At a meeting of eight hundred in the Odeon Theatre, one speaker observed that 'economic pressures arising from growing unemployment, high prices and lower wages [are] leading to greater dependence and degradation. . . . Nobody can deny . . . that women are more subject to abuses, sexual harassment on the job, and at home more and more of the frustrations are being taken out on us.'[24]

IMF-supported economic plans as pursued by the Seaga government entail decreased government at the expense of social welfare; health and educational services deteriorate and the poor are the main losers. Evidence of the social cost of the IMF/Seaga programmes comes from several sources. The *Miami Herald* reported in late 1981 that, while the rate of violent crime had gone down in Jamaica, burglaries and robberies had risen. Manley implied that due to its cut-backs in health services, the government was responsible for a polio outbreak that occurred in mid–1982. Scarce government housing, distributed on a non-partisan basis by the Manley government through the National Housing Trust established in 1975, was threatened by JLP partisanship in its allocation.[25] By 1983 poor people in Kingston's Coronation Market were saying that they were worse off than under Manley in the '70s.[26] Official legal emigration for the first three Seaga years amounted to 61,905 Jamaicans, approximately 4,000 more than during the last three troubled years of the Manley era.[27]

In her diary, Edna Manley documented both sweeping and mundane aspects of social deterioration. In early 1981 she sensed 'an attitude of loss of faith in all the areas that had a clear concept of nationalism—now it is every man for himself. . . . I think the teachers are the most notable example.' She later wrote of being shattered by 'all the alcoholism around me'. In mid-1982 she wrote of no light and power from seven in the evening until one o'clock next afternoon, ten days without either bread or crackers, and 'no water for days on end—wiping down with Limacol—looking askance at someone drinking the last drop of water instead of drinking a Coke!'[28]

New Cold War, Neocolonialism and Grenada

As early as December 1980 Manley warned his party that United States President-elect Ronald Reagan, finding the USSR and its allies entrenched in Eastern Europe, would be tempted to bully the Caribbean.[29] Superpower détente had begun to erode in 1979. Events in Jamaica in the '70s prefigured the collapse of détente, and in the early '80s the new Cold War added international pressure to economic and social problems.

Edward Seaga was the first foreign head of state to be invited to the White House by President Reagan. In April 1982 Reagan returned the favour with an eighteen-hour visit to Jamaica on his way to Barbados. The *Gleaner* observed that Reagan's stopover 'has not stirred much enthusiasm among the public.' Manley commented that United States foreign policy in the Caribbean was 'divisive, dangerous, warlike, and retrogressive and we view the visit of President Reagan in that light.'[30] The PNP's *Rising Sun* for April 1982 carried a cover drawing of an evil Reagan hoisting a Caribbean cupcake with Grenada, Cuba, and Nicaragua highlighted and the United States flag harpooning Jamaica. Edna Manley recorded in her diary that 'the great Mr. Reagan has come and gone with Government and some of the people genuflecting passionately to the Saviour of Jamaica.' She noted that Reagan had made a 'gaffe' in referring to the previous 'Communist government': 'Michael caught him on that one, and elements in the *Gleaner* obviously resented it too.'[31]

A visitor more to Manley's liking was Harry Belafonte, who was 'like a breath of fresh air' at the 6 July 1982 Norman Manley Awards ceremony. Manley defended Belafonte's right as a United States citizen to criticize Reagan, who had referred to Jamaica as his 'Caribbean showcase' and who had spread 'dishonest references to recent Jamaican history' by calling Manley's government communist.[32] Manley agreed with Belafonte that under Seaga Jamaica had become a 'pawn' and a 'stooge' of the United States.[33] Manley also approved of a speech at the PNP's annual conference in 1982 by Richard Hatcher, the Mayor of Gary, Indiana, which attacked Reagan's policies. Hatcher's address generated tremendous applause and the spontaneous singing of 'We Shall Overcome.' Edna Manley recorded that 'I've never seen Doug so uninhibitedly with excitement thrilled—and we all shot up onto the platform to shake hands and be joyful!'[34]

At Florida State University in April 1983 Manley said that he would like to improve his relations with the United States.[35] Doing that in the face of resurgent Reagan/Seaga neocolonialism and Manley's deep-seated anti-imperialism would be difficult. Some Jamaicans were willing partners in neocolonialism, as the Queen's visit in February 1983 showed. The Mayor of Montego Bay generously praised the British Commonwealth and noted blandly that Jamaica came under British 'influence' (read, imperialism and slavery) in 1655 and 'this was the beginning of a colorful and exciting process which provided a whole spectrum of experiences.'[36] Manley's mother noted that 'all this neo-colonialism—it leaves me a little stunned. The flooding, the swamping the country with America—anything that "really matters"—money economic growth etc.'[37] In an article reflecting Manley's views on the matter, Carl Rattray called for Jamaica to guard her own sovereignty rather than remain restricted by a monarchy with a foreign head of state in which final judicial appeal was to a committee of the English Privy Council.[38]

Some of Manley's activities conflicted with the historic hegemonic goals of United States foreign policy. In the Spring of 1983 he sent an open letter to the CARICOM heads of government urging them 'to speak out against US involvement in the armed invasion of a country [Nicaragua] that has shown nothing but friendship towards the territories that comprise our Community.' Manley also played a major role in sensitizing the Socialist International (SI) to the plight of the Third World countries. In 1981 an SI committee was established for that purpose as the result of Manley's efforts. At the SI congress in Portugal in April 1983, Manley was chosen Chairman of the SI's Permanent Economic Planning and Review Committee which had a mandate for formulating policies for the First World, the Third World, and relations between the two.[39] Manley also continued to be an outspoken advocate of the Non-Aligned Movement as crucial to Third World aspirations, defending Fidel Castro's selection as the movement's Chairman in 1979.[40] Manley's positions did not endear him to the United States government.

The most significant clash between United States policy and Manley's international approach came with the superpower's invasion of Grenada in October 1983. Manley had supported the revolutionary government of Maurice Bishop and his New Jewel Movement (NJM).[41] The invasion was popular in Grenada and in Jamaica for reasons ranging from conservative glee over the dispatch of a leftist regime to mass revulsion against the perceived killers of the popular Bishop. While condemning the madness that led to the death of Bishop and his colleagues, Manley took the unpopular route of censuring the invasion. On the day of the invasion Manley wrote to a young American that 'the tragedy of America is that a society dedicated to freedom and individual self-expression will not extend that principle to its neighbours.'[42] Later he wrote Julius Nyerere that 'the invasion of Grenada is a cynically inspired outrage. It was unnecessary; it was unprincipled[;] and compromises Caribbean sovereignty in a most dangerous way.'[43]

Manley regarded the collapse of the revolutionary government and the invasion as being linked historically to the struggle between neocolonialism and national independence. He applauded Bishop for his domestic policies and noted Bishop's controversial relations with Cuba, which provoked US hostility to Grenada. Bishop's execution was 'a very shocking and terrible chapter in Caribbean history', about which many facts were unknown as yet. Manley opposed the invasion, but he said his greatest concern was with the Caribbean leaders who invited the United States to invade. He questioned the legality of the Organization of Eastern Caribbean States' call for intervention, and regretted how the United States had begun to relate to the English-speaking Caribbean according to the historic pattern of US invasions of Latin America.[44] Manley's placement of the Grenada tragedy in the context of the struggle between neocolonialism and independence, highlighted the difference between himself and Seaga, who strongly supported the invasion. The Grenada invasion had a momentous impact on Jamaican politics.

Politics

In the early '80s Manley faced the task of helping rebuild a shattered party. Out of sixty seats in the House of Representatives, the PNP held only nine occupied by Manley, Ralph Brown, Dudley Thompson, D. K. Duncan, Portia Simpson, Anthony Spaulding, Terrence Gilette, Horace Clark, and Seymour Mullings. PNP members held eight seats in the less important Senate.

The defeat in 1980 naturally raised the question of Manley's leadership. On 8 February 1981 he offered to resign, stating that his leadership could be effective only in the context of 'certain principles, policies, and practices'. The PNP's National Executive Council (NEC) voted on the offer; eighty-three members rejected his resignation, four accepted it and twenty-five abstained. On 9 February Manley stated that the vote was not a viable basis for his leadership. On 15 February the NEC unanimously passed a resolution of confidence in Manley, except for three abstentions.[45] This did not end calls for Manley's resignation. In late 1982 the party newspaper published a letter from W. G. McFarlane, who called for a change in PNP leadership and observed that 'it would be a terrible impression to give Jamaica and the rest of the world that there is only one family and one man in Jamaica who is capable of being the Prime Minister of Jamaica.'[46] Most PNP members continued to support Manley in the absence of obvious alternatives, but succession posed a problem for the party.

There were more immediate concerns. Intra-party feuds continued. A PNP committee meeting on 5 July 1981 recognized three tendencies in the party: left, right and moderate. Ideological differences over such issues as 'imperialism' persisted. Manley proposed unifying the party around its *Principles and Objectives*. A more general crisis existed within the party as its grassroots groups were inactive, the NEC was not meeting regularly and many regional and constituency members failed to attend meetings. Achieving real internal democracy also required renewed, stronger commitment:

> Maximum effort must be concentrated on reviving the groups, the REC [Regional Executive Committees], making the NEC a systematic decision-making body and ensuring that the Executive has the built-in capability to address and resolve political issues which arise. In the absence of this genuine democracy within [the party] can never be a reality.[47]

The party did not lack self-criticism as this sensitive time. An unsigned article by Manley asserted that the PNP had little sense of its own history. 'No Party in modern times commits more to writing and few parties spend less time reading what has been written. The result is that the PNP tends to reinvent the wheel about every ten years or so.'[48]

The PNP committed itself anew to political education, supported by Manley and the PNP left, resisted by the party right. The rationale for political education was based on some lessons from the '70s. If the PNP was to be a party of 'transformation' it needed to develop cadres with a clear understanding of society. The cadres needed to understand the likely reaction of various classes to strategies of change so that the population could be educated to implement change without chaos. The party needed to be united in speech and action. Unity could only proceed from a shared analysis of history and the present situation.[49]

Manley frequently contributed articles to the political education supplement of *Rising Sun*, commenting on party matters and the larger issues of politics and economics. He explained the necessity of political education and of party criteria for candidates. 'Experience shows that not everybody should be a successful candidate. Many have the popularity to carry the vote but cannot represent a constituency and cannot form part of the effective Parliamentary leadership. . . .'[50] In an essay marking Jamaica's twenty-first anniversary of 'independence' Manley noted the positive side of Jamaica's recent history: great strides in education, productive development, trade unionism, and the evolution of the political system. The negative side was that a quarter of Jamaicans were chronically unemployed and that the distribution of work and income was grossly unequal and getting worse. The central challenge of the time, Manley wrote, was 'the mobilisation of the Jamaican people along a path of national self-reliance and involving a willingness to bear the necessary sacrifices while heroic efforts at developing production are made. . . .'[51] While moderating his own speaking style, he warned against demagoguery which 'might seem attractive in the short run, but will return to haunt us with a vengeance.'[52]

While confronting substantial internal problems, the PNP also had to concern itself with external threats. Edna Manley expressed a more widely-held view when she wrote that 'the thought of a one-party state gives me shivers.'[53] Earlier in 1982 two PNP supporters had been shot by police who justified their actions as moves against a 'communist terrorist camp'. The victims were Alguinan Williams, a member of the West Portland PNP executive since 1974, and Harvey Dwyer, a suspended policeman who 'had been beaten up by a top JLP activist after the 1980 elections.' Dwyer had been suspended for allegedly making insulting remarks about the Queen.[54]

The new Prime Minister, Edward Seaga, also faced political problems. Seaga had inherited Manley's political legacy, increased aspirations for social justice. Seaga's election was based on the premise that his administrative competence could ease the burdens of the poor. Yet the middle class and the private sector from which Seaga drew his support were too weak and demoralized for him to rely on, necessitating his turn to United States capital. This raised the key question as to the compatibility of United States economic interests and the economic and social needs of Jamaica's masses.[55] Carl Stone's polls showed declining support for the JLP, the recovery of the PNP, and a large bloc of uncommitted voters.[56] The trends in party support brought the issue of electoral reform to the fore.

In June 1982, when his party still lagged behind the JLP in popular support, Manley recalled the 'solemn agreement' made by the PNP and JLP in 1979 to reform the electoral system. An Electoral Advisory Committee had been set up, consisting of three independent members and two each from the JLP and PNP. The committee agreed unanimously on installing a universal photo identification system to prevent voter fraud by the first election after 1980 and making the Electoral Committee an independent self-funding commission deriving its powers from the constitution; Manley criticized the JLP government for inaction on the reforms.[57] Moreover, existing voter lists were outdated. Manley said in October 1983 that they contained a hundred thousand names of Jamaicans who had either died or emigrated, increasing the possibilities for fraud. Additionally, they disenfranchised 180 thousand who had reached majority but who had not been registered (and who tended to be strongly PNP).[58] The PNP continued to push the electoral reform issue throughout 1982 and 1983, while the JLP dragged its feet.

At the same time, the PNP recovered its position as the most popular party. In December 1982 Manley wrote of a 53 percent to 47 percent PNP lead over the JLP in the Stone Poll (leaving out uncommitted voters), but saw two dangers: that JLP propaganda and increased consumer spending might fool the people and that the JLP might fail to carry out the electoral reform.[59] In May and June 1983 the PNP held its first mass meetings since October 1980, turning out, by its own estimates, fifteen thousand in Savana-la-Mar, twenty thousand in Mandeville, and forty thousand in Kingston.[60]

In addition to declining popularity, Seaga faced the failure of an IMF test in September (which he refused to acknowledge) and the possibility of a devaluation of the Jamaican dollar which would kill his political fortunes. Then came the invasion of Grenada on 25 October, allowing Seaga the opportunity to play Reagan's junior partner, and temporarily boosting his popularity. No better opportunity was likely to present itself. On 25 November, in violation of the parties' electoral reform agreement, Seaga announced general elections for 15 December, with nominations to occur on 29 November.[61]

The PNP faced a dilemma, whether or not to participate in the elections. Pragmatism argued that the PNP would probably increase its seats in the sixty-member House of Representatives to some twenty-four or twenty-five (based on an election held on the outdated voter lists), though control of the government would probably remain narrowly in JLP hands. On the other hand, principle dictated to Manley and the PNP leadership the necessity of refusing to participate in elections in the absence of electoral reform and a new voter list.[62] Edna Manley recorded that Michael had told her that the PNP was ready for elections with a new list, but could not participate in a sham.[63] The PNP's National Executive Council voted 128–14 on 27 November in favour of Manley's position not to contest the 'bogus election'.

The result was that the JLP took all 60 seats, giving Jamaica a one-party House of Representatives and Seaga another five years as Prime Minister. Alfred Rattray, former Ambassador to the United States under Manley, regretted that Jamaican traditions and conventions were 'kicked aside' by expedience in the election of 1983. Jamaica's constitutional system 'was made to be run by men of honour. . . . We have to begin to devise a Constitution by taking into account the existence of persons whom I will not name.'[64] Carl Stone noted in 1985 that the 1983 election method was very unpopular, and that, but for Grenada, Seaga would have certainly been Jamaica's first one-term Prime Minister.[65] Edna Manley set the 'election' in the context of Jamaica's historical rivalries of partisan personalities:

> Busta and Norman in spite of their faults never broke faith over agreements. Shearer and Michael all through the union struggle and politically—never broke faith with each other. But this—son of a gun—will do anything to retain a place of power at whatever the spiritual cost to Jamaica.[66]

Had the fratricidal destruction of the Grenadian revolution and the United States invasion not occurred, it is likely that Seaga would have clung to power until 1985. Then, in the midst of economic crisis, he probably would have been voted out of office and Manley would have returned to power. However, Manley would have faced Reagan's Washington, a dubious prospect at best. Events in tiny Grenada certainly had a large impact in Jamaica, both in terms of what might have been and what was. The wheel of history is, at times, indeed enigmatic.

CHAPTER 22

New Consciousness and Responsibility

1984-8

> [Norman Manley] always said that when catastrophe overtook him, it always seemed as if he were being guided into new worlds of consciousness—actually even new worlds of responsibility.
>
> *Edna Manley*[1]

In the mid-'80s political and personal catastrophes dogged Michael Manley. The bogus election of 1983 meant another five years out of power for the PNP, which witnessed the further erosion of its programmes of the '70s. The economy shrank in 1984 and especially 1985, bringing down with it the health and education of the Jamaican people. Manley himself became estranged from his wife Beverley and experienced life-threatening illnesses in 1985 and 1987. In the latter year Edna Manley died after a lifetime's contribution to the national artistic and political causes.

But Manley grew by facing such difficulties. His enforced sabbatical from office gave him time to reflect on the past and plan for the future. The economic and social suffering of Jamaicans under Seaga made Manley's return to power possible, even after economic growth resumed in 1986. In that year, the PNP swept local elections. By 1988, it was widely predicted that the PNP would win general elections. The pall cast over the PNP's prospects by Manley's illnesses was dispelled by his strong recoveries. For Manley and his party the period from 1984 to 1988 was thus one of reflection on the past and present and preparation for the future.

Economic Crisis

In October 1984 Carl Stone observed that the average Jamaican was totally pre-occupied by economic survival. An angry worker told him that 'there is no seat for poor people on this bus being driven by Seaga.'[2] Bauxite was troubled. Consumer prices increased 28.9 per cent in 1984 and 23 percent in 1985. The Jamaican dollar underwent a nearly three-fold devaluation which made exports more competitive but imports more costly. The trade balance was persistently negative. Public debt steadily increased. Unemployment continued to be a major problem.[3] A great many economic problems had their roots in Jamaica's dependent, vulnerable position in the world economy.

The Seaga government doggedly implemented IMF policies. These required slashing budget deficits through cutting of government employment, social services, and unprofitable state enterprises, and adopting monetary policies (including devaluation) to restrict imports and promote exports. In mid–1984, the government laid off six thousand employees to conform to IMF policies; Manley condemned 'the arrogant decision taken in the IMF Board room in Washington to destroy the lives of people in Jamaica.'[4] He also accused the Governor of the Bank of Jamaica, nominally a neutral civil servant, of becoming 'an apologist for the Labour Party' because he refused to say whether Jamaica had passed the September IMF 1984 tests.[5] Following anti-government riots in January 1985, Carl Stone referred to 'the experts of Third World economic torture who preside over IMF policies here in Jamaica' who had led Jamaica beyond a threshold where pain exceeded benefits: 'If the Government and its battery of experts have no idea of what that threshold is, they all deserve to be flogged.'[6]

The decision of ALCOA to close its Halse Hall, Clarendon refinery on 20 February 1985 with twenty days notice dramatized the role of multinational corporations in Jamaica, regardless of which party was in power. Seaga decried ALCOA's decision as 'arrogant and offensive' and 'shocking and reprehensible'. He noted that the economy was driven over the past ten years by two main forces: the falling price of bauxite (which produced revenues for Jamaica of US$85 million in 1984) and the rising price of oil (which cost US$200 million in the same year).[7] In April the Seaga government reached an agreement with ALCOA whereby the government would acquire the closed plant and operate it as Clarendon Alumina Production (CAP), which ALCOA would manage.[8] This scarcely represented the private enterprise model favoured by the Reagan Administration, but the ALCOA controversy demonstrated why the state must play a major role in Jamaica's economy.

An aspect of Seaga's economic policies that particularly concerned Manley as a trade unionist were the 'free zones.' They were created to encourage foreign investment: investors were exempted from import licensing procedures and taxes on profits and subjected to minimal customs procedures. The zones generated about US$2.5 million in foreign exchange in 1983, but about US$15.5 million left the country. In 1985 the free zones employed about three thousand workers, ninety percent women, who frequently complained about low pay, high work quotas, and hot, stuffy working conditions.[9] Manley criticized the free zone 'sweatshops' as crowded with more machines than people, as based on the promotion of the 'cheapness' of the Jamaican worker, and as reminiscent of pre–1938 conditions.[10]

In 1986, the Jamaican economy began to grow again. However, one is struck by the similarity of the difficulties encountered by Seaga and by Manley with the IMF, the bauxite multinationals, and the private sector. Though the approaches of the two leaders had important differences, they faced international and national conditions that transcended ideology. Each had to confront the worsening social conditions caused by Jamaica's chronically deformed and crisis-ridden economy as well.

Social Deterioration

In July, 1984, Father Richard HoLung, a community social worker, addressed the Kingston Rotary Club on the plight of the poor street people of the capital city. He referred to the

> naked people, gone stark mad, the hungry that stagger through our streets cursing obscenities, and using so many fragmental words and sentences; the hands waving angrily and aimlessly at passersby, the screechy mad voices crying out to the heavens; our mad naked poor people caked with dirt, male or female genitals hanging out of dirty rags—our poor without shame or dignity. . . .

Father HoLung spoke of the massive numbers of poor children, about half the population of 'this great slum which Kingston is.'[11] Garnet Brown, a Jamaican consultant to the World Bank, saw Jamaica's problems from a loftier perspective: 'We are almost devoid of a national philosophy and we have become locked into imported norms, unrealistic tastes and ambitions, and an unquestioned reliance on foreign technical aid to solve our problems'.[12] There were many aspects to Jamaica's social deterioration in the mid-'80s, but the two with the greatest impact on most Jamaicans were the crisis in health services and the rising cost of food. These were among the social costs paid by the Jamaican people for the Seaga/IMF economic strategy.

One facet of the health crisis was the closing of three wards of Kingston Public Hospital because of a shortage of nurses. The Nurses Association of Jamaica complained that the government had failed to respond to their grievances, and in mid-July 1984 thirty-five nurses demonstrated to dramatize their complaints.[13] In August the Health Ministry announced the layoff of fifteen hundred employees.[14] The *Gleaner* began an investigation of the health crisis with a story headlined 'Health Services on Verge of Collapse.' Its investigator reported dilapidated clinics and inadequate supplies. It was announced that the university hospital was closing wards.[15] By October 1984, layoffs were said to be having a 'disastrous effect' at Kingston Public Hospital.[16] Carl Stone commented that the root of the health crisis was the IMF's insistence that the government budget deficit be cut by one-half in one year. He recalled Seaga's criticism of health conditions in the '70s and the JLP Prime Minister's 'promising to put the health services in order'.[17] Stone saw a 'racist side' to the crisis. 'Overseas experts' were advising the government that Jamaica was living above its means by having life expectancy and infant mortality levels comparable to those of the United States.[18] Matthew Beaubrun, Manley's personal physician, said in December 1987 that Jamaica's health services had reached their lowest level in the twentieth century.[19]

Poor and elderly Jamaicans' health was jeopardized by the rising cost of food. The government introduced a 'food security plan' in mid–1984, but professional nutritionists voiced serious reservations about it. Half the population was poor enough to qualify for food stamps, and by December 1984, 230 thousand elderly and very poor people had signed up. The JLP MP for Southern St Andrew reported that attendance at the Hugh Sherlock School dropped fifty percent when the price of a school lunch (a 'nutribun' and a half-pint of milk) went up from five to twenty cents. In April 1984 *Gleaner* columnist Franklin McKnight reported that food relief 'has almost completely evaporated in the fury of the dazzling pace of food price increases.' School meals, free at first, now required payment. The elderly poor received J$20 in food aid for two months; a ten-pound bag of rice cost J$17.50. McKnight said a return of the 'bad days' of 1980 would be 'very welcome indeed'.[20] The severe economic and social crisis of the mid-'80s contributed strongly to the unpopularity of Seaga and rising political fortunes of Manley and the PNP from 1984 on.

Politics, 1984

The economic and social crisis intensified the political environment soon after the bogus election of 1983. Political violence was one symptom. On 8 and 9 May 1984 feuding factions of the JLP clashed in the Rema neighborhood of Kingston, resulting in 8 deaths. Eleven JLP gang members were charged with crimes and scheduled to face a preliminary Gun Court inquiry in September.[21] In July the *Gleaner* reported that two people were shot and the police stoned as JLP supporters tried to break up a PNP conference in North Trelawny; PNP sources said a car belonging to the JLP MP Keith Russell was used in the attack.[22] At about the same time the PNP constituency office in Buff Bay was destroyed in a fire-bombing.[23]

Police abuse also became an issue. The Police Federation denounced those police who engaged in criminal acts.[24] Americas Watch, a human rights group, reported the high figure of 288 killed by the police in 1984, including incidents such as the following:

> February 1984. Six men at a work site were confronted by some 35 policemen, part of mobile reserve, who shot up the place, killing four of the six. An attorney explained that a JLP representative had complained that only PNP men where being hired. Two survivor eyewitnesses. No public account.[25]

In August, 1984, Prime Minister Seaga pointed to the serious indiscipline in Jamaican society and singled out the army as 'an oasis of discipline'.[26]

Social and economic crisis, political violence, and his earlier attacks on Manley's leadership naturally brought Seaga's leadership into question. George Eaton, Jamaican-born professor of economics and politics at Toronto's York University and Bustamante's biographer, believed Seaga's administrative performance was a great disappointment: 'He has allowed his Administration to become too centralized; he has taken over ministry after ministry for himself.'[27] A Stone Poll in September 1984 revealed that fifteen percent of the respondents regarded Seaga as doing a good job, forty-five percent believed he was 'trying but could be better', and thirty-eight percent believed he was performing poorly.[28]

In October 1984 the JLP government dissolved the Kingston and St Andrew Council (KSAC), a move reminiscent of the JLP's action in 1964. The KSAC, in a party-line vote of its fifteen PNP and ten JLP members, had called for the resignation of JLP Minister of Local Government Neville Lewis. The JLP gave KSAC financial laxity and mismanagement as its reasons for dissolving the KSAC, but PNP councillors said it was a political move. Manley denounced the dissolution as a 'cheap political ploy' to deny eight hundred thousand citizens in the city their right to choose their local representatives, and pointed out that the first recommendation of a management audit of the KSAC was that the body should *not* be dissolved.[29]

Meanwhile, Manley urged the PNP to prepare with seriousness and humility for possible elections. He announced that the party was working to resume 'the forward march of small business in Jamaica'.[30] Like the JLP, however, the PNP had lingering internal problems. Tony Spaulding, controversial Minister of Housing in the '70s, resigned as PNP caretaker of the crucial, embattled Southwestern St Andrew constituency, which he had represented in Parliament from 1972 to 1982. Manley later said that Spaulding had learned little since the defeat of 1980 and was thus unprepared to make the changes the PNP required should it return to power.[31]

Manley's political style in 1984 reflected what he had learned about excessive rhetoric. He projected the seriousness and humility he urged on his party. To the Montego Bay Chamber of Commerce and Industry in August he said that Jamaica would have constant crisis until Jamaicans learned that no one party or policy could solve the nation's problems. Both Seaga and he had tried very hard for twelve years 'and in the end of that it is on common ground to say that the country is on very serious ground economically. If there is one thing that we have learnt in the last twelve years, it is that there is no magic solution to our problems.' Both parties had made mistakes in concentrating too much on single programmes—the JLP amassing foreign exchange and the PNP in agricultural self-reliance—and neglecting others.[32] Carl Stone gave 'full marks to Manley' for the speech, calling it 'statesmanlike' and 'a rare display of political honesty'. He said the PNP had been a responsible opposition, despite 'unjustified provocation by the JLP over alleged trips to Cuba and Grenada by PNP persons and lame JLP efforts to revive the now dormant issue of anti-communism'.[33]

With Commonwealth leaders including Margaret Thatcher, President Kaunda, Julius Nyrere, Banda Hastings and Joe Clarke: Lusaka Summit, 1979.

Julius Nyerere, Michael Manley and Dudley Thompson (Jamaica Foreign Minister) in Tanzania, 1970's.

With Jesse Jackson and Edna Manley at a PNP Founders Day dinner, 1985.

With Vivian Richards, in London, 1988.

Campaigning.

Jamaica's popular leader, 1960's.

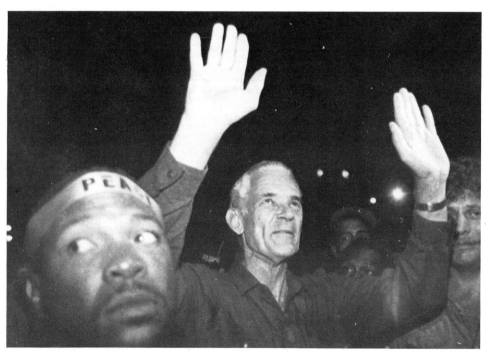

Victory night, 10 February 1989.

1985: Year of the Shadows

On 15 and 16 January 1985 riots against the Seaga government erupted, provoked by increases in the price of petrol and kerosene. A large number of roadblocks were thrown up by demonstrators, many of whom were without political connection, although some were PNP and WPJ activists. Women set up many roadblocks. Stone found the protests overwhelmingly popular.[34]

The PNP's political activity increased in early 1985, but was curbed when Manley was ill and underwent surgery first in February then in April after a severe attack of diverticulitis. He was also operated on for a benign thyroid growth. He complained later that political opponents tried to take advantage of the situation by spreading false rumours about his illness.[35]

While Manley was ill, two shocks hit the Jamaican political system. The first involved local government. In February 1985 the JLP Minister of Local Government casually announced that several functions of local government would be assumed by the central government. Stone commented that the merits of the action were debateable but 'how the government has gone about it is simply insulting to democracy and to the political traditions of this country.'[36] On 12 April, Seaga declared that the number of Parish Council seats was to be cut from 275 to 120 to save money on services duplicated by central government. A disgruntled JLP parish councillor said Seaga made the decision autocratically, without even a warning to his own party. The PNP protested that cutting the parish council seats threatened democracy and informed Governor-General Glasspole of their 'grave concern'. In early May Seaga retreated, proposing that the Electoral Advisory Committee decide the issue. In October the committee decided to cut the seats to 189, which both parties accepted.[37] The matter demonstrated the essentially authoritarian way Seaga made some major decisions.

The other blow involved an interview of PNP leftist D. K. Duncan with the *Gleaner* in late April. Duncan was highly critical of the PNP and of Manley's leadership. Duncan said in the '80s the PNP had become, as it was in the '60s, 'almost a mirror image of the JLP', 'a Social Democratic Party capable of mild reforms which definitely cannot meet the aspirations of the people.' Duncan's 'certain knowledge' was that the PNP fought the 1980 elections expecting to lose, but anticipating that Seaga's policies would lead to mass suffering, allowing the PNP to return to power with a programme to transform Jamaica radically. By 1982, however, 'some of the elements in the Party put pressure on the leadership and Michael Manley just lost his nerves.' Duncan blamed this not on Manley personally, but on the Social Democratic nature of the PNP which, he said, was never on the track toward radical transformation in the '70s.[38] Duncan was eventually censured by the party for these comments. His situation, like that of Tony Spaulding, indicated the decline of the PNP left.

Toward the end of summer Manley resumed political action. In

August he told a PNP party school graduation that central to the PNP was the goal of 'transformation'—'the serious re-organisation of the nature of the ownership structure of the resources in our society'[39]—and said that he could not be involved in political leadership lacking such a strategy. A lesson from the '70s was evident in Manley's warning to avoid the 'trap' of trying to transform everything at once.[40] With the knowledge that his party enjoyed majority support, Manley addressed enthusiastic crowds in settlements such as Black River on the southwest coast and Darliston in the western mountains. The people were anxious to see that 'Michael fit.' He explained the 'madness' of Jamaica's rapidly spiralling debt, which had more than doubled under Seaga in five years, and for what? Imported 'satellite dishes, caviar and tinned mushroom at US$20 a tin' which the vast majority of Jamaicans could not afford, while local farmers were displaced from local markets.[41]

At the PNP annual conference in September the party presented its alternative to Seaga's policies. It listed the priorities of a future PNP administration: (1) expanding production, especially of exports; (2) earning and conserving foreign exchange; (3) expanding employment opportunities, especially for the young; (4) guaranteeing minimum nutritional levels to all Jamaicans; and (5) restoring major social services, especially education and health.[42] As usual Manley delivered the main address on Sunday. The atmosphere was almost celebratory, and Manley spoke under a large banner displaying his words:

> We are against Capitalism but we are not Communist. We do not quarrel with any country which chooses to live under Capitalism or Communism. That is their business. We prefer to build Democratic Socialism here in Jamaica.

The PNP's 1985 policy alternative indicated that the party's and Manley's thoughts in many areas had changed. Ideological pronouncements were replaced by a problem-oriented pragmatism. The emphasis in the '70s on state ownership was replaced by a new accent on joint ventures between the state and the private sector. The private sector was given greater encouragement. There was more emphasis on producing wealth than redistributing it. The PNP retained its commitments to a non-aligned foreign policy, to cooperative and community-based enterprises, economic planning, social services, participatory democracy, and political mobilization. The big question was where to find the resources to fund the PNP's alternative.[43]

On the night of October 30, Manley addressed a meeting of an estimated eighty thousand demonstrators, calling for new elections and an end to the foreign exchange auction. The next day, he led a march to the Bank of Jamaica to protest the foreign exchange auction system. The crowds were disciplined, but Manley invoked the memory of the JBC strike of 1964 to imply what might happen 'if the powers that be won't hear us.'[44] He later reassured the private sector, much of which was hurt by Seaga's policies, in an informal exchange at the luxurious Pegasus Hotel; his comments received prolonged applause.[45]

In late November, Manley's approach to international economic issues was endorsed by Jesse Jackson, a United States presidential candidate in 1984 and 1988. Jackson criticized the Reagan Administration's Caribbean Basin Initiative for being planned in Washington and not by Caribbean people. He told the audience that as it demanded 'the right of sovereignty, the right to say "no" to a blind, often merciless International Monetary Fund that will choke you to death,' it was marching to the beat of a liberation drummer.[46]

Manley called 1985 'the year of the shadows'.[47] He was referring to the global arms race and the debt crisis, but maybe he also had in mind the fatal riots of January, his own illnesses, and Seaga's tenacious clinging on to power despite his unpopularity and the socio-economic crisis. But, as it turned out, a better new year was dawning.

1986: Referendum on Seaga

The central event of 1986 was the big PNP victory in the local elections of late July, widely interpreted as a referendum on Seaga's policies, though the Prime Minister denied that view. Before and after the vote Jamaicans were agitated by controversial new taxes, a debate over how to spend a governmental windfall from lower oil prices, increased higher education fees, and allegations of governmental human rights violations and renewed racism. Although the economy grew, this seemed to have little effect on most Jamaicans' political convictions.

Manley addressed a 'Save the People' rally and all-night vigil at National Heroes Park on 10 March 1986. He decried government policies toward students, teachers, doctors and nurses. The next day he led a march on Jamaica House, the Prime Minister's official residence, to present a petition bearing ninety-two thousand signatures on how the J$450 million oil windfall should be spent. Manley left after a few minutes, but thousands of protesters stayed, hurling insults at Jamaica House.[48] Following the rally the protesters were attacked, probably by JLP supporters. Stone called the violence a 'reign of terror' and suggested that 'it's as if someone was determined to telegraph a message that presence at a PNP rally was a risky business.'[49]

At about this time, the Chairman of the Jamaica Council of Human Rights, Dennis Daly, denounced the 'flagrant infringement of civil liberties' in Jamaica. Daly mentioned arbitrary arrest and detention, 'police brutalities and killings', limitations on freedom of expression in the media, the Seaga government's attempts to 'undermine' free and fair elections, and economic policies with 'devastating effects on civil and political rights'.[50]

Manley continued to criticize Seaga's economic policies. In May the Prime Minister introduced price cuts, price subsidies, price controls and increased public spending, all in defiance of IMF dictates, and in response to pressure from the PNP, the WPJ, organized labour, 'the man in the street', and the increasingly numerous critics of his government.[51] Manley observed that the moves were an 'insulting bribe' to voters, since the price cuts were much smaller than price increases which had occurred since 1980. He noted that no expenditures were planned for health and education from the oil price windfall. Manley reckoned that per capita Gross Domestic Product had fallen four percent between 1980 and 1985, while wages had declined from fifty-one to forty-eight percent of GDP. He also said that the JLP government's Ministry of Health had admitted that while the minimum wage could about cover the weekly food needs of a family in the late '70s, in 1985 it could cover only one-third of those needs.[52] Manley criticized the *Gleaner* for praising Seaga's budget deficit reduction, demonstrating that the deficit in the year 1983/4 was significantly higher than in the last year of Manley's administration (1980/1), and not that different in the intervening years.[53]

Seaga set the local elections for 29 July 1986, two years late. Manley said that, despite government statements to the contrary, the PNP and the people saw the elections as a referendum on Seaga's policies. The PNP campaigned on the economy (the effects of deregulation, devaluation, and tight credit); the destruction of local government and the impairment of democratic participation; deteriorating social services; and corruption.[54]

Both major parties acted to minimize the 'tribal' political violence that had so marred elections in 1976 and 1980, and indeed had so retarded Jamaican development and democracy.[55] Rock-throwing occurred between PNP and JLP members in Falmouth.[56] Manley's motorcade was fired upon in mid-July in Spanish Town, and on another occasion his party was attacked in a 'rock stone ambush' in which one person was wounded.[57] Stone reported that 'Mr Manley's speech at the Spanish Town meeting which I attended was a model of responsible leadership as he tried to cool down the passions of PNP activists who were ready if he had just given the signal to resume the political war. A new politics is being attempted with far-reaching implications.'[58] Two were killed on election day, but violence was confined mainly to strongholds of either the PNP or JLP in Kingston and St Andrew.[59]

The PNP won the elections impressively. Sixty-four percent of the registered voters turned out, a high figure for local elections. Fifty-seven percent voted for the PNP, which gained control of twelve of Jamaica's thirteen parish councils. If the vote had been for Parliament, the PNP would have won forty-four seats to the JLP's sixteen.[60] Stone wrote that in effect Seaga was Jamaica's first one-term prime minister because 'if he had to face the polls in free and fair elections he would have lost.' He added that 'within the next nine to twelve months, fresh parliamentary elections must be called if the government has any respect for democracy and for the will of the people.'[61] Seaga clung to power early into 1989, as he repeatedly vowed he would.[62]

In 1986 Manley wrote essays and delivered addresses that illustrated his values and concerns. An article on Julius K. Nyerere described the Tanzanian leader as modest and unassuming, but possessing an iron will and exceptional intellect which saw to the heart of things. Manley listed two of Nyerere's characteristics as indispensable to greatness: 'absolute personal integrity' and a sense of humour that made greatness 'more accessible to the rest of us'.[63] General Augusto Pinochet of Chile was Nyerere's polar opposite. Pinochet had gained power 'by treason and murder, and has made murder an instrument of state policy.' Although Manley described himself as 'not a person who easily comes to the defence of violence,' he admitted that 'I would be a hypocrite if I did not register my profound disappointment at the failure' of an attempt on Pinochet's life.[64] In a third article Manley continued his long opposition to the apartheid regime in South Africa by criticizing the opposition of Reagan, Thatcher, and West Germany's Helmut Kohl to sanctions against the racist government.[65]

Despite his desire to improve relations with the United States, Manley remained ambivalent about 'the most powerful, the most complex and in many ways the most extraordinary nation in history.' He deplored the decline of Martin Luther King's dream. 'Now the dream is in retreat, as social programmes are slashed every year in the world's most powerful economy.' Third World victims of 'a structurally created debt crisis' were told by the United States 'to export even if it kills them.' Still Manley dared hope that the contradictions in the world economy 'might in due course create a politics [in the United States] in which the sense of international cooperation is seen as both the extension and ultimately the best guarantee of U.S. interest[.]'[66]

A reason for hope was the existence in the United States of groups such as Trans Africa, which Manley addressed in June. He attributed the rise of radical right, in the United States and Europe to the inflation crisis of the '70s, described the social costs of reducing inflation on both First World and Third World countries, and suggested an alternative approach. Manley explained why he led the Socialist International to change its economic strategy and reviewed SI's programmes, elaborated in *Global Challenge*. He encouraged private initiative and entrepreneurship, but warned against trampling on the weak and rejected the notion that productivity and social conscience are incompatible.[67]

Towards the end of the year, Manley delivered the second Manzur Qadir Memorial Lecture at Lahore, Pakistan. He tried to detect the 'logic of history' and turn it to Third World use. In the face of multinational corporate organization of the world economy, he argued for multilateral public institutions to overcome the limits of the nation-state. He renewed his call for a North-South alliance based in equity to replace past and present patterns of dominance.[68] It was characteristic of Manley that while dealing, as opposition leader, with Jamaica's awesome socioeconomic and political problems, he continued to articulate a compelling vision of global challenge and global cooperation.

1987: Personal Loss and Illness

In early 1987 Manley experienced personal setbacks that added to the challenges presented by the '80s. On 9 February Edna Manley died in her sleep, three weeks short of her eighty-seventh birthday. She had remained an active artist practically until her death. Her last finished work, the painting 'Worship' (December 1986), was a marvellously expressive portrait of two women representing complementary sides of the religious experience—the one accepting and believing, the other doubting and sceptical—perhaps two sides of Edna Manley's own character. Her influence on Michael was greater than that of Norman Manley, but Michael did not immediately allow himself to explore the depths of personal loss. Instead, he threw himself into preparations for Edna's funeral. That public event surpassed the funerals for Alexander Bustamante and Norman Manley, and equalled that of Bob Marley in 1981. Rex Nettleford suggested that the outpouring of feeling for Edna Manley was because, in a sense, four people were buried in her: herself, Bustamante's cousin, Norman Manley's wife, and Michael Manley's mother.[69] Mourners came from all classes and from both major political parties, a hopeful augury of the political détente between Manley and Seaga that began later in the year.

In April Manley's diverticulitis recurred, complicated by a severe viral infection that nearly cost him his life. His colon was removed, but he recovered fully and was able to resume his normal activities with slight adjustments. His close brush with death and the other political and personal events of the 1980s forced him to grow emotionally, a dimension of personal development that had lagged behind his political and intellectual maturation. In his convalescence, he hastened to repair personal friendships damaged by politics. His new perspective on life added a useful element of detachment from his politics, which had sometimes been hurt by excessive passion.[70] Inevitably, Manley's health would become a political issue, but his vigorous recovery quashed it. Manley's personal physician reported in December 1987 that Manley's recovery was excellent.[71] Typically, Manley used a nearly-catastrophic event as the chance to explore himself for new possibilities.

1987–8: The Economy, Crime, and Politics

In the last two years of Seaga's tenure, Jamaicans continued to worry about the crises in health and education and the suffering of the poor majority. Three issues stood out, however: the economy, crime, and politics.

Prime Minister Seaga argued that his harsh policies had worked and that the time had come to address social and infrastructural needs. The 1986 elections and polls through mid–1988 had attested his government's unpopularity. In 1986 he had defied a joint commission of the United States Agency for International Development, the World Bank, and the IMF that argued that still more economic austerity was required in Jamaica. Carl Stone, a frequent critic of Seaga's economic strategy before 1986, wrote in May 1988 that Seaga's priorities were correct and that the economy had performed robustly since 1986.[72]

Naturally 'Seaganomics' was under constant criticism from the left, but more damaging was the attack from the right. One of Seaga's most persistent critics was the *Gleaner*'s conservative financial editor Basil Buck, a former JLP official who had criticized Manley's economics in the '70s. Buck complained constantly about the trade gap and argued that the economic growth of 1986 'came about primarily by some fancy accounting foot work between the Government and the Bank of Jamaica. . . .'[73] Buck also pointed to the continuing high costs of IMF support: 'our health services are in shambles,' 'our educational system is creaking at the joints,' 'our roads have become pot holes,' 'our transportation system is a mess,' 'our housing is getting scarcer and scarcer, with some living in old cars and caves. . . .'[74] A colleague of Buck's even claimed, most ironically, that Seaga's government was just as 'socialistic and interventionist' as Manley's.[75] Buck did find the economy of 1987 more solid than that of 1986, although the recovery was still fragile and dependent on good luck.[76]

Seaga designated 1988 'The Year of the Worker' to commemorate the fiftieth anniversary of the labour revolt of 1938. Some observers noted the cynicism since under him the working class was being pushed deeper and deeper into poverty: 'Just what are we celebrating in this so-called year of the worker?' asked Stone.[77] The 'Year of the Worker' highlighted Manley's trade union credentials and Seaga's lack of them. A November 1987 survey showed that Jamaican workers regarded Manley as the most outstanding trade unionist in Jamaica (twenty-four percent of responses, as compared to twenty percent for Shearer), as the most admired politician (forty-one percent, as compared to nineteen percent for Seaga), and as the politician most committed to the workers' cause (fifty percent as compared to twenty-two percent for Shearer and six percent for Seaga).[78]

A second issue troubling Jamaicans in 1987 and 1988 was violent and drug-related crime. Symptomatic were the events surrounding the acquittal of Lester Lloyd Coke in July 1987. Coke was one of four men charged with murder in the JLP internal warfare in Rema in May 1984. When he was acquitted, his supporters, lining King and Tower streets in downtown Kingston, fired gunshots into the air that sent police and bystanders scurrying for cover. The *Gleaner* responded with an editorial entitled 'Anarchy', which stated that it was 'alarming that the behaviour smacks of political celebration, in particular since the incident which led to the murder trial arose from a dispute between rival JLP gangs.' 'We are left to assume,' the editorial continued, 'that the very guardians of the peace fear some parallel authority of ghetto power which thumbs its nose even at the highest courts of the land.'[79]

In September 1987 Jamaicans were shocked by the murder of reggae superstar Peter Tosh. Tosh was one of the more political reggae musicians, and identified with the PNP although he could also be critical of it. Manley described Tosh's murder as 'a blot of shame on all Jamaicans.'[80] In his main speech at the PNP's annual conference Manley said that in the past three years crime, mostly linked to drugs, had become the most deadly threat in the nation's history. The PNP leader called for unity to fight crime, for an end to political tribalism, since no one party could fight crime alone.[81]

But crime was a political issue. JLP Minister of National Security Errol Anderson said Paul Burke, Secretary of the PNP's Human Rights Commission, was a link between the party and criminal gangs. Responding, Manley questioned the role of 'certain elements in Tivoli Gardens' in the cocaine business and accused Anderson of 'a new low in the political life of this country' for trying to smear the PNP by citing unsubstantiated claims. Burke denied the claim, and—as frequently happened in such circumstances—no charges were brought against him.[82]

Drugs and crime in Jamaica acquired a crucial international dimension with the development of the 'posses', violent, drug-dealing Jamaican gangs in the United States. In November 1987 there was a nationwide crackdown on the Jamaican posses in the United States, and their violence threatened the tourist trade and Jamaica-United States relations. In New York, Manley denounced the posses, but urged police not to 'blur the edges of justice' by confusing the posses with poor Jamaicans and other blacks.[83] Sensational, dubious news reports from the United States in early 1988 attempted to tie both parties to the drug trade. A CBS-TV programme showed a reported drug dealer making a large contribution at the 1987 PNP annual conference. Knowledgeable, independent observers debunked the programme.[84] Prime Minister Seaga's undisguised pleasure at the CBS report was short-lived. In April, a convicted American drug dealer, in uncorroborated testimony before a United States Senate subcommittee, tied Seaga to the drug trade. Seaga said he was gravely concerned that 'rumours in a bar are given the dignity of testimony in a committee of the U.S. Senate' and he protested formally to the US State Department. The harm in such reports, Carl Stone argued, was that they diverted attention from the real issues. He pointed out that both the Manley and Seaga governments had moved against the drug trade to the extent of their resources.[85]

The controversies over the economy and crime and drugs fuelled the slow drift toward general elections. Foreign journalists featured Manley's more moderate posture and his acceptance of responsibility for the failures of the '70s.[86] The return of David Coore as PNP spokesman for foreign affairs symbolized the victory of PNP moderates, and led to criticism of Manley and the party by Jamaican leftists.[87] The trend toward moderation—what Manley called 'dynamic Social Democracy'—was not unmixed. Political education, an important project avidly supported by the left and by Manley, but an anathema to the party right, continued under the direction of young leftist Tony Bogues. In Kingston, the party school graduated 179 students on 6 September 1987.[88] In early 1988, Hugh Small, a leftist government minister in the '70s, became a candidate for the general elections, apparently accepting the new, more moderate thrust of PNP policies. Manley still regarded the leftists of the PNPYO as among the most intellectually dynamic and creative elements of the party.[89] His attempts to maintain the PNP as a broad majority class alliance committed to serious change was bound to produce internal party tensions, especially if the party came to power.

Improved relations between the *Gleaner* and Manley resulted in a lengthy interview under the headline 'Democracy Cannot Survive Politically-Induced Violence—says Michael Manley.' Manley stated that leftist D. K. Duncan had no post in the party and would have none in any future PNP government. He affirmed that a PNP government would negotiate with the IMF. He referred to the PNP as Social Democratic, avoiding the more controversial 'Democratic Socialist'. Manley stated that a future PNP government would restore relations with Cuba, although it would avoid doing things that in the '70s had been misinterpreted by and caused problems with some Jamaicans and with Washington. He said the PNP had worked very hard in the '80s on improving relations with Washington out of 'genuine feelings of friendship and admiration', although 'we will never undertake to agree with everything the USA does in its foreign policy.' On political violence, Manley stated that 'no political party could dare to say to you in honesty that there are no gunmen who may not attach themselves somewhere to the PNP or the JLP in a constituency organisation,' but that 'as a categorical absolute assurance, they are not under our control, they are no part of our strategy and we have no intention to exploit anything of that sort in the politics of Jamaica.' Manley endorsed the idea of supplying large security forces during elections to those areas which 'the political parties cannot or will not control.'[90] Manley could not have hoped for a much better forum to get his current thinking across to Jamaicans outside the PNP.

The most promising political development of the late '80s was the détente between Manley and Seaga that might, with much hard work, diminish political tribalism and violence. On 6 November 1986 Manley had written to Seaga proposing a bipartisan approach to crime and violence. A second letter in October 1987 led to extensive talks. Seaga commented that the climate between the parties was better than it had been in fifteen years as the result of his government's policy and the restraint of the PNP in opposition. Manley praised Seaga's response, and the *Gleaner* carried pictures of the two shaking hands at a Jamaica

House State Dinner and at a National Prayer Breakfast.[91] At a PNP meeting on 29 March 1988 in Spanish Town, Manley said he believed that there was a 'genuine agreement' between the parties to stop electoral violence. He threatened to denounce publicly those who hadn't got the message.[92]

Carl Stone predicted for years that Manley would win the next general elections, likely to be waged on the issue of Seaga's managerialism versus Manley's populism. Stone believed that Seaga managed the economy in the '80s better than Manley had in the '70s. But 'Manley's political image is that of the man in today's Jamaica most committed to sharing up the cake equitably and taking care of the needs of the poor.' The rub for Manley would be after his return to power when he would have about eighteen months to show that the good populist and trade unionist could also be a good manager. Failing that, the effect on the PNP could be far more disastrous than 1980.[93]

CHAPTER 23

Michael Manley in Jamaican History

The Leader in History

In the last 150 years, wildly different views on the influence of leaders in history have been expressed. In 1840, Thomas Carlyle proposed his 'Great Man' theory.

> . . . Universal History, the history of what man has accomplished in the world, is at bottom the History of the Great Men, who have worked here. They were the leaders of men, these great ones; the modellers, patterns, and in a wide sense creators, of whatsoever the general mass of men contrived to do or to attain. . . .[1]

A generation later, Leo Tolstoy argued a nearly opposite point of view in *War and Peace* (1869): 'One only has to penetrate to the essence of any historical event, that is, to the activity of the mass of men who take part in it, to be convinced that the will of the historic hero does not control the actions of the mass but is itself controlled.'[2] Not too long ago an American political scientist, Murray Edelman, argued along modified Tolstoyan lines:

> Recent research on leadership suggests that 'leadership' may be the wrong word. The word connotes an individual free to choose his course of action and able to induce others to follow his lead because of his superior intelligence, knowledge, skill, and the force of his personality. The emphasis in modern leadership theory is rather upon the willingness of followers to follow.

Edelman believed that leadership served a largely symbolic purpose.

> Because it is apparently intolerable for men to admit the key role of accident, of ignorance, and of unplanned processes in their affairs, the leader serves a vital function by personifying and reifying the processes. As an individual, he can be praised and blamed and given 'responsibility' in a way that processes cannot. Incumbents of high public office therefore become objects of acclaim for the satisfied, scapegoats for the unsatisfied, and symbols of aspiration or of whatever is opposed.[3]

Edelman studied primarily leadership in the United States, where surely the role of individual leaders is diminished by the size of the society, the magnitude of the bureaucracy, the institutionalized 'balance of powers', the complexity of interest groups, and the sheer magnitude of problems. In this tradition, a journalist recently argued the impotence of American leaders faced with the waxing and waning of vast demographic forces and the price of imported oil.

> [Jimmy] Carter was battered by the same forces that bolstered Ronald Reagan. Yet neither had any real control over these forces. Nothing the White House has done, whether in style or in substance, has significantly affected the ebb and flow of our national economy in the last decade.[4]

Jamaica is neither Tolstoy's Russia nor the recent United States. A view that seems to fit the Jamaican case better is that of the Marxist historian George Plekhanov, dating from 1886.

> . . . By virtue of particular traits of their character, individuals can influence the fate of society. Sometimes this influence is very considerable; but the possibility of exercising this influence, and its extent, are determined by the form of organization of society, by the relationship of forces within it. The character of the individual is a 'factor' in social development only where, when, and to the extent that social relations permit it to be such.[5]

As we review Michael Manley's role in Jamaican history it is well to keep these perspectives in mind. Carlyle's beliefs may in fact have contributed to an English heroic tradition—very evident in Manley's own attitudes about heroes and leaders—that was internalized in Jamaican social relations and history along with the Afro-Caribbean 'man of words' tradition and helped prepare the way for leaders like Garvey, Bustamante, and Michael Manley. Tolstoy reminds us, however, that history is ultimately made by *all* the people, in ways which still defy the collective understanding of social scientists and historians. Manley's supporters and opponents, and the uncommitted, the more than two million Jamaican people, created the possibilities and set the limits on what he and they could do. Among other things, Jamaican leaders do play a symbolic and dramaturgical role, freeing most from the wearying business of trying to understand economics, society, politics, and international affairs. It would be too much to say, however, that in the small, intimate island-state of Jamaica, with its underdeveloped institutions and commanding personalities, that the role of leaders is only symbolic. Manley did wield significant power, but only within the constantly shifting limits that Jamaican social relations—understood as encompassing economics and politics—enabled him to do.

Early in this work it was suggested that the role of Michael Manley in Jamaican history has been shrouded by two myths. The positive, heroic myth of 'Joshua' yielding his 'rod of correction' against the oppressors of the 'sufferers', reflected real suffering and a heroic, messianic Jamaican tradition. The negative 'Muddler/Marionette Myth', crafted by Manley's political enemies, mirrored the interests, fears, and anxieties of Jamaica's upper and middle classes. It is not that 'history' starkly contrasts to 'myth', but rather that the historical record suggests a less partisan, more complex, subtle, and nuanced interpretation than 'Joshua' or the 'Muddler/Marionette'. Such an explanation needs to consider the formative influences on Manley, the complex legacy of his years in power, and his growth and development in the '80s.

Formative Influences

Michael Manley's life up to 1989 has spanned the last thirty-eight years of British imperialism in Jamaica and, since 1962, twenty-seven years of neocolonial national 'independence' under the shadow of the American eagle. Imperialism and neocolonialism, with their historic byproduct of Third World underdevelopment, thus form the broad background to Manley's emergence as a leader. The more proximate backdrop was Jamaican society as it evolved from the '20s to the '80s: a stagnant economy being partly vitalized by the addition of dynamic bauxite, tourist, and manufacturing enterprises, but retaining its structural dependence and gross inequality; a society gradually yielding some of the rigidity of its colour/class stratification, but unable to remedy many major problems afflicting the black majority; a political system, intimately linked to Manley family leadership, achieving formal 'independence', building national institutions and formal democracy, but leaving the majority substantially limited in its power to control its economic, social, and political existence.

Manley comes from a dissident, progressive sector of the small 'brown' middle class. Historically the Jamaican bourgeoisie has been a problematic class, jealous of the tiny white 'oligarchy' and fearful and disdaining of the black 'masses', earning its keep more through administration than productivity. Manley's earliest training for leadership came, as he puts it, 'sitting at the knee of Gamaliel', his father Norman Washington Manley. 'N. W.' provided many elements of his son's leadership portfolio: belief in democratic method; intellectualism and openness to new ideas; faith in the 'logic of history' as a guide to politics; devotion (at times abstract) to a mild form of socialism in the Fabian tradition of the British Labour Party; belief in planning, organization, and technocracy; and breadth of interests. However, Manley has said more than once that his mother, Edna Manley, was an even more profound influence. From her he received a measure of the artist's humane temperament and sensibility, indifference to distinctions of colour and class, support and encouragement to keep his own counsel, and perhaps even a scepticism of 'isms', despite his frequent invocation of them. Both parents endowed Manley with nationalism in its twin political and cultural spheres. They also gave their son a sense of duty to society, along with a bit of *noblesse oblige*, and a legacy of being at odds with the reactionary Jamaican oligarchy.

The family connection of the Manleys to Alexander Bustamante and

Hugh Shearer of the rival JLP is a primary fact of Jamaican political sociology. Also linking the familial and political is that fact that in its fifty years, the PNP has had only two leaders, father and son. Family pride has motivated Manley to defend and extend the political inheritance of Norman Manley, whom the son still deifies. However, to be a Manley is a mixed blessing. Manley probably would not have become a leader without the paternal connection, but on the other hand he wrestled for years to establish his own persona, free from the long shadow cast by Norman Manley.

Manley is also a product of colonial secondary education. Prestigious Jamaica College enhanced his already privileged status. But bullying by older students gave him a taste of oppression which later helped him to identify with Jamaica's 'sufferers'. The curriculum was intended to produce good colonial subjects and intermediaries between Crown and 'masses', but the system was marked by crosscurrents which enabled Manley to become critical of the educational establishment. His conflict with Jamaica College authorities symbolized the muted rise of Jamaican nationalism.

The London School of Economics continued the ambivalent role of formal education in Manley's life. He found it difficult to conform to requirements and for a while vacillated between the arts and political economy, choosing the latter. Harold Laski was a very important influence. Manley still regards the English professor/politician as the leading theorist on the marriage of liberal democracy and socialism, the performance of which was Manley's chosen task. London exposed him to other nationalistic students from various Caribbean and African colonies.

Absent from Jamaica from 1943 to 1951, Manley reimmersed himself in the country's life via journalism and, more significantly, trade unionism. *Public Opinion* gave him the opportunity to probe for 'The Root of the Matter' in partisan issues of politics, economics and foreign affairs. Trade unionism was a profound experience, second only to leading Jamaica in the '70s. Unionism sensitized Manley to the hard lives of men and women in sugar field and bauxite quarry. Through it he staked out a career and a persona ('Young Boy', 'Mike') distinct from his father's. Trade unionism helped narrow the chasm between a middle-class, brown leader and the black rank-and-file suspicious of middle-class browns. It honed Manley's superior skills as advocate and negotiator, talents he would use as politician and statesman seeking, more often than not, to harmonize conflicting classes, parties and nations. Trade unionism also meant that Manley would bring to 'liberal democracy' the insights of one who had studied the gross inequalities of economic classes under Jamaica's dependent capitalism. Above all, Manley's trade unionism established him as the premier worker's advocate in Jamaica, an invaluable asset where eighty percent are either workers, peasants, or poor.

In the '60s Manley continued his trade unionism while embracing other causes of that tumultuous time. 'Joshua' was born in the JBC strike of 1964, in which the issues were mostly civil libertarian and political. Middle-aged, Manley was receptive to younger Jamaicans' critiques of the establishment expressed in Rastafarianism, Black Power, the Walter Rodney riots, the New World Movement, and *Abeng*. He broadened his conceptions of politics and trade unionism and considered the economic, social and cultural dimensions of 'democracy' neglected by the Westminster model. The 1972 campaign drew on the disaffections of the '60s, including those of alienated capitalists and middle-class groups, and showed Manley's mastery of 'charismatic', 'messianic', 'populist' or 'social democratic' politics and modern campaign techniques.

The Complex Legacy of the 1970s

The legacy of Manley's years of leadership, 1972–80, is complex, but has been frequently oversimplified and negatively distorted beyond recognition.[6] During the '70s, Jamaica experienced a variant of the more general crisis in the Caribbean and Latin America. Throughout the hemisphere, economic growth from 1945 to 1970 masked economies that remained essentially weak, vulnerable, underdeveloped and neo-colonial; in the '70s and '80s, those economies succumbed to unprecedented and devastating world economic trends. Meanwhile social changes such as industrialization, urbanization, the spread of mass media, trade unionism, improved education, and political mobilization led to increasingly massive demands, often by grievously neglected or repressed groups, for a better life. Social ferment and new classes and groups led to new political parties which were, in some measure, populist, nationalist, statist, and sometimes 'socialist'. These parties aggregated and stimulated the demands of their followers in return for votes, but could not reshape national economies sufficiently to reward the increased demand for higher living standards. In countries such as Argentina, Brazil, Chile, and Uruguay (some of the most 'modern' of Latin America), military coups and brutal dictatorships arose to put the genies of populism and nationalism back in the bottle. Watching nervously over all this, helping when the need arose to repress the forces of change, was the United States government, concerned with the threat which populism and nationalism (more than 'communism') might pose to its strategic and economic interests. Although it avoided the military coup, Jamaica under Manley was an acute example of the more general syndrome.

Assessing Manley's complicated legacy requires understanding its four parts: the economy, social reform, politics, and foreign policy. The economy constitutes the most basic and most negative aspect of the Manley legacy, although it developed under exceedingly difficult extenuating circumstances and contained substantial achievements that critics often minimize or ignore. Manley's government set out to radically reform the economy in the direction of national sovereignty, redistribution of wealth and economic power, self-reliance, more productive use of national resources, and promotion of social justice. He inherited a seriously flawed economy, and the reform effort was soon swamped by a long list of circumstances beyond his government's control; the twin oil shocks of 1973 and 1979 (Seaga benefited from the windfall of 1986); international inflation and its effects on imports; stagflation; the flight of managers, technicians, and capital; the organized violence of the opposition; the hostility of the US and the role of the IMF, and so on.[7]

Critics point out that economies comparable to Jamaica's, such as those of Barbados, the Dominican Republic, and Costa Rica fared better (using the very limited measure of per capita GDP growth) in the hostile economic climate of the '70s. Jamaica started the '70s well above the average per capita GDP for Latin America and the Caribbean and finished below the average.[8] Such comparisons can, of course, be very misleading. In 1973, Oscar Arias Sánchez of Costa Rica, one of the 'successful' countries, observed: 'We have learned that a modern economy, growing urbanization and industrialization, and increased education and health benefits do not reduce socio-economic inequality. On the contrary, the inequality today is more marked than it was twenty-five years ago.'[9] Part of the difference in Jamaica's experience and that of comparable countries was timing. In 1981 Costa Rica suffered a severe economic crisis, enhancing its vulnerability to United States pressure and threatening what many observers uncritically regard as its model 'democracy'. In the early '80s Barbados experienced a major emergency based on threats to sugar and tourism. In 1984 the Dominican Republic was torn by riots against an IMF programme adopted by its centre-left government; the army killed over seventy persons, wounded another two hundred and arrested more than four thousand. The Dominican Republic, despite its record of economic growth, remains a nation of massive poverty and desperation.[10] Moreover, per capita GDP continued to decline in Jamaica under Seaga until at least 1986.[11]

Nonetheless, the Jamaican people could not see into the future and did not care very much about events in other countries. The economy was central to the PNP debacle in 1980, as Manley has admitted.

> Basic to everything that happened was the question of the economy. There had been negative growth. There were shortages. Unemployment had crept up from 24% to 26.8%. . . .
>
> World economic crisis and its effects placed unprecedented strain on Jamaican management. It took us a long time to respond adequately to this challenge. Unquestionably we overestimated the capacity of the bureaucracy and the management sector to run the wide range of new institutions. In a sense we overloaded the circuit.[12]

Although Manley bristled at Seaga's charges of mismanagement, the

JLP leader had a point which scored with some of the electorate. Sugar cooperatives, food farms, and the Special Employment Programme for the hard-core unemployed, were—on balance—failures, and the Manley government lost ground on the crucial problem of unemployment. The goal of economic sovereignty and self-reliance was sacrificed on the IMF altar.

It should be noted, however, that there *were* some economic successes amidst the general failure. Domestic agriculture grew strongly in response to the Manley government's policies, notably Land Lease. Douglas Manley reported that in his rural constituency of South Manchester ordinary people commonly said in the '70s that 'black man make money,' meaning that small farmers were profiting.[13] A number of successful programmes were continued by the Seaga regime in the '80s, including the renamed State Trading Corporation, the export promotions agency, the Bauxite Levy (at a slightly reduced rate), and trade pacts with the Soviet Union.

Manley was also more successful than Seaga subsequently in controlling the trade deficit, which was serious enough in the '70s, but skyrocketed under Seaga. Also, Manley was criticized heavily for borrowing from abroad, but Seaga borrowed even more, running up the national debt. During the first five years of Seaga's rule, from 1981 to 1985, with massive assistance from Washington, all major economic sectors either stagnated or declined, while social services plummeted.[14] Whether the resumed growth of 1986 and 1987 is a temporary or permanent trend remains to be seen, but in any case most Jamaicans have failed to benefit from it.

Nonetheless, despite large extenuating factors and some solid successes, the 'Manley' economic legacy is on balance a negative one, though largely beyond his government's control. Manley's efforts in the '80s to pursue better relations with the private sector and with the United States: his recognition of the importance of implementing and administering a few carefully chosen reforms, and the PNP's emphasis on small business, foreign exchange, and productivity are evidence of a desire to remedy the negative economic experiences of the '70s.

A second Manley legacy was social reform, which was heroic in its pursuit of social justice, but necessarily constrained by economic and political realities. Manley inherited an explosive situation as shown by the unrest of the late '60s and most fundamentally by the growth of unemployed, untrained, desperate, and rebellious ghetto youth, a new and largely anarchic force which, in its intensity, none of his predecessors had faced. Despite economic limits, Manley's party sponsored a wide variety of social reforms, aimed especially at youth, women, and the poor. The literacy programme JAMAL educated some two hundred thousand adults, one-tenth of the total population, and resulted in the internationally-acclaimed working-class women's theatre group, 'Sistren'.[15] Although the Special Employment Programme was an economic failure, it helped thousands of poor women and men survive. The Manley government also sponsored the Status of Children Act which removed the legal stigma of 'illegitimacy' and made law conform to majority practice as opposed to minority mores. The Family Court provided an improved venue for resolution of domestic problems. The Maternity Leave Law of 1979 also brought potential benefits to women. Modest increases in pensions and relief and a new Mental Health Law aided the poor. Nutrition programmes and community health aides helped in reducing infant mortality, which rose again in the '80s under Seaga. Educational reforms increased poor people's access to a key element of social mobility. The social ownership of RJR radio broke with the tradition of mass media controlled by narrow special interests. The National Housing Trust removed a scarce benefit, public housing, from partisan distribution at the cost of some ill-will by the PNP's own supporters, whose opposition to victimization was limited to victimization of *them*. A particularly unsuccessful reform was the National Youth Service, which was poorly planned and beyond the financial capacity of the government.

In addition to economic resources a committed government is needed to sustain social reforms designed to meet the majority's basic needs. In the '80s, many social reforms have been eroded. Even so, a lasting legacy of the Manley government's social reforms, was the expectation of the Jamaican poor that government, despite its limited resources, should seek social justice. This expectation was tempered considerably by the continued misery of the '80s.

Manley's Political Legacy

A third Manley influence was the political legacy which was also mixed, but on balance positive. The most negative aspect of that legacy was Manley's role as party leader, which cannot be separated from the nature of the PNP. Manley's leadership was flawed in that it could not ultimately resolve intra-party fighting, which had implications for economic strategy and for democracy itself. Manley's inherited position is in the moderate party centre. His dissatisfaction with the traditional politics of the PNP right, with its philistinism and lack of vision, and his desire for radical change, led him to move the party leftward in the '70s. He enlisted the support of D.K. Duncan, Arnold Bertram, Hugh Small, and Beverley Manley. Ferment on the left led to indiscipline and exaggerated rhetoric which Manley could not control, in part because of his own guilt about his privileged background and position. Thus Manley's attempt at 'democratic socialism', plagued by many forms of opposition, also failed through lack of party unity. In the PNP, Manley inherited an imperfect tool for serious change, but as party leader Manley must bear some responsibility for failing to repair it.

On the positive side, Manley has been the most democratic leader Jamaica has ever known. The leaders of the Jamaican Labour Party have enjoyed popular support, but their's has been a totally conventional, uninspired approach to 'democracy', emphasizing formalities and masking authoritarianism. Alexander Bustamante never understood the give-and-take, compromise, and tolerance essential to the democratic process. Hugh Shearer, as Prime Minister from 1967 to 1972, was prone to use police force against manifestations of popular resentment and even mere exercise of routine civil liberties. In the '80s, Edward Seaga has ruled in a highly individualistic, authoritarian manner, making many major decisions privately, certainly not as the result of any systematic consultation with interest groups or the people.

The PNP, since its origin in 1938, was rhetorically committed to 'democratic' rule, although its early brown middle-class leadership limited its willingness and ability to gain black majority support and stimulate majority decision-making power. Norman Manley, like most of his contemporaries, accepted uncritically the British Westminster model of parliamentary democracy, to which he had an unswerving loyalty. Notwithstanding the more radical approach he envisioned in the '60s, his brand of 'democracy' was unintentionally limited by his adoption of the Puerto Rican economic model with its grave socioeconomic inequality, clearer—to be sure—in retrospect than at the time.

By comparison even with the democracy of Norman Manley, Michael Manley's own beliefs and actions have been more profoundly democratic. Manley knows the limitations of the Westminster model, which restricts the popular mobilization necessary for a more genuine democracy. His commitment to democracy recognizes the necessity of economic and social democracy, as expressed in experimental community councils and economic enterprises, cooperatives, and worker participation in the organization and management of work. His vision of democracy transcends the concrete concerns that plague the poor Jamaican men, women and children in their struggle for survival. Manley's concept of democracy also outstrips the sometimes petty concerns of politicians of both parties interested mostly in winning elections, power, and distributing rewards to clients.

When Manley became Prime Minister in 1972, Jamaican democracy was moribund. Supported by many helpers, he stimulated more interest and participation in the democratic process. He also strove to identify democracy with majority interest (without which it is a fraud), with social justice and with change. This led to the historic class realignment of 1976, with workers and the poor increasingly attaching themselves to the PNP, while the upper and middle classes (and some remaining workers, peasants, and poor people) aligned with the JLP. Ultimately, the internal party conflicts of the PNP and above all, the economic crisis and dealings with the IMF undermined democratization, but a permanent result of Manley's approach to democracy was a massive consciousness-raising, to which the JLP had to adapt. Once again, the extent to which that new consciousness has survived the '80s only time will tell.

The Foreign Policy Legacy

The United Nations gold medal Manley received in 1978 placed him in the company of such leaders as Kwame Nkrumah, Jawaharlal Nehru, Paul Robeson, and Olof Palme. Although the medal was awarded for Manley's decades of opposition to apartheid, it is a fitting symbol of Manley's strongest legacy, his global vision and foreign policy, deeply marked by his personal touch.

Manley's foreign policy reflected a determination to understand the flow of world history and to reach a more just future for Third World nations. Characteristically, Manley begins a treatment of 'the politics of development' with 'the search for justice and equality.'[16] Because of Jamaica's small size and limited resources, and because of its profound dependency on alien economic, political, and cultural influences, its history has responded to global tides beyond its control which have restricted its development and sovereignty. Rooted in personal and local experience, Manley's foreign policy transcended those grounds through his advocacy of regional integration, pan-Caribbean nationalism, and Third World solidarity and the NIEO. To implement this policy he and his government—notably Foreign Ministers Dudley Thompson and P. J. Patterson—consistently and energetically pursued complementary approaches through the Caribbean Common Market (CARICOM), British Commonwealth, African, Caribbean, and Pacific (ACP) group, Non-Aligned Movement, Socialist International (SI), and the UN, compiling an impressive diplomatic record for little Jamaica.

Manley's vision and foreign policy drew on deep traditions from the nationalist and progressive history of the Third World. He regularly cited the Bolivian dream of a unified Latin America and the heroic struggles of Afro-Asian decolonization. Closer to home he refused to renounce the dream of Caribbean integration despite the ill-starred precedent of the West Indies Federation (1958–62).

Michael Manley's role has been to keep the flame of regional integration and Third World solidarity alive in the face of economic crisis and political strife which have compelled Third World countries to struggle separately for survival. Particularly notable were his government's regional initiatives toward the Spanish-speaking regional powers Venezuela and Mexico and toward progressive parties in Puerto Rico and the Dominican Republic. Manley has been concerned not solely with national sovereignty, but with regional, Caribbean sovereignty as well. This helps explain his opposition to the United States invasion of Grenada, which Manley rightly perceived as a fateful blow to Caribbean sovereignty even though the invasion was popular both in that tragic ministate and in Jamaica. It also helps account for his solidarity with the Nicaraguan revolution in the '80s. In the Western Hemisphere, no country's sovereignty is secure, unless every country's sovereignty is safe. Manley's conclusion that the Jamaican future rests on broad regional economic and political solidarity is correct.

To change the unjust and troubled global economy, however, even a unified Caribbean or Latin America is not sufficient. There is thus a practical side to Manley's genuinely idealistic commitments to African liberation: the search for potential allies in a global reform project, which expands to truly global significance with his vigorous promotion of the Non-Aligned Movement. Given the disparity of wealth and power, even a unified Third World might not prevail against the First World. The logic of the situation led to Manley's role in SI, the British Commonwealth,[17] the UN and other forums where First World and Third World interests are mediated. His chairmanship of the Economic Committee of the SI has provided European social democracy with a Third World plan based on the interrelatedness of well-being in the First and Third Worlds. The work of SI under the leadership of Willy Brandt, the late Olof Palme, and Manley has been called 'a perfect example' of what is needed in response to the Third World debt crisis and 'the global structure of inequality that undergirds it.'[18]

Manley has long shared the common Third World view that East-West rivalry perpetuates North-South inequality and Third World underdevelopment. As Mexican novelist and diplomat Carlos Fuentes has observed, injecting Third World struggles for social justice into the Cold War denaturalizes and dehumanizes them.[19] Manley's foreign policy perhaps inevitably confronted the power of the United States and its intolerance of the real independence of Western Hemisphere nations perceived to tread upon US strategic and economic interests.

In retrospect, the most questionable part of Manley's foreign policy was the close relationship with Cuba. Manley rightly saw friendship with Cuba as natural and essential to Jamaican sovereignty and independence. Cuba-Jamaica relationships were principled, fraternal and valuable to Jamaica, and advanced regional solidarity. But Jamaican 'socialism' had long been weakened by the 'Red Scare'. The extremely close ties with Cuba played into the hands of the JLP and alienated moderate Jamaicans. Moreover they scratched a raw nerve of the United States. In local politics and global *realpolitik* the Cuba policy cost Manley's government dearly. The overplayed 'Cuba card' does emphasize, however, that out of Jamaica's leaders Manley has been the most committed to true national sovereignty and independence. Acknowledging reality, a future Manley government's relations with Cuba would attempt to maximize regional solidarity while minimizing the costs.

In terms of practical achievement, the record of Manley's foreign policy was mixed. Some projects failed to fulfil hopes because of unfavourable economic developments and lack of political will in other countries: the aluminium scheme with Mexico and Venezuela and the International Bauxite Association, for example. More successful were the Bauxite Levy—which was an act of foreign policy—and diversified trade relations with the Soviet Union and Eastern Europe.

One serious problem with Manley's foreign policy and global vision is its failure to connect to Jamaicans' daily reality. Manley admits that he has not succeeded in convincing most Jamaicans that their local destiny is the result of global structures and dynamics. Most Jamaicans are friendly to the United States, many would like to migrate there, and many oppose attacks on it.[20] A recent poll of workers showed that Ronald Reagan was the most admired foreign head of state, considerably more so than Fidel Castro.[21] Pro-Americanism seems much stronger in Jamaica than in Latin American countries, and constitutes a strong obstacle to Manley's global vision.

In 1979, almost exactly a year before he was voted out of power, Manley assessed his government's goals and achievements. In economics, it had tried to increase control of basic resources, but was unable to augment 'returns to us' because foreign forces were more powerful than Jamaica; thus overall production declined. The government enacted numerous social reforms, but they were limited by the decline in production which left less to share. The government had widened the democratic process and deepened political consciousness, but that had provoked extreme opposition. Manley admitted that many mistakes were made, but the international environment was also overwhelming. 'In terms of the heroic aspirations of 1972 the Jamaican drama contains a real element of tragedy,' he said.[22]

The 1980s: Opposition and Anti-'Tribalism'

Manley has continued to grow as thinker, activist, and leader since the defeat of 1980. He has reflected at length on the '70s and has elaborated revised strategies for change.[23] Equally as important have been Manley's actions as Leader of the Opposition and his efforts to end the PNP-JLP 'tribalism' that has so wounded and enfeebled the working classes, poor people and Jamaica generally.

In the '50s, as noted earlier, Manley was 'product and perpetuator' of the bitter partisan rivalry of the PNP/NWU and JLP/BITU. Although there were important episodes of Manley-Shearer cooperation in the union struggle, Manley's confrontations with Edward Seaga and Pearnel Charles in the '60s and '70s continued the pattern of antagonism. However, early in his tenure as Prime Minister, developing a more truly national perspective, Manley began to speak to his own party and to the wider society of the madness of dogmatic loyalty to party and equally unthinking hatred of the other party. Manley's 'democratic socialist' project was severely wounded by an opposition which knew virtually no bounds. To its great credit, the PNP left realized clearly that 'tribalism' was antithetical to the achievement of 'democratic socialism', based as it must be on the unity of workers, peasants, urban 'masses', and progressive sectors of the middle class.

Victimized in the '70s by extremist opposition, Manley in the '80s has provided the society with a practical lesson on 'responsible' opposition. Given Manley's tendency toward competition and partisan confrontation (which conflicts with his equal tendency toward compassion and conciliation), it required great discipline to forge a new style of opposition politics. It also necessitated setting aside deeply personal wounds in the interest of what Jamaica wants and needs, political peace so that partisan differences may be contested in politics, but minimized in the shared task of national development. Thus Manley identified the need for bi-partisan approaches to *national* problems such as the bauxite industry, crime, violence, and drugs, and promised shared responsibilities with the JLP in the event of a return to power by the PNP. Manley's 'new politics' has been praised by independent observers such as Carl Stone and eventually by Edward Seaga as well. On 26 August 1988 Manley and Seaga signed a peace agreement in anticipation of the forthcoming elections. No one thinks 'tribalism' will be easily ended, but Prime Minister Seaga has encouraged this détente and it is supported by important business, civic, and church groups. Manley deserves major credit for paving the way towards realization of this fundamental hope.

Conclusion

Michael Manley is a complex man whose profound impact on recent Jamaican history defies simple myths and one-paragraph summaries. The overarching theme of his life has been the struggle against colonialism and neocolonialism, within a Jamaica that has changed significantly but which still denies a decent life to most of its citizens. The titles of his first books suggest his significance: *The Politics of Change, A Voice at the Workplace, Struggle in the Periphery*. Manley's role has been largely determined for him by his status as a privileged but dissident member of the middle class, by his extraordinary parents Norman Washington Manley and Edna Manley, by his formal education at Jamaica College and the London School of Economics, and by Jamaica's history, social relations, economic forces, and political dynamics. He has shaped that inheritance through thought and action, most significantly in trade union and political struggles motivated by his desire for social justice, economic self-reliance, democratization, national sovereignty, and development. His legacy to date is, like the man, complex: weakest in its economic record; stronger but mixed in its political aspects; just and compassionate in the promotion of social reform though limited by circumstances; and of world-wide importance in foreign policy and his own personal stature as Third World spokesman. The legacy would be stronger yet without the determined opposition of domestic and foreign foes and the implacable rush of economic forces. In time this legacy may emerge with greater clarity; but Manley has been, like Norman Manley, Edna Manley, and Alexander Bustamante, one of the initiators of the process of Jamaican national development. We are learning just how agonizing and protracted that struggle is in the neocolonial Third World. In time, it may also emerge that Manley's main significance, validated by union and political struggle, has been his visionary comprehension of global economics and politics.

Epilogue: Victory 1989

On 9 February 1989 Michael Manley and the People's National Party were returned to power with a national vote of near-landslide proportions. Despite consistently favourable Stone polls from the mid–'80s, the results of the election had been thrown in doubt the previous September by Hurricane Gilbert. As the PNP was preparing to celebrate its fiftieth anniversary, the storm destroyed homes, schools, farms, businesses, utilities and roads. The hurricane's massive destruction added to Jamaica's chronic economic woes and a general pre-election atmosphere of uncertainty. Manley's lead in Carl Stone's polls fell from fifteen to six points, but his hefty edge was restored by December, and Stone's polls very accurately predicted the eventual election result.

Manley and the PNP won fifty-seven percent of the popular vote and forty-five of the sixty seats in the elective House of Representatives. Although thirteen people were killed in political violence, the election contrasted sharply with the awful bloodletting of 1980. Both parties and their leaders deserved credit for lowering the level of political tribalism. There was little ideological content to the election. In its aftermath Manley stressed national unity, a theme echoed in an ecumenical religious service attended by him and Edward Seaga. A cabinet of moderates was appointed, with Portia Simpson and P. J. Patterson as 'super-ministers' (of Labour, Welfare and Sports; and Production, Development, and Planning, respectively). Other important appointments included David Coore as Foreign Minister, Hugh Small (the only prominent leftist of the '70s among the appointees) as Minister of Mining and Energy, Seymour Mullings in the crucial post of Finance Minister, K. D. Knight as Minister of National Security, and Douglas Manley as Minister of Youth, Culture, and Community Development. Manley named successful, respected businessmen Barclay Ewart and O. K. Melhado to head JAMPRO (a development agency) and the Jamaica Tourist Board, respectively; both are close personal friends of Manley's, and their appointments were designed to reassure the private sector. As Prime Minister, Michael Manley originally intended to concentrate on monitoring and coordinating the government as a whole. In the immediate aftermath of victory, Manley stressed production and education as government priorities. In foreign policy, he planned to build good relations with the United States and Western Europe and to attempt to strengthen CARICOM. Diplomatic relations with Cuba would be restored, but at a lower profile than had existed in the '70s.[1]

Reaction to the Manley/PNP victory fell out along predictable lines. Julio A. Muriente Pérez, in Puerto Rico's socialist weekly, *Claridad*, observed that Manley's victory was a defeat for the 'Puerto Rican model' of economic development which Edward Seaga and Ronald Reagan had attempted to revive.[2] *Caribbean Contact*, a monthly located in Bridgetown, Barbados, editorialized along similar lines:

> For the past ten years the Caribbean has undergone the equivalent of a counter-revolution from right-wing and conservative groups and parties. During that period the interests of Caribbean societies and Caribbean peoples were subordinated to the interests of powers and institutions outside the Caribbean.
>
> The ideologies of 'privatisation' and 'restructuring' were implemented in a manner in which the infrastructure, the health services and the educational services in particular were severely neglected and in some cases run down and the implication made that the peoples of the Caribbean are for sale.
>
> The importance of the results of Jamaica's election is that that implication has been clearly and decisively defeated.[3]

More cautiously, *The New York Times* observed that Manley 'and his struggling country merit Washington's support.' The *Times* continued: 'Now it's Mr Manley's turn to cope with a $4 billion foreign debt and to reassure the private sector while trying to deliver on campaign promises to help the poor.'[4] The United States government, struggling under new president George Bush to define its foreign policy, delivered a US$30 million instalment of post-hurricane aid to the new Jamaican government on 15 February, an event Manley termed 'an interesting example of continuity'.[5] The JLP's Edward Seaga promised vigorous opposition in the House of Representatives, and warned of the dangers of departing from his government's economic strategy.[6]

One of Michael Manley's most notable traits is his seemingly incurable optimism. His many successes have been punctuated with painful personal, professional, and political defeats, especially the electoral loss of 1980. Always in the public eye, as a private soul Manley has felt what very few of us have, the agony of responsibility for guiding a society toward a more just future combined with a violent reversal of that quest. Perhaps then, it was in part a collective act of optimism when the Jamaican people voted strongly to return the government of their country to Michael Manley and the People's National Party.[7] The Jamaicans did so in full light of the traumas of the '70s and the '80s, and consistent with the tradition, since 1944, that has allowed each party two terms before it is voted out of office. Manley and the PNP had the chance to redeem some of the unrealized promise of the '70s.

The story is incomplete, but barring unforeseen developments Michael Manley will continue to evolve as a Caribbean leader. His deeds, his thought, and his fundamental humanism will undoubtedly continue to contribute much to Jamaican, Caribbean and Third World history. Hopefully, Manley will inspire an increased appreciation in the First World—especially the United States—of the just needs of Third World people. Hopefully he will help motivate actions to assist the birth of a more equitable world. Michael Manley's life, in all its complexity, has to date been a compelling witness of one 'struggle in the periphery', testimony of the kinds of feeling, thought, and action necessary if we are all to meet the global challenge of the present and future.

APPENDIX A

Principal Policies and Programmes of the PNP Government

1972
Promotion of New International Economic Order (NIEO)
Special Employment Programme (for chronically unemployed)
Skill training programme
Workers' Bank
Jamaica Adult Literary (JAMAL)
Lowered Voting Age (to 18)
Community health aides
Operation GROW
Land Lease (land reform)
Civil Service reclassification
Youth training increased
Public housing programme

1973
Cultural training centre
Food subsidies on flour and condensed milk
Uniforms for primary school children
Free secondary education
Free university education
National Youth Service
Rent restriction act revised
Equal pay for women
Women's Bureau
Jamaica Nutrition Holdings
Property taxes increased
Initiated founding of International Bauxite Association
Home Guard
Promotion of Non-Aligned Movement

1974
Self-supporting farmers development loan programme
Family Court
National minimum wage
Gun Court
Suppression of Crime Act
National Insurance Scheme pensions increased
Poor relief increased

Agricultural Marketing Corporation outlets in low income areas
New Mental Health law
Free education for the handicapped
Construction of small industries complexes
Sugar cooperatives
Bauxite Production Levy
Nationalization of multinational bauxite companies
Development venture capital financing loans
Jamaica Public Service (electricity) Company established
Jamaica Merchant Marine established
Jamaica Omnibus Service established

1975
Capital gains and inheritance taxes increased
Compulsory recognition of trade unions
Worker participation in management introduced

1976
National Housing Trust

1977
Small enterprise development company
State Trading Corporation
National Commercial Bank

1979
Maternity leave with pay

Note: Adapted from Evelyne Huber Stephens and John D. Stephens, *Democratic Socialism in Jamaica* (Princeton, New Jersey: Princeton University Press, 1986), pp. 70–71. Used with permission of Princeton University Press.

APPENDIX B

The Elections of 1976 and 1980

Although the distribution of popular vote between the two parties changed less than one percent from 1972 to 1976, the PNP increased its margin of House seats from 36–17 to 47–13 in the expanded house. The premier analyst of Jamaican politics, Carl Stone explained this as a result of several factors: (1) the single member district/simple majority system: (2) consituency boundary changes, i.e. gerrymandering, which cost the JLP two seats; (3) the more even geographical spread of votes for both parties, which in a single member district system enhances the seat majority of the party winning the popular vote: and (4) 'illegal voting and registration manipulation by the PNP' that 'won at least two or three seats for the PNP that probably would have been won by the JLP candidates in rural parishes.' If Stone is correct the PNP's 47–13 majority in the House would have been reduced to a still powerful 43–17 or 42–18.[1]

A study of overvoting (more votes than registered voters) in the elections of 1976 and 1980 concluded that overvoting had not changed either the election's overall result, but if institutionalized, could, like violence, become a threat to the right to vote. The study showed three constituencies had, in the words of the Chief Electoral Officer, 'a high percentage of overvoting': Western Kingston (held by the JLP's Edward Seaga), St Andrew Southern (held by the PNP's Anthony Spaulding), and St Andrew Southwestern (held by the PNP's Portia Simpson). Overvoting in Manley's East Central Kingston constituency in 1976 was almost as marked as that in Seaga's constituency; that tendency disappeared in Manley's constituency in 1980 (when the JLP benefited more than the PNP from overvoting in Manley's constituency), though not in Seaga's. What was perhaps most disturbing to the author of the study was the tendency of overvoting to occur in the constituencies of leaders in both parties as a part of the 'turf politics' of urban Kingston. This would seem to diminish the possibilities of reform.[2]

1. Carl Stone, *Democracy and Clientelism in Jamaica* (New Brunswick, New Jersey: Transaction Books, 1980), p. 188.
2. Mark Figueroa, 'An Assesment of Overvoting in Jamaica', *Social and Economic Studies* (September, 1985), pp. 71, 77, 94 and 96.

APPENDIX C

The PNP's 'Principles of Democratic Socialism'

The Democratic Socialist process will provide for:

1. The right of every Jamaican to form or join any political party of his/her choice, and to compete for State power in democratically contested elections.

2. The right of every Jamaican to meaningful employment and economic and social security.

3. The right of every Jamaican to own private, personal and productive property and to save to achieve these ends, subject to, and consistent with, the social and economic needs of the working people.

4. The right of every Jamaican to equality before the law, and consequently, the right to equality of opportunity, to equality of rights and to entitlement of security and social justice.

5. The right of the Jamaican people to establish sovereignty over their natural resources and to order their social, economic and political affairs wihtout interference from any other nation.

6. The patriotic duty of every Jamaican to cooperate with the majority in building the nation, to respect our national heritage, to contribute to the development and welfare of the nation as a whole.

7. The deepening of the democratic process so that the collective wisdom and experience of the Jamaican working people can become the decisive factor in the decision-making process at all levels. This will undoubtedly represent considerable advance in the excercising of the democratic rights and responsibilities which in our capitalist stage of development is largely limited to voting every five years. An important consequence of this advance will be raising the level of involvement and participation of the vast majority—the working class and small farmers—in national life, in production as well as in cultural and political life.

8. Obviously the trade union movement as well as the movement for worker participation will have to play leading roles in this deepening of the democratic process. Hence, as the entire socialist process deepens, the working class and small farmers will naturally play a more active role in national life and within the class alliance than they had in the past or were capable of playing under capitalism.

Note: From *Principles and Objectives of the People's National Party* (Kingston: PNP, 1979), pp. 10–11.

APPENDIX D

Excerpts from the PNP's 'Glossary of Terms'

Capitalism: 'a system of economic organization based on exploitation, in which the major means of production, distribution and exchange are privately owned by a small minority, constituting the capitalist class, and in which workers, constituting the working class, sell their labour power for wages. The surplus generated by the workers is appropriated by the capitalists as profit. Resources are allocated, and production determined primarily by the play of market forces and the profit motive. As a result, social needs are often neglected in favour of more profitable, but unnecessary luxury production.'

Scientific Socialism: 'as distinct from Democratic Socialism, is a political philosophy based on the teachings of Marx and Engels as developed by Lenin. It proclaims the inevitability of class struggle in a capitalist society, believes in the accentuation of that struggle to the end that the working class will overthrow and destroy the capitalist state leading to the setting up of the dictatorship of the working class. The dictatorship of the working class is seen as the instrument through which socialism will be established.'

Communism: 'describes the form of society which it is believed will evolve from scientific socialism. It is ideally conceived as a social and economic system in which all the means of production, distribution and exchange will be collectively owned and consumer goods will be distributed according to the needs of each person. Finally, it is envisaged that classes will disappear and the state will wither away.'

Imperialism: 'in the capitalist era, it is the process by which capitalist organizations and institutions of a more developed country come to own and/or control the raw materials and/or the basic productive processes of another, less developed country. The purpose is exploitation and the effect is the creation of a dominant influence upon the life of the exploited country. Generally proceeding by military conquest in its earliest stages, it is now often characterized by international corporations which operate beyond the reach of any one country. Also to be noted is the tendency of developed capitalist states to use the levers of power and influence available to them to support and protect their international corporations at the expense of the exploited States.'

Nationalism: 'describes the emotion associated with a highly developed sense of collective unity among the members of a Nation State. It is the force that drives colonial societies to achieve political independence and all societies to defend themselves from attack.

'It can also express itself in a tendency to aggressiveness in dealings with other nations. It is sometimes exploited to keep disadvantaged members of a society from protesting against their condition.'

Democratic Socialism: 'a political and economic theory under which the means of production and exchange are owned and/or controlled by the people. It is a system in which political power is used to ensure that exploitation is abolished, that the opportunities of society are equally available to all and that the wealth of the community is fairly distributed. A process rather than a rigid dogma, its application must depend on the particular conditions which obtain from time to time in each country. It emphasizes co-operation rather than competition, and service rather than self-interest as the basic motive forces for personal, group and communal action. Its ultimate objective is the building of a classless society by removing the element of entrenched economic privilege which is the basis of class divisions. As distinct from scientific socialism, its method is based on the alliance of classes around clear objectives.'

Social Democracy: 'describes a political process employing broad reforms of and controls over a capitalist system to create a more just and equitable society without changing the system itself fundamentally.'

Note: From *Principles and Objectivities of the People's National Party* (Kingston: PNP, 1979), 61–7. Also defined in the glossary are class, state capitalism, colonialism, neocolonialism, democracy, egalitarianism, fascism, feudalism, patriotism, social ownership, social control, totalitarianism, and 'working people'.

NOTES

See Index for note on abbreviations

Introduction

1. In this book the unmodified name 'Manley' is used exclusively to refer to Michael Manley to avoid confusion with other members of the family.
2. Under the Jamaican constitution, the government may call elections within five years of its election to office and may do so with very little notice at any time.
3. Excellent studies include the several studies of Carl Stone, listed in the Bibliography; Michael Kaufman, *Jamaica Under Manley: Dilemmas of Socialism and Democracy* (Westport: Connecticut: Lawrence Hill, 1985): and Evelyn Huber Stephens and John D. Stephens, *Democratic Socialism in Jamaica: The Political Movement and Social Transformation in Dependent Capitalism* (Princeton, New Jersey: Princeton University Press, 1986).
4. MM, *The Politics of Change: A Jamaican Testament* (London: André Deutsch, 1974); *A Voice at the Workplace: Reflections on Colonialism and the Jamaican Worker* (London: André Deutsch, 1975); *Jamaica: Struggle in the Periphery* (London: Third World Media, 1982); *Up the Down Escalator: Development and the International Economy, A Jamaican Case Study* (Washington, D.C.: Howard University Press, 1987); and *A History of West Indies Cricket* (London: André Deutsch. 1988). See also: John Hearne (ed.), *The Search for Solutions: Selections from the Speeches and Writings of Michael Manley* (Oshawa, Ontario, Canada: Maple House, 1976), and Socialist International Committee on Economic Policy (chairman, Michael Manley), *Global Challenge: From Crisis to Cooperation: Breaking the North-South Stalemate* (London: Pan Books, 1985). Many of Manley's shorter writings and speeches are listed in the Notes and Bibliography.
5. Broadly speaking, 'democratic socialism' is the attempt to unite socialist economics with truly democratic politics. It occupies a middle ground between liberal democracy—with its emphasis on formal political practices and absence of social and economic democracy and mass participation in decision-making—and Marxism-Leninism, with its tendency toward the one-pary state and centralized decision-making. For the PNP's 'Principles of Democratic Socialism', see Appendix C.
6. Capitalism in the Third World is fundamentally different from First World capitalism because of its dependent nature. It depends on foreign markets, resources, capital, technology, and expertise which it ultimately does not control. Critics argue that this dependency precludes authentic development.
7. 'A Historia do Brasil, Escrita nos EUA', *Veja*, 24 November 1971.
8. Eric Williams, *British Historians and the West Indies* (New York: Scribner's, 1966). See also the *Bulletin of Eastern Caribbean Affairs* (May/June 1982), special issue devoted to Caribbean Historiography; and Derek Walcott's 'The Muse of History' and Gorden Rohlehr's 'History as Absurdity' in: Orde Coombs

(ed.), *Is Massa Day Done? Black Moods in the Caribbean* (Garden City, New York: Doubleday, 1974).

CHAPTER 1 **Confrontation and Compassion**

1. This chapter is largely based on the author's interview with Manley of 15 August 1984; unless otherwise notes, quotes are from that interview. Another version of the events is in Manley's *A Voice at the Workplace*. pp. 18–19.
2. Coore was Manley's closest friend at the time and later served as Deputy Prime Minister in Manley's government during the '70s. In an interview on 22 December 1987 Coore corroborated many of the events discussed in this chapter.
3. *Facety* is a Jamaicanism for 'impertinent'.
4. Manley noted that 'they say I organized this thing to this day in Jamaica, which is totally untrue.'
5. On 13 March 1972, Hardie wrote from England to congratulate Manley on his election as Prime Minister: 'Our acquaintance some thirty years or so ago was brief and stormy and I guess we were both unnecessarily bruised and burnt by the flames of [illegible word] irreconcilable idealisms, our immaturity and what we believed we represent.' Manley replied on 14 April 1972: 'I was fascinated to have you confirm that you, too, were badly burnt in 1943.' Manley wrote that Hardie's letter 'through its comprehension and humanity laid the hurt finally to rest.' Letters in MPF.
6. Donat Crichton, letter to the editor published under the title 'Thanks, Michael Manley,' *DG* 10 December 1987. Crichton was an unsucessful JLP candidate for parliament in 1967 and heard the story directly from Esme Grant.

CHAPTER 2 **What Sort of Man?**

1. EM, Diary, 29 November 1975.
2. EM, Diary, 23 February 1983.
   This was written after Edna found out that the American Ambassador had described her to FBI director J. Edgar Hoover, presumably in the '40s, as a 'committed communist'. She was stunned by the false claim.
3. Letter to Beverley Anderson-Manley, 25 June 1980, PNP file 'Citizens' Mail [1980].'
4. Christopher Arawak [pseud.] *Jamaica's Michael Manley: Messiah, Muddler, or Marionette?* (Miami: Sir Henry Morgan Press, 1980).
5. William H. McNeill, *Mythistory and Other Essays* (Chicago: University of Chicago Press, 1986), p.3. McNeill's comments were part of his presidential address for the American Historical Association's annual meeting in 1985.
6. MM, Interview with author, December 1987.
7. On the importance of speaking and eloquence in the West Indies, see Roger D. Abrahams, *The Man-of-Words in the West Indies: Performance and the Emergence of Creole Culture* (Baltimore and London: Johns Hopkins University Press, 1983).
8. Interview with author, 12 September 1985.
9. For examples, see *Jamaica: Struggle in the Periphery*, p. 15. and *Voice at the Workplace*, pp. 148–9.
10. Interviews, Thelma Manley, 24 September 1985, and Rex Nettleford, 23 December 1987.
11. Pamela O'Gorman, interview with author, 24 December 1987. Miss O'Gorman, a friend of the Manleys since 1958, has served as a music critic on the Jamaican Broadcasting Corporation (JBC) and the *Gleaner*, was Director of Music at the University of the West Indies from 1969 to 1975, and since then has headed the Jamaica School of Music.

12. In at least one case an important friendship became a victim of the politics of change. John Hearne, a novelist, was a very close friend for decades who broke with Manley in the mid–'70s and ended up writing vicious, and in one case libelous, attacks on him in the *Gleaner*. For the views of Hearne and other Manley critics, see chapters 16–20.
13. Interview with author, December 1987.
14. Author's interview with Dr. Matthew Beaubrun, 19 December 1987.
15. MM untitled manuscript, 12 September 1974, in MPF.
16. For what it is worth, my initial impression of Manley after interviewing him three times in 1984 was of an open and generous man. My opinion has not altered in that respect since. Admittedly, Manley and I have a somewhat artificial relationship.
17. John Dorschner, *Tropic Magazine (Miami Herald)*, 11 April 1976, reprinted in *JWG*, 27 April 1976, p. 7.
18. *JWG*. 10 April 1978, p. 31.
19. *JWG*, 14 April 1978, p. 13. Additional negative comment is presented below in chapters dealing with the '70s in sections treating the media.
20. Hans J. Massaquoi, 'Jamaica: A New Beginning Under New Leadership,' *Ebony* (October 1973), p. 38.
21. *JWG*, 6 August 1974, p. 2.
22. *The Guardian*, 17 April 1974.
23. Max Lerner, 'A Talk in Jamaica', *The New York Post*, 19 February 1975.
24. The *Star*, 28 June 1976.
25. *JWG*, 21 December 1976.
26. Andrew Kopkind, 'Trouble in Paradise', *Columbia Journalism Review* (March/-April 1980), p. 42.
27. Michael Massing, 'The Jamaica Experiment', *The Atlantic Monthly* (September 1983), p. 49.
28. MM, letter dated 31 July 1974, MPF.
29. MM, letter of 12 November 1984, MPF.
30. MM to Norman Washington Manley, 1 October 1944, BAM collection.
31. MM, letter dated 7 June 1978, MPF.
32. MM, letter of 20 January 1977, PNP file 'Executive 1977', PNP library.
33. MM, taped interview with author, 12 September 1985. Questioned about the degree to which his emotions and enthusiasms outstripped his intellectual judgments, Manley admitted that it might not be only 'slightly' but 'maybe "considerably".'
34. Confidential interview.
35. Author's interview, Beverley Anderson-Manley, 5 September 1985.
36. Wayne Brown, *Edna Manley, The Private Years: 1900–1938* (London: André Deutsch, 1975).

CHAPTER 3 **Roots**

1. MM, *Jamaica: Struggle in the Periphery* pp. 16–23.
2. On creolization, see Edward Brathwaite, *The Development of Creole Society in Jamaica, 1770–1820* (Oxford, England: Clarendon Press, 1971).
3. R. Renny, *A History of Jamaica* (London, 1807), p. 241n, cited in Brathwaite, p. 208.
4. Charles Campbell, *Memoirs of Charles Campbell* (Glasgow, 1828), p. 19, quoted in Brathwaite, pp. 207–8. Punctuation as in original.
5. In 1835, the traveller Richard R. Madden was moved to write of Jamaica: 'In a word, where all things, both buckras and blacks/ Are in fits and by starts whether rigid or lax/ And in faith and in politics, never it seems/ Are content if their notions are not in extremes!' R. R. Madden, *Twelve Months' Residence*

in the West Indies During the Transition from Slavery to Apprenticeship, (2 Vols, London, 1835), 1: 96, quoted in Philip Curtin, *Two Jamaicas: The Role of Ideas in a Tropical Colony, 1830–1865* (Cambridge: Harvard U.P, 1955), p. 52.

6. Birth Certificate, file 'Manley, N. W.—Birth Certificate', NMP. Two personal remembrances of Norman Manley exist: Philip M. Sherlock, *Norman Manley* (London: Macmillan, 1980) and Victor Stafford Reid, *The Horses of the Morning: About the Rt. Excellent N. W. Manley, Q.C. M.M., National Hero of Jamaica: An Understanding* (Kingston: Caribbean Authors Publishing Co. 1985). Also of interest is the long introduction by Rex Nettleford to the collection of speeches *Norman Washington Manley and the New Jamaica: Selected Speeches and Writings 1938–1968* (New York: Africana Publishing, 1971). Two somewhat different autobiographical fragments are: 'My Early Years: Fragment of an Autobiography' in Nettleford (ed.) *Norman Washington Manley and the New Jamaica*, pp. xcv-cxii, and 'The Autobiography of Norman Washington Manley', *Jamaica Journal*, 7 (March/June 1973), pp. 4–19, 92. A definitive scholarly biography fully utilizing the sources in the NMP remains to be written.
7. NWM, 'Autobiography', *Jamaica Journal*, 7 (March/June 1973), pp. 4–5.
8. George E. Eaton, *Alexander Bustamante and Modern Jamaica* (Kingston: Kingston Publishers, 1975), p. 1. In an interview with the author in December 1987, Manley named Alexander Shearer as the key ancestor of his family. Alexander Shearer was also the ancestor of Manley's long-term friendly adversary, Hugh Shearer, Jamaica's Prime Minister from 1967 to 1972.
9. Philip Sherlock, *Norman Manley*, p. 7.
10. NWM, 'My Early Years', p. xcviii.
11. *Jamaica College Magazine*, Easter Term 1911, pp. 5, 8, 14; Midsummer Term 1911, pp. 3 and 13; and Midsummer Term 1912, *passim*.
12. Sherlock, *Norman Manley*, p. 57.
13. Victor Stafford Reid, *Horses of the Morning*, p. 38.
14. This section relies heavily on the sensitive recreation of EM's early decades in Wayne Brown, *Edna Manley*, especially pp. 1–45. Brown's book also contains much on Norman and is based on the voluminous correspondence of the two. Useful sources on EM's art include David Boxer, 'Edna Manley: Sculptor,' *Jamaica Journal*, 18 (February/April 1985), pp. 25–40; Boxer, *Jamaican Art, 1922–1982* (Washington, D.C.: Smithsonian Institution and Kingston: National Gallery of Jamaica, 1983); 'Edna Manley: Selected Sculpture and Drawings, 1922–1976,' (Kingston: National Gallery of Jamaica, 1980); and 'Edna Manley: The Seventies', (Kingston: National Gallery of Jamaica, 1980.
15. Brown, *Edna Manley*, p. 19.
16. Quoted in Brown, *Edna Manley*, p. 21.
17. Brown says 29 February 1900, but under the Gregorian calendar reforms 'century years' such as 1900 are leap years only if divisable by 400. Thus there was no 29 February in 1900. EM apparently preferred to celebrate the more exotic date as her birthday. Much was later made of the fact that the JLP government chose 29 February, 1972 for the elections which brought down the JLP and elevated Manley to Prime Minister.
18. Brown, *Edna Manley*, p. 45.
19. Brown, *Edna Manley*, pp. 49–58.
20. Brown, *Edna Manley*, p. 71.
21. Brown, *Edna Manley*, p. 72.
22. William McCulloch to NM, 31 March and 17 April 1921, NMP, original emphasis.
23. William McCulloch to NM, 19 May, 3 August, and 30 August 1921, NMP.
24. Brown, *Edna Manley*, p. 18 notes the contradictory evidence on this score.
25. With the rise of pride in blackness and the African heritage in Jamaica in the

'60s EM was sometimes subjected to malicious reverse discrimination. Because of her extreme fairness of skin, gossipers said, she could not possibly be Michael's mother.

26. Brown, *Edna Manley*, p. 115.
27. Brown, *Edna Manley*, p. 116.
28. NM, 'My Early Years', pp. civ-cv. In *Edna Manley*, Brown notes that NM's greatest weakness may have been the tendency, at some critical times, to suspend the rational analysis of society, as evidenced here in his words 'I did not study it.'
29. Brown *Edna Manley*, p. 134.

CHAPTER 4 **Youth 1924–43**

1. EM, Diary, 13 March 1977.
2. EM, undated note [postmarked 9 May 1929], on stationary of *SS Patuca*, NMP.
3. Letter, EM to NM, Orpington, Kent, England, undated [c. 1937], NMP.
4. On the conflicts between art and family in Edna's work at this time see David Boxer, 'Edna Manley: Sculptor', *Jamaica Journal* (February-April 1985), pp. 27–8.
5. Quoted in Sherlock, *Norman Manley*, pp. 66–7.
6. *DG*, 3 October 1932.
7. William McCulloch to NM, 12 August 1933, NMP.
8. NM 'The Autobiography of Norman Manley', *Jamaica Journal* (March/June 1973), p. 14.
9. Sherlock, *Norman Manley*, p. 71.
10. NM, 'Autobiography', p. 15. Reid, Norman Manley's chosen biographer, records that NM and Garvey 'squared off and nearly came to blows. Garvey, a beefy energetic man, chuckled at his slight but sinewy challenger's invitation to "step outside," and declined. Thus was lost an incomparably hilarious footnote to history.' (Reid, *Horses of the Morning*, p. 66.)
11. NM, 'Autobiography', p. 15. Brown (*Edna Manley*, p. 175) reports that NM's client asked for £1000 in damages and received a mere £30.
12. NM, 'Autobiography', p. 15, and Reid, *Horses of the Morning*, p. 66.
13. Brown, *Edna Manley*, p. 175.
14. NM, 'My Early Years', p. cvi.
15. Sherlock, *Norman Manley*, p. 77.
16. MM, interview, 12 August 1984.
17. MM, interview, 12 August 1984.
18. Letter to MM, undated [April-May 1983?], MPF.
19. EM, Ms Diary, 23 October 1982, on the occasion of Miss Boyd's ninetieth birthday.
20. Jamaica, Office of the Prime Minister, 'Michael Norman Manley: Prime Minister of Jamaica, A Political Biography', ([Kingston], 1975), p. 1.
21. T. A. Marryshow, 'Edna Manley, Artist, Inspirer of Youth, and What a Politician's Wife Should Be,' (undated typescript copy, c. 1950s?), MPF
22. George Campbell, Interview, 13 September 1985.
23. MM, 'The Root of the Matter,' *Public Opinion*, 20 November 1954.
24. George Campbell, interview, 13 September 1985.
25. DM, interview, 4 September 1985.
26. EM, interview, September 1985.
27. MM, interview, 12 August 1984. A somewhat different version of this story is found in Sherlock, *Norman Manley*, p. 15. Questioned about the differences, Manley insisted his version is correct.
28. MM, interview, 12 August 1984.
29. EM to NM, Arthur's Seat, undated [1930s], NMP.

30. EM to NM, envelope postmarked 12 June 1936, NMP.
31. EM to NM, postmarked Kingston 15 June 1936, NMP.
32. EM to NM, undated [c. June or July 1936], NMP.
33. EM to NM, Orpington, Kent, England, 2 letters [c. 1937], NMP.
34. EM to NM, [c. 1937], NMP. 50 years later, Corina Meeks, a former special assistant to Manley when he was PM described him to me as a racehorse who needed to be reined in at times (interview, 21 December 1987).
35. EM to NM, [early 1940s], NMP.
36. EM to NM, Drumblair, [c. 1940], NMP. In this letter, Edna gave her age as 40.
37. Jean Fairweather, 'Michael Manley, Who is Leading Jamaica Into Democratic Socialism, Is Many Things to Many People,' *Commonwealth* (April-May 1975), p. 5.
38. DM, interview, 6 September 1985.
39. George Campbell, *First Poems* (New York: Garland, 1981), p. 107. Reprinted with permission
40. MM to NM, 3 November 1940, BAM.
41. MM, 'Poem,' 24 November 1940, BAM.
42. MM to NM, 31 January 1941, BAM.

CHAPTER 5 **Jamaica College (1935–43)**

1. Quoted in A. W. Singham, *Hero and the Crowd*, pp. 77–8. C. O. Cutteridge was the English author of textbooks used in the West Indies.
2. MM, 'The Prime Minister. . . . at U. W. I.,' 22 November 1976, PNP Archives, Speeches files.
3. Adam Kuper, *Changing Jamaica* (London: Routledge and Kegan Paul, 1976), pp. 68–9. Kuper's book is an adaptation of a report commissioned by the government headed by MM. Except as noted, much of this chapter, including unattributed quotations, is based on interviews with Manley on 12 and 15 August 1984.
4. Eric Eustace Williams, *Inward Hunger: The Education of a Prime Minister* ([Chicago]: University of Chicago Press, [1971]), pp. 33–4.
5. MM, *The Politics of Change*, p. 141.
6. Morris Cargill, *Jamaica Farewell* (Secaucus, New Jersey: Lyle Stuart, 1978), pp. 50–73. Cargill, writing for the *Jamaican Daily Gleaner*, was a strident and often unfair critic of the government headed by MM.
7. DM, interview, 4 September 1985.
8. *Jamaica College Magazine*, December 1941.
9. D. K. Duncan, interview, 17 September 1985.
10. John Maxwell, interview, 7 September 1985.
11. Quotations from Wayne Brown, *Edna Manley*, p. 232.
12. *Jamaica College Magazine* (December 1941), p. 10.
13. MM to JM, 11 February 1974, MPF. In this letter, Manley compared his record to that of his son Joseph who won two academic prizes in one year at Munro College. In this way Manley attempted to spare *his* son the anxiety that he experienced in dealing with a gifted father.
14. A. W. Singham, *Hero and the Crowd*, p. 81.
15. MM, *The Politics of Change*, p.138.
16. MM to NM, 31 January 1941, BAM.

CHAPTER 6 **Canada (1943–5)**

1. Trevor Munroe, *The Politics of Constitutional Decolonization: Jamaica, 1944–62* (Mona, Jamaica: Institute of Social and Economic Research, University of the West Indies, 1972), p. 32.

2. BAM, in her study of the early PNP, comments that 'there is no doubt that to both the Americans and the British—two opposing "tribes" – political parties . . . would make an ideal situation of "divide and rule".' BAM cites a memo from FBI Director J. Edgar Hoover to Assistant Secretary of State Adolf A. Berle (1 September 1942, U.S. National Archives Document 844D 00/47): '. . . The real danger is thought to lie in the possibility of a combination of the two parties which would place in the hands of Manley, the PNP leader and Busta, complete control of the island. Such a combination is not considered probable because of the wide gulf which separates the two men and their organizations.' (BAM, 'The Ideological and Programmatic Platforms of the People's National Party (1938–44).' M.Sc. Thesis, University of the West Indies, Mona, Jamaica, 1985, pp. 122–3)
3. Monroe, *Politics of Constitutional Decolonization*, p. 36.
4. Gordon Lewis, *The Growth of the Modern West Indies*, p. 179. By the late '80s many Jamaicans saw MM as the most important political figure of independent Jamaica.
5. Munroe, *Politics of Constitutional Decolonization*, p. 36.
6. David Coore, interview, 22 December 1987.
7. MM, interview, 15 August 1984.
8. *Ibid.*
9. MM, letter of 24 March 1975, MPF.
10. MM to NM, Lachine, Quebec, 8 October 1943. The remainder of this chapter is based largely on nineteen letters Manley wrote to his father during this period. In 1985 the letters were in the possession of BAM who made them available to me. Subsequent citations in this chapter from these will be made by date only.
11. 16 October 1943.
12. 11 February 1944.
13. Single letter, 20 and 21 February 1944.
14. 7 March 1944.
15. 4 April 1944. Manley's scores improved when he was diagnosed as myopic and fitted with glasses.
16. 1 October 1944.
17. 29 October 1944.
18. 28 January and 8 February 1945.
19. 8 October 1943.
20. 7 March 1944. The word 'excommunicate' recalls the unpleasant time Manley spent in 'Coventry' at Jamaica College.
21. 29 May 1944.
22. 20 November 1944.
23. 6 October 1943.
24. 8 October 1943. In 1987, in the aftermath of EM's death. Manley gave the 'Horses of the Morning' to the 'National Art Museum for the people of Jamaica.
25. 28 April 1944.
26. 29 May [1944].
27. 6 October 1943.
28. The 'West end' of Kingston was then Bustamante's stronghold and personal constituency. PNP gains later forced him out to a safe rural seat. The West Kingston ghetto, through a violent process in the mid–'60s, became the political fortress of Edward Seaga, MM's bitter rival of the '70s and '80s.
29. 8 October 1943.
30. 11 February 1944.

31. MM, letter dated only 1975 in MPF, 1975. Manley observed in the letter that 'All memories of hunger and cold die hard.'
32. 6 October 1943. Manley reckoned later that he was in good enough shape to run a fifty-second quarter mile (Manley's letter of 24 March 1975.)
33. 7 March 1944.
34. 1 October 1944.
35. MM to EM, 18 November 1943.
36. 16 October 1943.
37. EM to NM, undated [early '40s]. NMP.
38. 28 April 1944.
39. DM interview, 4 September 1985.
40. 20 November 1944.
41. 1 October 1944. Manley's relationship with intellectuals has been contradictory. Possessed of great intelligence and wit, he performed only indifferently in academic settings. He clearly enjoys the company of intellectuals, but is also often critical of them. He can refer to himself as an intellectual, although he disdains the sustained, concentrated, detailed scholarship characteristic of the academic intellectual. His main model of the intellectual was Harold Laski, a brilliant thinker and writer whose intellectual activity may have suffered because of his political activism.
42. 8 February 1945.
43. Manley's Canadian letters are interesting, in some respects, for what they don't contain, or treat only in passing: race relations and anti-imperialism, for example.

CHAPTER 7 **London, 1945–51**

1. Richard Ellman and Robert O'Clair (eds.), *The Norton Anthology of Modern Poetry* (New York: Norton, 1973, p.490.
2. Singham, *Hero and the Crowd*, p. 95.Singham goes on to argue that 'one of the by-products of marginality is the capacity to accommodate to anyone and everyone with whom the individual comes into contact.' a trait that has often been observed in MM.
3. Kwame Nkrumah, *Ghana: The Autobiography of Kwame Nkrumah*, (New York: Thomas Nelson, 1957) pp. 49–61. In their study of West Indians in Britain, Clarence Senior and DM cited the work of the British sociologist A. H. Richmond, *The Colour Problem*, as providing conclusive evidence that one-third of the British were tolerant, one-third were mildly prejudiced, and one-third 'extremely prejudiced against coloured people'. Prejudice stemmed from three sources: class, colour, and xenophobia. A light-skinned student (such as a Manley) would have fewer problems than a dark-skinned worker (Senior and Manley, *The West Indian in Britain* [London: Fabian Colonial Bureau, 1956], pp. 8, 10).
4. George Campbell, interview, 13 September 1985.
5. DM, interview, 6 September 1985.
6. MM, 'Root of the Matter', *Public Opinion*, 16 May 1953.
7. MM to NM, 17 December 1945, collection of letters in the possession of BAM. Except as noted, letters cited in this chapter come from this collection.
8. MM, 'Root of the Matter', *Public Opinion*, 25 September 1954.
9. MM, 'Root of the Matter', *Public Opinion*, 5 March 1955.
10. MM to NM, 5 April [1946?].
11. MM to NM, single letter dated 23 and 26 January 1946.
12. MM to NM, 8 July 1946.
13. MM to NM, 22 October 1951.
14. All quotations in this paragraph from MM to NM, 20 November 1945.

15. David Coore, interview, 22 December 1987.
16. MM to NM, 17 December 1945. In this letter Manley was characteristically concerned about his father's possible reactions to his lengthy comments on art, religion and science: 'I hope you're not smiling too ironically—I believe too in every man having a job and getting on with it., It's just that I feel that the artist, the Philosopher & man of religion are not doing there's [*sic*].'
17. MM to NM, 22 December 1946.
18. MM to NM, 17 December 1945.
19. MM to NM, single letter dated 23 and 26 January 1946.
20. MM to NM, 5 April [1946?].
21. MM, *Voice at the Workplace*, p. 16.
22. *Ibid.*, 16–17.
23. MM to NM, St Agnes, Cornwall, 21 March 1947 and 31 July [1947]; quotation from the second letter.
24. MM to NM, 11 November 1947.
25. *Current Biography (1941)*, p. 493.
26. *Ibid.* Eric Williams, long time Prime Minister of Trinidad and Tobago and recipient of a PhD in history from Oxford in the '30s, regarded D. W. Brogan of Oxford and later Cambridge as a less popular but more solid authority on democracy in the US than Laski; see Williams, *Inward Hunger*, p.48.
27. In the '70s MM was similarly labelled by those to the right and left of him, respectively.
28. Christian H. Eismann, 'Laski', *The McGraw-Hill Encyclopedia of World Biography* (New York: McGraw-Hill, 1973) v 6:pp.341–2.
29. Herbert Deane. *The Political Ideas of Harold J. Laski*, New York: Columbia University Press, 1955.)
30. MM to NM, 23 January 1948. In 1988 Manley could not recall Laski's reaction to his paper. Not all West Indians were so enamoured by Laski. In June 1944, the radical Trinidadian intellectual C. L. R. James attacked Laski for the falseness of his politics, which he said was not really Marxism, but 'petty bourgeois radicalism', and for seeking modern values in the totalitarian state in Russia. See James' critique of Laski's *Faith, Reason and Civilization* (1944) in James, *The Future in the Present* (Westport Connecticut: Lawrence Hill, 1977), pp.95–105.
31. MM, statement about LSE, submitted at the request of its director, 21 August 1979, MPF.
32. MM, interview, 6 September 1985. Manley recalls that Laski said relatively little about how democratic socialism might be established in the 'Third World'. As noted above, Laski himself had become increasingly pessimistic about instituting socialism via constitutional means except in such societies as Britain and the US. In the '80s of course, those societies moved in the opposite direction.
33. MM to NM, 8 July 1946.
34. MM to NM, 6 July [1948?]
35. NM to MM, 13 July 1948, NMP 4/60/2B/13(35).
36. MM to NM, 23 January 1948. The literature on the 'Puerto Rican model' is vast and controversial; useful introductions may be found in Gordon K. Lewis, *Puerto Rico: Freedom and Power in the Caribbean* (New York: Monthly Review, 1963), Chs 10–11 and Raymond Carr, *Puerto Rico: A Colonial Experiment* (New York: Vintage, 1984), Ch 8.

    For background on Arthur Lewis, see Tony Johnson, 'Getting Out of the Poverty Cycle: An Interview with Sir Arthur Lewis,' *JWG* 22 March 1977, p. 13).
37. Interview, MM, 9 September 1985. As early as 1948, NM was much impressed

by the Puerto Rican experience. He wrote to his son Douglas that the Puerto Ricans were working on 'a very intelligent and well-thought out [*sic*] plan.' He noted that the Puerto Ricans 'pretty well control their own Civil Service . . . staffed by a very enthusiastic lot of Puerto Ricans who work in complete sympathy and harmony with the basic plans of the Government. It is extraordinary what a good atmosphere this creates by comparison with our own. I feel we have a lot to learn from them both as to method and as to spirit.' NM went on to comment: 'Muñoz Marin and his popular party [The Popular Democratic Party] have enormous power. Too much really to be good for them but they steadily keep in mind the fundamental job of doing things the right way and creating a basic tradition of clean politics, honest tactics and democratic procedures.' NM to DM, 2 July 1948 (copy), NMP, 4/60/2B(36).

38. The cover of the paperback edition of Manley's *Jamaica: Struggle on the Periphery* depicts an aggressive eagle about to descend on Jamaica.
39. MM to NM, single letter dated 23 and 26 January 1946.
40. MM, interview, 6 September 1985.
41. MM to NM, 23 January 1948.
42. 'Michael Norman Manley: Political Biolgraphy', (1975), p.5; MM 'Root of the Matter', *Public Opinion*, 2 May 1953.
43. Dorothy Butler Gilliam, *Paul Robeson: All-American* (Washington, DC: New Republic, 1976), pp 72–3.
44. MM, interview, 17 December 1987. The actual origins of 'Joshua' are discussed in Chapter 11.
45. MM to NM, 17 December 1945. In this letter Manley referred, with the supreme confidence of a 21 year old, to Picasso as 'a cynical caricaturist' who 'doesn't admit it.'
46. MM to NM, single letter, 23 and 26 January 1946.
47. MM to NM, 23 January 1948.
48. MM to NM, 20 November 1945.
49. MM to EM, 11 December 1945. Of the analysis—his 'first attempt at analysing, a group of things entirely on my own'—Manley wrote: 'I know it's cheek to send *you* this at all, but I think you'll understand.'
50. MM to NM, 20 November 1945.
51. MM, interviews, 12 August 1984 and 6 September 1985.
52. MM to NM, 8 July 1946.
53. MM to NM, 10 October 1946.
54. MM to NM, 23 January, 1948.

CHAPTER 8 **'Root of the Matter', 1952–5**

1. MM to NM, 19 June 1951, BAM.
2. MM, *Voice at the Workplace*, pp. 21–22.
3. MM, 'Root of the Matter,' *Public Opinion*, 15 November 1952. The context of this observation was Manley's observation of the trials of a poor old man on the streets of Kingston. Future citations of 'Root of the Matter' in this chapter will be by date only.
4. 7 March 1953.
5. 17 September 1955.
6. 9 April 1955.
7. 11 June 1955.
8. 25 June 1955. Manley depicted Mais's *Hills Were Joyful Together* as 'the most authentic account of West Indian Life to come from the pen of a West Indian'.
9. 27 August 1955.
10. 25 October 1952.

11. 28 June 1952.
12. 6 February 1954.
13. 21 February 1953.
14. 8 March and 29 November 1952.
15. 15 November 1952. Manley noted that the care of children was little better than that of the elderly, 'so the children remain on the streets,' as, indeed, thousands of them did in the '80s.
16. 8 December 1952.
17. 11 June 1955.
18. 7 March 1953. In the '70s the Manley government enacted laws to remove the stigma of 'illegitimacy' from children.
19. 25 June 1955, 27 November 1954, 28 November 1953.
20. 5 February 1955.
21. 3 May 1952.
22. 29 November 1952. For British Governor Sir Hugh Foot, however, Manley had high regard: see his column of 13 September 1952.
23. 21 November 1953.
24. 1 October 1955.
25. 26 February 1955, Margaret, Manley wrote, was 'never so poised as to seem aloof, yet never suffering from the gaucherie of over-enthusiasm'.
26. 29 March 1952. Other columns documented specific instances of British racism: see columns of 3 January and 12 December 1953.
27. 18 October 1955. This column accurately predicts some problems the Manley administration had with 'nationalism' in the '70s. As Manley admitted, he placed great hope in internationalism and Third World solidarity, and worked relatively little at inculcating Jamaican nationalism per se. Meanwhile, the more conservative JLP tried to advance its agenda under the rubric of a more parochial 'nationalism', and used the word extensively in its political propaganda.
28. 12 July 1952. See also the column of 17 July 1952, entirely devoted to nationalism, where Manley argued that Jamaicans must not attempt to manage the country before gaining experience in managing its institutions, that Jamaica's highest priority was development and not national pride, and that Jamaica must become economically self-supporting.
29. 18 October 1955. Here Manley quoted the Irish poet and mystic George Russell: 'A nation exists primarily because of its own imagination of itself.'
30. 16 August 1952. The 'custos' was a colonial official.
31. 30 August 1952.
32. 2 August 1952.
33. In 1952, the quartet set a world record of 3:3.9 in the four times 440 metre event. Wint had won the four hundred metre event in the 1948 Olympics and Rhoden had triumphed in the same event in the 1952 Olympics. Years later, Manley spoke of Jamaica's record in these and other events with great pride.
34. 2 August 1952.
35. 17 May 1952.
36. 6 September 1952.
37. 18 April 1953.
38. 16 April 1955.
39. 21 February 1953. During the '70s, Manley himself would often be accused of the kind of 'personality politics' he deplored earlier in Bustamante.
40. 20 September 1952. The word 'democracy' is used so uncritically so often as to have lost much real meaning. Although widely referred to as a 'democracy', subject ultimately to British authority, Jamaica's political system conformed

more closely to a liberal political system legitimized by the *appearance* of 'democracy', as discussed by Peeler in his *Latin American Democracies*. Peeler's critique applies equally to the United States and Britain, which, he believed were not genuine 'democracies' in the sense that democracy means direct popular participation in and control of decisions re. allocation of national resources. Therefore, the US and Britain were hardly entitled to promote 'democracy' abroad, except as a rather hollow slogan. Manley himself, while appreciating 'democracy' for other features, would become critical of its weakness when mobilizing a Third World country for national development.

41. 7 November 1953.
42. 31 December 1955.
43. 13 September 1952.
44. 19 April 1952. When I read these words to Manley in September 1985, he responded: 'I still don't lie.' For additional comments in this vein see the column of 19 September 1953.
45. 23 July 1955. The context of this was that NM was in England conducting negotiations for the marketing of Jamaican bananas and citrus. In his son's opinion, he was being unfairly attacked by JLP journalists in England.
46. 16 April 1955.
47. 17 January 1953, 28 February 1953, 24 May 1952. When we first met in New York City in 1984, Manley warned me that there were three versions of Jamaican reality: the PNP version, the JLP version, and that version that manifests itself (rarely) when partisan politics are put aside. In the '80s, he made major efforts to diminish the political 'tribalism' that had brought much grief to Jamaica.
48. 12 March 1955.
49. 8 December 1952.
50. 26 September 1953. In his column of the following 10 October, Manley referred to 'those who *nearly* lost their hurricane houses because they belonged to the Minority Party or no party at all' [italics added]. A particular target of Manley's was the JLP's Rose Leon, an MP. According to Manley, Leon initiated the affair by shamefully insulting EM in Parliament. Manley attacked Leon in print repeatedly throughout 1953, transgressing the tolerance he said was necessary to democracy ('Root of the Matter', 28 February, 9 May, 3 and 31 October, and 12 December 1953). However, Jamaica is a small society and in the '70s, having switched parties, Leon became a valued minister in Manley's government. She wrote him on 7 February 1977 that 'I know you are a principled person who always stood for justice': (PNP file 'PNP Executive' 1977).
51. 24 September 1955. The image of Jericho here is striking, in the light of Manley's later adoption of the political sobriquet, Joshua (see Chapter 11 below.)
52. 12 March 1955.
53. 8 March 1952. One of Rose Leon's misrepresentations that Manley attempted to correct was her claim that Richard Hart and Ken Hill had been expelled from the PNP because they would not accept the supposed communist views of the party. On the contrary, Manley said, Hart attacked the PNP 'every night' because of its anticommunism ('Root of the Matter', 31 October 1953). Manley's exposure to the teachings of Harold Laski would have given him an intelligent scepticism of communism, especially in the Stalin era.

    George Padmore, a West Indian and once the foremost black in the Communist International until he decided that Stalin was manipulating blacks in the interest of the USSR alone, subsequently provided an interesting perspective on the events in Jamaica. Accoding to Padmore, the communists had attempted to infiltrate trade unions under the leadership of the West Indian representative of the World Federation of Trade Unions, Ferdinand Smith. Smith was a Jamaican trade union boss who lived in the US for 30 years, until expelled for communist activities by the FBI. Although the communists were unable to capture the Caribbean Labour Congress, established in 1945 to coordinate regional unions and socialist parties, they succeeded in disrupting it. Padmore also claimed that the communists did in fact make an attempt to capture the PNP, leading to the ouster of the 'four H's'. Padmore claimed that after his ouster from the PNP, 'Mr. Hill tried to form an alliance with . . . Sir Alexander Bustamante, who made use of him against his political rival and cousin, Mr. Norman Manley, but dropped Hill as a political liability on the approach of the 1955 general election. For Sir Alexander is as big a "red-baiter" as Senator Joe McCarthy.' (George Padmore, *Pan-Africanism or Communism* [Garden City, New York: Doubleday, 1971], pp 321–2.

According to William Knowles, in the '50s West Indian socialism—as distinct from communism—was mild and intellectual. It was introduced by West Indians returning from English universities and emphasized anticolonialism, self-government and economic planning. Knowles saw it as contributing to the missionary zeal and honesty of politicians. Some identified with 'socialism' simply because 'capitalism' was identified with colonialism which 'has such a bad reputation in a colonial area.' 'Capitalism is synonymous with lethargy and the status quo and is, therefore, unacceptable to many West Indians.' Socialism meant 'an enthusiastic determination to improve social, economic and political conditions.' (William H. Knowles, *Trade Union Development and Industrial Relations in the British West Indies* [Berkeley: University of California Press, 1959], pp 134–5.

54. 29 January 1955.
55. 5 February 1955.
56. 31 December 1955.
57. 12 January 1952.
58. 2 February 1952.
59. The provisions were included in Puerto Rico's Industrial Incentives Act of 1947. The quotation is from 'Root of the Matter', 31 May 1952.
60. 8 December 1952. In fact, high rates of unemployment continued to plague Puerto Rico, even during the boom of the '60s, and remain one of the principal limitations of the model. This is true despite the fact that Puerto Ricans, as United States citizens, enjoy unlimited rights of legal migration to the mainland, which is not true of non-US possessions.
61. 31 January 1953.
62. 8 October 1955. Manley noted that British tax laws gave no tax holiday for investments in Jamaica, although American tax law did. This was one factor pushing Jamaica out of the British economic orbit and into the American sphere.
63. 22 August 1953.
64. 10 October 1953. These comments were made in response to criticism of the union organization effort made in the publication *Spotlight*, of which Hunter, the manager of Stresscrete, was also a manager (see 'Root of the Matter', 17 October 1953).
65. 23 January 1954.
66. 12 April 1952.
67. 17 October 1953.
68. 30 April 1955. For a mocking sarcastic article on 'bourgeois hysteria' over 'flying teacups' in the USSR, see Manley's column of 10 May 1952.
69. 14 June 1952.

70. 5 July 1952.
71. 9 August 1952.
72. 26 April 1952. Manley went on: 'When are the people of the world going to decide that it is themselves as individuals who alone have inalienable rights. Capitalism elevated property above people. Hitler and Mussolini persuaded a generation that the State was greater than the people for whom it existed to serve. Communism seeks now to invest the Party and its disciples with a mystique, an omniscience, that entitles it to blind obedience. Only liberalism which is, by itself, ineffectual[,] and socialism seem any longer to give a damn for human beings as such.'
73. 8 November 1952. Manley's comment on 'hero-worshipping emotionalism' is interesting in the light of the 'messianic' hero-worship said to be characteristic of Jamaican politics and viewed by many as playing no small role in Manley's own subsequent political career.
74. 7 June 1952. In the '70s, as Prime Minister, Manley would volunteer Jamaican manpower and resources for African national liberation.
75. 3 April 1954.

CHAPTER 9 **'Young Boy': Trade Unionist, 1962–72**

1. Gordon Lewis, *Growth of the Modern West Indies*, pp. 180–6.
2. For the origins of the Jamaican labour movement see Knowles, *Trade Union Development*, pp. 70–73.
3. Trade union membership represented only 11.9% of the work force in 1946, 14% in 1960, and 18% in 1974, according to Ralph Gonsalves. 'The Trade Union Movement in Jamaica: Its Growth and Some Resultant Problems', in Carl Stone and Aggrey Brown (eds), *Essays on Power and Change in Jamaica*, ([Kingston?]: Jamaica Publishing House, 1977), p. 93. According to Gonsalves, the NWU had three times as many non-dues-paying members in 1974 as dues-payers.
4. For 'political unionism' see: Knowles, *Trade Union Development* and Caswell L. Johnson's articles: 'Emergence of Political Unionism in Economics of British Colonial Origin: The Cases of Jamaica and Trinidad'. *American Journal of Economics and Sociology* (April 1980), pp. 151–64: 'Political Unionism and Autonomy in Economies of British Colonial Origin: The Cases of Jamaica and Trinidad'. *ibid.* (July 1980), pp. 237–48; and 'Political Unionism and the Collective Objective in Economies of British Colonial Origin: The Cases of Jamaica and Trinidad', *Ibid.* (October 1975), pp. 365–80.

   'Political unionism' tended to break down in the '70s, to be replaced by 'business unionism'. An example of this was that bauxite workers, although loyal to the NWU, voted heavily against Manley's PNP in the 1980 election.
5. MM, *A Voice at the Workplace*. p. 53.
6. Knowles, *Trade Union Development*, p. 74. Knowles reported that the BITU was losing membership to the NWU for the reasons cited and because its leaders' involvement in politics led to the neglect of union business.
7. MM, 'The Angry Egalitarian: Roger Mais and Human Rights,' ([Kingston: M. Manley], 1968) p. 2.
8. Manley, *Voice at the Workplace*. p. 64.
9. *Ibid.*, p. 69.
10. *Ibid.*, p. 29.
11. *Ibid.*, p. 32.
12. Except as noted, the account of Manley's activities in the sugar field follows his account in *Voice at the Workplace*, pp. 81–113. Knowles, *Trade Union Development*, p. 73 says that the NWU received funds from the Cuban sugar workers' unions.

13. Manley, 'Root of the Matter', *Public Opinion*. 12 December 1953.
14. Thelma Manley, interview with author, 24 September 1985.
15. *Ibid.*
16. Terry Smith, 'Union and Political Party Go Hand in Hand', *Caribbean Business News*, September 1970.
17. *DG* 20, 24, and 26 June 1958.
18. MM, *Voice at the Workplace*, p. 105. Two things about this speech stand out. One is the phrase 'the word is love,' which Manley would use in the '70s as a political slogan. The other is his almost mystical attention to 'inspiration' and 'the inner essence of the moment', which recalls his mother's self-characterization as 'a damned mystic' and her artistic search for the essence of things.
19. *Ibid*. pp. 114–152.
20. MM, 'Root of the Matter', *Public Opinion*, 4 July 1953.
21. MM, 'Root of the Matter', *Public Opinion*, 11 July 1953.
22. MM, *Voice at the Workplace*, p. 122.
23. Quotations in this paragraph from Manley, *Voice at the Workplace*, 126–8.
24. Knowles, *Trade Union Development*, p. 73. Knowles adds that the bauxite workers became the backbone of the NWU.
25. MM, 'Root of the Matter', *Public Opinion*, 28 March 1953; see also *Voice at the Workplace*, pp. 130–2.
26. MM, *Voice at the Workplace*, p. 134.
27. The Vickers incident is described in *Voice at the Workplace*, pp. 136–41.
28. *DG*, 21 October 1957.
29. Knowles, *Trade Union Development*, pp. 74 and 98 Knowles was apparently unaware that MM held positions on the PNP's National Executive Council and Central Executive Committee.
30. *DG*, 14 December 1959.
31. NM to MM, 19 January 1961 (copy), NMP 4/60/2B/25(146).
32. NM to MM, 25 January 1961, NMP 4/60/2B/25(147).
33. NM to MM, 6 February 1961, NMP 4/60/2B/25(148).
34. *DG*, 22 October 1957.
35. *DG*, 27 and 29 November 1961.
36. Lewis, *Growth of the Modern West Indies*, p. 179.
37. *JWG*, 6 August 1974, p. 2.
38. Terry Smith, 'Union and Political Party Go Hand in Hand', *Caribbean Business News* (September 1970). Smith also attributed the NWU's rise to a wealth of expertise 'drawn from numerous and powerful international affiliations,' that represented 'veiled subsidies and grants', and the efforts of Thossy Kelly, a 'tough, burly trade unionist' acceptable to all factions.
39. Manley, *A Voice at the Workplace*, p. 69.
40. MM, Budget Speech, 1969, ms., p. 30.
41. MM, untitled transcript of address to PNP, 12 September 1974, pp. 30–1, PNP speeches files.
42. 'The Prime Minister at U.W.I.', 22 November 1976, PNP speeches files, pp. 11, 43.

CHAPTER 10 **The Political Context**

1. Terry Lacey, *Violence and Politics in Jamaica, 1960–70* (Totowa. New Jersey: F. Cass. 1977), pp. 45 and 54. See also Carl Stone, *Democracy and Clientelism in Jamaica* (New Brunswick, New Jersey: Transaction Books, 1980). Kuper (*Changing Jamaica*, p. 113) notes that both the PNP and the JLP have done some gerrymandering, but believes the election results have been a fair reflection of popular support. In 1967, with only 50.7% of the votes, the

JLP won 33 out of 53 seats in Parliament; in 1972 after its own 'energetic gerrymandering' the JLP lost 8% of the vote and its number of seats dropped to 16.

2. Vera Rubin, 'His Vision of Caribbean Man', in: Ken I. Boodhoo (ed.), *Eric Williams: The Man and the Leader* (Lanham, Maryland: University Press of America, 1986), p. 95.
3. Paul Sutton, 'The Scholar: A Personal Appreciation', in Boodhoo (ed.), *Eric Williams*, p. 34, citing Trinidad and Tobago, *Report of the Constution Commission*, (22 January 1974), paragraph 27.
4. Although Jamaica's political system retains its formally 'democratic' character, it is a strongly authoritarian system in which the PM exercises great power, a tendency which has increased rather than diminished over time. A fascinating trend is the increasingly one-sided allocation of seats in the House of Representatives between 1967 and 1983, irrespective of which party was in power.

Seats in House of Representatives, 1967–83

	1967	1972	1976	1980	1983
JLP	33	16	13	51	60
PNP	20	37	47	9	0

5. Ramesh Deosaran, 'A Psychological Portrait of Power', in Boodhoo (ed.), *Eric Williams*, p. 17.
6. Gordon Lewis, review of Cheddi Jagan's *The West on Trial: My Fight for Guyana's Freedom*, in: *Caribbean Studies* (January 1968), pp. 59–61.
7. Wendell Bell, *Jamaican Leaders: Political Attitudes in a New Nation* (Berkeley: University of California Press, 1964).
8. A. W. Singham, *The Hero and the Crowd*, p. 96. Although NM enjoyed playing the parliamentary game, MM has had limited interest in it. MM interview with author, 6 September 1985.
9. Stone, *Electoral Behaviour and Public Opinion in Jamaica* ([Mona, Jamaica]: Institute of Social and Economic Research, 1974). pp. 35–6. The party-line vote has diminished recently.
10. PNP, 'Man of Destiny: Norman Washington Manley,' (Kingston, 1962).
11. Singham, *Hero and the Crowd*, pp. 9–10, 88–9, 152, 199.
12. Kuper, *Changing Jamaica*, p. 136.
13. On the 'twenty-one families', see Stanley Reid, 'An Introductory Approach to the Concentration of Power in the Jamaican Corporate Economy and Notes on Its Origin', in: Carl Stone and Aggrey Brown (eds.). *Essays on Power and Change in Jamaica* ([Kingston]: Jamaica Publishing House, 1977), pp. 15–44.
14. Lewis, *Growth of the Modern West Indies*, p. 189. Lewis notes that middle-class politicians must appear to attack the middle class since it is scarcely a popular class with the poor majority, but that is a small price for 'the real control that the middle-class Jamaican civilization now [late '60s] finally exercises over both parties.
15. For details and analysis see Stone, *Democracy and Clientelism in Jamaica*.
16. Kaufman, *Jamaica Under Manley*, p. 52.
17. Manley denies as 'absolutely untrue' the allegations by Kuper (*Changing Jamaica*, p. 144) that Manley won his parliamentary seat in 1967 only through Bustamante's intervention and that Shearer narrowly held his seat in 1972 because Bustamante persuaded Manley, as a *quid pro quo*, that Shearer should remain in Parliament.
18. Kuper, *Changing Jamaica*, p. 120.
19. Stone, *Race, Class and Political Behaviour in Urban Jamaica* (Mona, Jamaica: Institute of Social and Economic Research, 1973), ch. 3.
20. Stone, review of Adam Kuper's *Changing Jamaica*, in *Caribbean Studies*, July 1976, pp. 246–9.

21. BAM, 'The Ideological and Programmatic Platforms of the People's National Party', pp. 105–6. In an interview with the author (5 September 1985) she said that there were many continuities between the early PNP and the party in the '70s and '80s.
22. *Ibid.*, p. 107.
23. Kaufman, *Jamaica Under Manley*, pp. 48–9.
24. George E. Eaton, *Alexander Bustamante*, p. 134. Eaton (p. 108) maintains that NM's definition of socialism was 'consistent with the Marxian formulation.' But Eaton goes on to ask a key question applicable to MM's 'socialism' of the '70s, as well: given a capitalist economy with a 'capitalist class' of considerable power and the will to resist, could Jamaica be transformed into a 'socialist' society without a highly coercive state apparatus? This remains a dilemma of Jamaican 'democratic socialism.'
25. Sherlock, *Norman Manley*, pp. 101–2.
26. BAM, 'Ideological and Programmantic Platforms of the People's National Party', p. 156.
27. NWM, 'A Message to My Constituents the Voters of Eastern St Andrew', (December 1949), Institute of Jamaica.
28. Kaufman, *Jamaica Under Manley*, p. 51 notes that the wealthy Matalon family, which backed the PNP, had extensive holdings in public utilities.
29. Gordon Lewis, 'The Decline and Fall of Manley's Socialism', *DG*, 22 October 1959. Lewis, intimately familiar with Puerto Rico, noted the influence on Manley of its 'Operation Bootstrap' industrialization-by-invitation programme and predicted that one day the Jamaican intelligentsia would find itself making the same 'agonising reappraisal' of the Puerto Rican model that its Puerto Rican counterpart had already started to do.
30. MM, *Jamaica: Struggle on the Periphery*, p. 21.
31. BAM, 'The Ideological and Programmatic Platforms of the People's National Party', p. 155.
32. *Ibid.*, citing letter of NM to [Harold] Stannard, British Home Council, 21 December 1942.
33. Memorandum, J. Edgar Hoover, Director, FBI, to Adolf A Berle, Jr, Assistant Secretary of State, 16 November 1942: U.S. National Archives Document 844D 00/48 (declassified), reproduced in BAM. *Ibid.*, Appendix IX, pp. 305–10. For further information on the early PNP leaders, see Louis Lindsay, 'The Myth of Independence', (Mona Jamaica: Institute of Social and Economic Research, 1975), pp. 28–32.
34. MM, *Jamaica: Struggle in the Periphery*, p. 121.
35. Kaufman, *Jamaica Under Manley*, p. 50.
36. MM, interview with author, December 1987.
37. BAM, 'Ideological and Programmatic Platforms of the People's National Party', pp. 166–7, 227–8, 246. Local PNP units are known as 'groups'.
38. Lindsay, 'The Myth of Independence', pp. 16–17.
39. BM, 'Ideological and Programmatic Platforms of the People's National Party', p. 113. BM notes also that NM was also held suspect, to some degree, by the colonial authorities and the defenders of the status quo (p. 117). These facts point to the intermediate role occupied by NM both within the PNP and between the colony and the metropolis and help explain some of the contradictions of his policies.
40. MM to NM, 29 October 1944, in possession of BAM.
41. Quoted in David Boxer, 'Edna Manley: Sculptor', pp. 35–6.
42. BAM, 'Ideological and Programmatic Platforms of the People's National Party', p. 247. In her general conclusion Manley concludes that the multi-

class nature of the PNP 'does not gear it for fundamental change' and requires the party leader to be a 'compromiser-balancer' (p. 245).

43. Kaufman, *Jamaica Under Manley*, p. 49.
44. Knowles, *Trade Union Development*, p. 74
45. Eaton, *Alexander Bustamante*, p. 154.
46. Lewis, *Growth of the Modern West Indies*, p. 180. More positively, Lewis notes that some of the redistribution of economic and social power in the 20 years after 1938 was traceable to the tremendous impact of the 'Busta' personality, with its 'raucous and bogus radicalism, its tremendous braggadocio, its personification of the Jamaican folk-hero, the spiderman Anansi, who survives in a hostile world by cunningly exploiting the weaknesses of his enemies.' Busta's appeal to 'the unlettered crowd' was that 'they glorified in him the poses and perquisites they felt they could never obtain for themselves.' 'Busta seemed, much more than his arch rival, to embody the self-image of the Jamaican populace,' although his politics, in Lewis's view, was essentially negative and sterile (pp. 180–1, 184).
47. Cargill, no friend of the Manleys, comments that NM's call for a referendum on the Federation was 'a most honorable and statesmanlike thing . . . bitterly opposed by his son' (Cargill, *Jamaica Farewell*, p. 162).

CHAPTER 11 **Joshua Awakened (1962–7)**

1. C. L. R. James, 'Parties, Politics and Economics in the Caribbean,' in: James, *Spheres of Existence: Selected Writings* (London: Allison and Busby, 1980), pp. 150–6.
2. M. G. Smith, 'The Plural Framework of Jamaican Society', in: Lambros Comitas and David Lowenthal (eds.) *Slaves, Free Men, Citizens: West Indies Perspectives* (Garden City, New York: Anchor, 1973), p. 188.
3. Lewis, *Growth of the Modern West Indies*, p. 273.
4. Lindsay, 'Myth of Independence', p. 4.
5. Cargill, *Jamaica Farewell*, pp. 96–7.
6. MM, *Budget Speech, 1969*, (mimeo), p. 1.
7. Eaton, *Alexander Bustamante*, pp. 196–206, 240–2.
8. Lacey, *Violence and Politics in Jamaica*, pp. 46–7.
9. Lewis, *Growth of the Modern West Indies*, p. 171. Writing 20 years later, CS called the 1964 dissolution of the KSAC 'entirely political' (*DG*, 22 October 1984). In the '80s the JLP regime of ES carried out a campaign to reduce the number of local councillors. This was done in the name of efficiency, but had undeniably negative implications for local democracy (see Ch. 22).
10. Lacey, *Violence and Politics in Jamaica*, p. 80.
11. The JLP's gerrymandering is reported in Eaton, *Alexander Bustamante*, p. 240; Lacey, *Violence and Politics in Jamaica*, p. 47; and Kuper, *Changing Jamaica*, p. 113. See also CS 'How do Electoral Boundaries Affect Results', *JWG*, 7 September 1976, p. 7; Kaufman, *Jamaica Under Manley*, p. 125; and *JWG*, 26 April 1979, p. 9. It is widely believed that both parties engage in the practices described in this paragraph, although each attributes such evils only to the other.
12. C. L. R. James, 'The Making of the Caribbean People' (1966), reprinted in James, *Spheres of Existence*, p. 184.
13. NM to DM, 27 September 1961, NMP 4/60/2B/25(145).
14. Quoted in Boxer, 'Edna Manley: Sculptor', p. 34.
15. Quoted in *Ibid.*, p. 35.
16. Sherlock, *Norman Manley*, pp. 188, 191.
17. *DG*, 26 August 1962. The PNP executive had given Manley the permission to do what he had done, and it was consistent with the party's constitution.

On the other hand, the selection process seems inconsistent with the professed internal democracy of the party.

18. *The Jamaica Times*, 22 September 1962, p. 10.
19. *DG*, 17 September and 4 December 1962.
20. MM, 'Economic Criteria by Which Arbitrators May be Guided', in G. E. Eaton (ed.) *Proceedings of Industrial Relations Conference on Theory & Practice of Arbitration, November 11–12, 1962* (Mimeo: Mona, Jamaica: Institute of Social and Economic Research, 1963), pp. 53–5.
21. John Patmos, 'Off the Cuff', *DG*, 30 August 1963.
22. *SG*. 22 September 1963, p. 8.
23. MM, *Voice at the Workplace*, p. 212. A lengthy chapter (pp. 153–86) of this book is devoted to the strike. All quotations about the strike, unless otherwise identified, are from this source.
24. Years later, 'Thomas Wright' (Morris Cargill), a frequently bitter critic of the Manleys, noted that when NM had led the government in the '50s he had insisted that 'Wright' be able to express his views on JBC, even though they were vastly different from Manley's and despite considerable pressure on the elder Manley not to do so. ('Thomas Wright', *JWG*, 29 June 1976.)
25. John Maxwell, undated typed notes accompanying a xerox copy of his 29 November 1973 broadcast review of Manley's *Politics of Change*, in MPF.
26. MM, *Voice at the Workplace*, p. 185. These words were eerily predictive of how the Manley government of the late '70s was made to do what it did not want to do by 'tremendous economic pressure'.
27. MM, interview with author, 9 September 1985.
28. Lacey, *Violence and Politics in Jamaica*, p. 58, and *SG*, 12 September 1964, p. 6. The *Gleaner's* reporter believed that the PNP's policies were being determined in an essentially democratic way, that they were not being imposed by a small cabal or by one man.
29. *DG*, 7 November 1964.
30. *SG*, 8 November 1964, and *DG*, 12 and 19 November 1964.
31. *DG*, 7 December 1964. MM's 1965 income, according to the income tax assessment in his personal files in the PNP archive, was £1560, on which he paid taxes of £285.8.4.
32. *DG*, 26 February 1966.
33. Undated memorandum in file titled 'Visit to Israel September 3–17, 1965', NMP 4/60/1. In this memo NM noted that 'we are planning radical changes if we get back into power including a ceiling on holdings and development of cooperative and national farms.'
34. MM, *Voice at the Workplace*, p. 212.
35. Frank Hill, 'Interview of Norman Washington Manley', 7 August 1964 Institute of Jamacia, Ms 1632

CHAPTER 12 **Joshua Prepares (1967–72)**

1. Lacey, *Violence and Politics in Jamaica*, pp. 82–5.
2. Barry Chevannes, 'The Rastafari and the Urban Youth', in: Carl Stone and Aggrey Brown (eds.), *Perspectives on Jamaica in the Seventies* (Kingston: Jamaica Publishing House, 1981), pp. 394–5.
3. *Ibid.*, p. 395.
4. NM, 'West Kingston. The Real Danger. Jamaica Beware: A Broadcast by the President of the People's National Party', (PNP Newsletter, July 1966), Institute of Jamaica. Kaufman, *Jamaica Under Manley* (p. 144) comments: 'The constituency represented by Edward Seaga, Member of Parliament and the future Prime Minister, was the scene of some of the first armed groups. As a result, many Jamaicans believe that Mr. Seaga bears a major responsibility

for introducing guns into Jamaican politics.' In an interview with me on 23 December 1977, the respected political scientist and journalist Carl Stone told me that he thought that the violence had started with JLP politician D. C. Tavares, that the PNP's Dudley Thompson's supporters had responded in kind, and that only then had Seaga's supporters become involved.

5. *DG*, 16 August 1966.
6. *SG*, 14 August 1966, p. 2.
7. *SG*, 21 August 1966 and *DG*, 22 August 1966.
8. 'Thomas Wright', 'Candidly Yours', *DG*, 24 August, 1966.
9. *DG*, 27 August 1966.
10. *SG*, 28 August 1966.
11. Lacey, *Violence and Politics in Jamaica*, pp. 49–50.
12. MM, *Jamaica: Struggle in the Periphery*, p. 170.
13. MM, interview with author, 16 September 1985.
14. MM, interview with author, 12 September 1985.
15. 'Michael Manley—1969 Political Man of the Year', *SG*, 28 December 1969. For a detailed account of the 1967 election see Anita M. Waters, *Race, Class and Political Symbols: Rastafari and Reggae in Jamaican Politics*, (New Brunswick, New Jersey: Transaction, 1985), Ch. 3.
16. MM, interview with author, 16 September 1985.
17. Two contrasting accounts of the Alpart conflict exist: Manley's *Voice at the Workplace*, pp. 142–50 and that of his rival, Pearnel Charles, *Detained* (Kingston: Kingston Publishers, 1978), pp. 88–90.
18. Charles' version (*Detained*, p. 89) is that Manley incited the crowd to attack him, that Cooke hit Charles first, and that he fired only in self-defence. Charles adds that he was acquitted of 'shooting with intent' by the Black River Circuit Court, despite Manley's testimony that Charles had hit Cooke first. Charles' account includes the curious detail that during the fighting Manley shouted to Charles: 'You damn fool! I am trying to help you.'
19. MM to NM, on NWU stationery, 11 October 1968. NMP 4/60/2B/39(55).
20. *DG*, 9 December 1968.
21. MM, 'Angry Egalitarian', pp. 10–16.
22. Waters, *Race, Class and Political Symbols*, p. 93.
23. Waters, *Race, Class, and Political Symbols*, pp. 93–5; Munroe, *Politics or Constitutional Decolonization*, p. 200; and Lacey, *Violence and Politics in Jamaica*, p. 52. Rodney was the author of *A History of the Upper Guinea Coast, 1545–1800* (London: Oxford University Press, 1970); *Groundings with My Brothers* (London: Bogle L'Ouverture, 1969) which deals with his Jamaican experiences; *How Europe Undeveloped Africa*, London: Bogle L'Ouverture, 1972; and *A History of the Guyanese Working People, 1881–1905* (Baltimore: Johns Hopkins University Press, 1981). An opponent of the dictatorship of Forbes Burnham in Guyana, Rodney was murdered there in 1980 under circumstances still not fully explained.
24. *DG*, 10 February 1969.
25. *DG*, 16 February 1969. The political reporter stated that no one had a better record as a trade unionist in Jamaica, although Shearer and Frank Hill had comparable records.
26. *Abeng*, 8 February 1969, quoted in Waters, *Race, Class, and Political Symbols*, pp. 96–7.
27. Lacey, *Violence and Politics in Jamaica*, p. 53.
28. MM, *Budget Speech, 1969*.
29. MM, interview with author, 6 September 1985.
30. *DG*, 27 October 1969.
31. Sherlock, *Norman Manley*, pp. 199–200.

32. *DG*, 27 October 1969.
33. *SG*, 28 December 1969. In my interviews with him in 1985, Manley bridled at the idea that he modelled his political career on that of his father, but by 1987 he seemed more comfortable with it.
34. Letters of 20 January, 5 and 16 March 1970, all in MPF.
35. MM, letter of 25 November 1969, MPF. The letter was written to a student writing a master's thesis on Jamaican foreign policy.
36. MM, 'Overcoming insularity in Jamaica', *Foreign Affairs* (October 1970), pp. 100–110.
37. Letters from Ethiopian Orthodox Church to MM, 10 December 1969 and 20 January 1970; Manley to Ethiopian Orthodox Church, 10 March 1970, in MPF.
38. DM to MM, 9 February and 9 May 1970; MM to DM 18 May 1970, in PNP archive, file 'Douglas Personal'.
39. *DG*, 1 March 1971.
40. MM to DM, 28 April 1970, PNP archive, file 'Douglas Personal'.
41. Olive Senior, *The Message is Change: A Perspective on the 1972 General Elections* (Kingston: Kingston Publishers, 1972), pp. 65–6. Some observers see in Joshua and his rod a cynical manipulation of the people's beliefs. Others disagree. See Kuper, *Changing Jamaica*, p. 106, and Stone, *Electoral Behaviour*, p. 35. For a lengthy treatment of the election see Waters, *Race, Class, and Political Symbols*, Ch. 5.
42. [Tony Verity], 'Interview [with] Michael Manley', *West Indies Chronicle* (May 1972), p. 188.
43. Stone, *Electoral Behaviour*, p. 19, and Waters, *Race, Class, and Political Symbols*, p. 169.
44. Stone, *Electoral Behaviour*, p. 41.
45. Stone, *Electoral Behaviour*, p. 19 and *Class, Race, and Political Behaviour*, pp. 169, 171.
46. Waters, *Race, Class, and Political Symbols*, p. 138, citing Carl Stone's analysis in the *DG*, 8 June 1980.
47. *JWG* 8 and 15 March 1972.
48. The Jamaican Supreme Court overturned the election in February 1974 because of ballot tampering by DM's supporters, without his knowledge or approval. DM's opponent, Arthur Henry Winnington Williams, brought no charges against DM personally, and the court made no finding against him (*JWG*, 27 February 1974), pp. 27 and 29). 'Thomas Wright', generally critical of the Manleys, commented:

    > Technically Dr. Douglas Manley is probably liable to certain penalties, now that he has lost his seat in the House. But no right-thinking person would like to see him penalised, under the circumstances in which he lost his seat. I have no doubt that by the time these words are printed the [PNP] government will have introduced a special Bill to exonerate and indemnify Dr. Manley, retrospectively from any liability—one of the very few instances where retrospective legislation is, in my view, morally permissible. The government will also no doubt pass legislation making all acts legal during the time he served as a Minister. I hope nobody will object to any such legislation. It is only right and proper that it should be passed—and passed with the full support of the [JLP] Opposition. (*JWG*, 6 March 1974, p. 34).

    Commentators noted that the disputed election was held under the JLP government's general, obviously faulty, control, and that it revealed serious weaknesses in Jamaica's electoral procedures. For additional details and commentary, see *JWG* 5 April 1972, p. 15; 19 April 1972, p. 9; 26 April 1972, p.9; 7 November 1973, p. 8; 6 March 1974, p. 3; 20 March 1974, p. 24 (*DG* editorial of 13 March); 24 September 1974, p. 4; 7 January 1975, p. 22; and *DG* editorial,

6 November 1975. Following the Supreme Court's action MM appointed his brother Senator. DM won the 1976 election in South Manchester, and represented the district until 1980.
49. *DG*, 4 March 1972.
50. Lacey, *Violence and Politics in Jamaica*, pp. 55–6.
51. John Maxwell, interview with author, 7 September 1985.

CHAPTER 13 **Populism, 1972–3**

1. Stone, 'The Long Road From 1938', *JWG*, 24 May 1978, pp. 15, 29.
2. However, severe nutrition problems remained as of 1972; see MM, *Budget Debate Speech, May 1st 1972* (Kingston; Government Printer, [1972]), p. 61.
3. A *Gleaner* editorial on 29 March 1972 acknowledged that Jamaica's income distribution was 'very uneven' and that 'there must be a more equitable distribution of the nation's wealth.' The editorial continued that 'we do not favour the entrepreneur who earns excessive profits at the expense of the consumer and worker,' although the entrepreneur should be able to earn reasonable profits. In November of 1973 it was reported that in 1972 one-half of income earners gained less than J$500 per year, and that income distribution was little changed from 1958, when it had been reported as among the most unequal in the world (*JWG*, 14 November 1973, p. 33).
4. Kaufman, *Jamaica Under Manley*, p. 209.
5. Answers on form, 'East Central Kingston Skill Survey', in file 'E. C. Kgn Socialist', PNP Archive.
6. *JWG*, 23 February and 5 March 1972.
7. [Tony Verity], 'Interview [with] Mr. Michael Manley', *West Indies Chronicle* (May 1972), p. 188. This conflicts with Manley's statement that in 1972 he met with the PNP's National Executive Council, reminded it of the Party's 'socialist history', and suggested the need to clarify the party's ideology, a move which caused 'immediate controversy' in the party (Manley, *Jamaica: Struggle in the Periphery*, p. 120).
8. MM, *Jamaica: Struggle in the Periphery*, p. 21.
9. *Ibid.*, Ch. 8.
10. For a brief discussion of populism, see Emilio Willems, *Latin American Culture: An Anthropological Synthesis* (New York: Harper & Row, 1975), pp. 290–3. Kaufman (*Jamaica Under Manley* pp. 78–81) prefers to call the early Manley government 'social democratic'. The Jamaican experience was dynamic, shifting, and—while it drew on imported models—to some extent *sui generis* and resistant to rigid ideological labelling.
11. *JWG*, 1 March 1972.
12. Abeng Group, 'The Elections. Where from Here?' (Kingston, [1972], p. [1].
13. *JWG*, 8 March 1972, p. 9.
14. *JWG*, 15 March 1972.
15. *DG*, 1 March 1972.
16. *JWG*, 15 March 1972, p. 37. The resulting law, however, was indifferently followed by parliamentarians of both parties.
17. *JWG*, 10 May 1972, p. 10.
18. Abeng Group, 'Joshua. Aaron. Moses. Eli: Two Months of Selling out Jamaica', (Kingston, [1972], p. [2].)
19. Letter to MM, 6 March 1972, in MPF.
20. In October 1972, Manley exchanged cordial letters with members of the US government and even a state governor; letters of 5 October 1972, MPF.
21. Letter to MM, 28 July [1972], MPF.
22. Letter to MM, [August 1972], MPF: original italics.

23. Letter to Manley, 11 September 1972. This lengthy letter is indicative of the enthusiasm Manley inspired in Jamaicans and non-Jamaicans alike.
24. 'Jamaican Joshua', *Time*, 2 August 1972; Manley letter, 25 August 1972, MPF.
25. For details and evaluation see Kaufman, *Jamaica Under Manley*, Ch. 4 and Stephens, *Democratic Socialism in Jamaica*, Ch. 3. On pp. 70–71 of the latter source is a handy year-by-year listing of major policies and initiatives of the PNP government between 1972 and 1979. (See Appendix A).
26. Manley, *Budget Debate Speech, May 1st 1972*: quotes from pp. 67, 16, 70. For a longer, retrospective account of the setting of the government's domestic and foreign agendas, see Manley, *Jamaica: Struggle in the Periphery*, Chs. 4 and 5.
27. Martin Ira Glassner, 'The Foreign Relations of Jamaica and Trinidad and Tobago, 1960–5'. *Caribbean Studies*, (October 1970), 126–8.
28. Kaufman, *Jamaica Under Manley*, p. 86.
29. Manley, letter of 18 October 1972. MPF.
30. Lord Caradon [Hugh Foot] to Manley, 18 March [1972], MPF.
31. MM, [Untitled transcript of taped address to PNP meeting], 12 September 1974, PNP archive, p. 27. Manley went on to say 'the trouble is the PNP cannot change Jamaica if it cannot also change the world.' Entered in handwriting after the preceding sentence were the words 'think that one through—think it through!'
32. MM, *Jamaica: Struggle in the Periphery*, pp. 59, 64–5.
33. *Ibid.*, p. 79.
34. Seymour M. Hersh, *The Price of Power: Kissinger in the Nixon White House* (New York: Summit Books, 1983), p. 136.

    A review of several of Kissinger's books revealed no references to Manley and only one brief (and misleading) reference to Jamaica. In an April 1980 address, 'The Future of American Foreign Policy', he included in a survey of what he considered a 'rockslide' of unfavourable events after the victory of the Sandinistas in Nicaragua in 1979 the statement that 'Castroite elements are already implanted in Jamaica' (Kissinger, *For the Record: Selected Statements, 1977–'80* [Boston: Little, Brown, 1981], p. 288.)
35. MM, *Jamaica Under Manley* pp. 65–8.
36. *JWG*, 4 October 1972, p. 1.
37. MM, 'The Prime Minister of Jamaica . . . 1972 Address to the United Nations', (Kingston: Jamaica Information Service, 1972.)
38. Carmelo Mesa-Lago, *Cuba in the 1970s: Pragmatism and Institutionalization* (Revised edition, Albuquerque: University of New Mexico Press, 1978), pp. 120–5.
39. Editorial, *DG*, 30 March 1973, in *JWG*, 4 April 1978, p. 4.
40. Mesa-Lago, *Cuba in the 1970s*, throughout.
41. Letter of Manley, 31 July 1972). MPF. Ulric Simmonds was the *Gleaner's* 'political reporter'.
42. Letter of 15 August 1972, MPF.
43. Letter to Manley, 15 September 1972, MPF. Manley's admiration for Tanzania's leader, Julius Nyerere, one of the most estimable of African leaders, is well-known. For a sampling of Nyerere's views, see his *Man and Development* (London: Oxford University Press, 1974).
44. Letter of Manley, 18 October 1972. Diverticulitis, in this case an infection of abnormal sacs in the intestines, plagued Manley for many years.
45. Unsigned, undated copy of a declaration by Manley, MPF. This document states that Manley first saw the pamphlets on 1 December 1972.
46. *JWG*, 6 December 1972.
47. MM, letter of 19 December 1972, MPF.

48. MM to Rachel Drummond, 19 December 1972, MPF.
49. Of the entire 1972–80 period Manley later wrote: 'In retrospect it does not seem that our programme was all that radical. Many of the schemes we tried to introduce already exist in most of Europe. But they were enough to play a part in a process that brought Jamaica to the verge of a civil war..to be fought between tribalised poor people in the name of the two parties.' (*Jamaica: Struggle in the Periphery*, p. 95).
50. Rex Nettleford, News analysis, 29 April 1973, file 'General Correspondence', PNP Archive. Nettleford is the editor of *Norman Washington Manley and the New Jamaica*, and author of such books as *Caribbean Cultural Identity* (Los Angeles: University of California at Los Angeles, 1979) and *Mirror, Mirror: Identity, Race and Protest in Jamaica* (New York: Morrow, 1972). He has also served as long-time director of Jamaica's National Dance Theatre Company and head of the Trade Union Education Institute at the University of the West Indies.
51. *JWG*, 15 August 1973, p. 12.
52. MM, *Jamaica: Struggle in the Periphery*, p. 44.
53. Adam Kuper, *Changing Jamaica* (London: Routledge, 1976), pp. 131–2. Kuper believed that ambiguity was characteristic of PNP reform efforts and noted: 'No outsider can hope to establish the crucial reason for this characteristic ambiguity. It may have arisen through disagreement in the cabinet, imperfectly resolved, or the obfuscation of civil servants, or sheer incompetence.'
   Crucial factors in this ambiguity would be the internal ideological split in the PNP, based in class relations and class conflict more generally characteristic of Jamaica, the sheer magnitude of what the PNP was trying to do (and the failure to prioritize), and the inexperience of the government.
54. MM, *Jamaica: Struggle in the Periphery*, p. 88. Manley had proposed in a PNP meeting to abolish private schools altogether and received a resounding 'no' from the party leadership, much of which, like Manley himself, was the product of such schools, with children enrolled in them.
55. MM, *Jamaica: Struggle in the Periphery*, p. 92.
56. *Ibid.*, p. 83.
57. *JWG*, 10 October 1973.
58. George Crile, 'Our Man in Jamaica', *Harper's* (October 1974), pp. 87–96; quote from p. 87. Crile's is the most complete account of the de Roulet Affair, based largely on interviews with the former ambassador. Unless otherwise noted, my account follows Crile. Curiously, Manley omits the de Roulet Affair from his *Jamaica: Stuggle in the Periphery*.
59. Hans J. Massaquoi, 'Jamaica, A New Beginning Under New Leadership,' *Ebony*, (October 1973), p. 38.
60. Quoted in Crile, 'Our Man in Jamaica', p. 94.
61. *Ibid.*
62. *Ibid.* Crile reported that de Roulet explained to the Senate subcommittee that 'Manley and his advisers were fearful that the CIA might intervene in the Jamaican elections as it had allegedly done several years before in Guyana.' Sometime later, without giving details, Pearnel Charles of the JLP suggested that the CIA did intervene in the 1972 election, without saying that if it had, it would have probably intervened on the side of the JLP, given the ambassador's leanings (*JWG*, 1 August 1973, p. 3).
63. *JWG* 27 July 1976, pp. 6, 29; see also the column by Malcolm Sharp, p. 28. Kaufman (*Jamaica Under Manley*, p. 194, n. 7) notes that the *Miami Herald* of 18 July 1976 attributed the fact of the PNP receiving $20,000 from ALCOA to William Isaacs, PNP Finance Chairman.
64. Editorial, *JWG*, 1 August 1973.

65. Crile, 'Our Man in Jamaica', p. 89. See also Percy Miller, 'De Roulet, Visa-seekers, and U.S. Foreign Aid Seekers'. *JWG* 8 August 1973, p. 9, where it is noted that de Roulet 'disowned any responsibility for the discomforts [visa applicants] suffered while standing up in the queues'. Miller calls this a revealing disclosure of de Roulet's attitudes toward United States-Jamaican relations.
66. *JWG*, 1 August 1973. p. 3.
67. Crile, pp. 87, 95. According to a confidential source in a position to know and interviewed by me, Manley sometimes drank heavily before 1972, but carried his liquor well so that only those who knew him could tell the difference. After election to Prime Minister he generally limited himself to moderate amounts of white wine.
68. Letter to Manley, 4 October 1973, file 'Miscellaneous 1975'. PNP Archive.
69. MM, 'Address . . . at the Inaugural Session of the Fourteenth Meeting of the Board of Governors of the Inter-American Development Bank', 7 May 1973, file 'Speeches, 1973,' PNP Archive.
70. MM, 'Speech . . . at Luncheon with Exporters and Trade Associations in Caracas, Venezuela', [March 1973], file 'Speeches, 1973.' PNP Archive. See also 'Prime Minister's Statement: Official Visit . . . to Venezuela, March 28–31, 1973, 31 March 1973, PNP Archive, and Rafael Caldera and MM, *Venezuela and Jamaica: Speeches* (Caracas: Central Information Office, 1973).
71. MM, *Jamaica: Struggle in the Periphery*, pp. 108–9.
72. MM, interview with author, 16 September 1985.
73. Editorial, *DG*, 11 September 1973, and news accounts, *JWG*, 19 September 1973.
74. MM, 'The Prime Minister Addresses Members of the University Community, University of Dar Es Salaam, Tanzania,' 12 September 1973, file 'Speeches, 1973', PNP Archive.
75. MM, 'Statement . . . at a Press Conference . . . Held on . . . Return from the Non-Aligned Summit Conference', 14 September 1973, file 'Speeches, 1973.' PNP Archive.
76. MM to Kenneth Kaunda, 26 September 1973, MPF (1978 [misfiled]). Kaunda's reply of 20 December 1973 stated that he was not surprised at the opposition: 'I can imagine that people who have no idea of what the situation is like here cannot react in any other way.'
77. On the US role in the overthrow of Allende, see Hersh, *The Price of Power*, Chs. 21 and 22.
78. *JWG*, 3 October 1973, p. 14.
79. Fidel Castro Ruz to MM, 2 October 1973, typed official translation of handwritten letter, MPF (1978 [misfiled]). Castro went on to express his country's desire to strengthen ties with Jamaica, and recognized the need for discretion. He also reminded Manley that it could count on Cuban support 'loyally and without any [selfish] interest at all, for your courageous and historic task of consolidating your country's independence, and leading the political process of your noble and great people through ways of progress and social justice.'
80. *DG*, 28 and 29 December 1973.
81. *JWG*, 16 January 1974, p. 31.

CHAPTER 14 **Return to 'Democratic Socialism' (1974)**

1. Kaufman, *Jamaica Under Manley*, p. 1.
2. Rachel Drummond to MM, undated [November, 1974?], MPF.
3. *JWG*, 5 June 1974, p. 1.
4. Editorial, *DG* 19 April 1974. The main criticisms this editorial made of the first two years of Manley's government were excessive 'rhetoric' emanating

from the more radical elements of the PNP and problems of implementing the numerous new programmes.

5. *JWG*, 6 February 1974.
6. MM, *Politics of Change*, p. 248
7. BBC 2 'News at 10' transcript (copy), 27 November 1974, in PNP file 'General Correspondence. 1973–1974.'
8. Rex Nettleford, '[JBC] News Analysis', 6 May 1974. PNP file 'General Correspondence, 1973–1974,
9. Manley, *Politics of Change*. p. 252.
10. Editorials. *DG*. 20 March and 5 April 1974.
11. MM, *Politics of Change*. p. 252.
12. *DG*, 8 March 1974. For evaluations see Stephens and Stephens, *Democratic Socialism in Jamaica*. p. 72 (who regard the programme as 'fairly successful') and Kaufman, *Jamaica Under Manley*. p. 97. (who regards the programme as 'very successful'. Manley believed that one of the most successful aspects of the literacy programme was the grass-roots women's theatre group, Sistren (interview with author, September 1985).
13. MM, 'Statement to the House of Representatives . . . November 20, 1974', PNP speeches files.
14. Editorial, *DG*. 30 November 1974.
15. Stephen, *Democratic Socialism in Jamaica*, p. 102.
16. MM, letter of 25 January 1974, MPF.
17. *JWG*. 30 January 1974. p. 1.
18. *JWG*. 13 August 1974. p. 18.
19. 'JMA President Calls for Ministry of Production and Economic Dynamism,' undated press release, stamped 17 June 1974 by the Office of the Prime Minister, PNP file 'General Correspondence, 1973–1974'.
20. Editorial, *DG*, 19 August 1974.
21. Editorial, *DG*, 14 November 1974.
22. Stephens, *Democratic Socialism in Jamaica*, pp. 94.99–100.
23. *Ibid.*, pp. 89–90.
24. MM, *Jamaica: Struggle in the Periphery*, p. 45.
25. Letter to Manley, 17 August 1972, in MPF.
26. MM, *Politics of Change*, pp. 113–14.
27. MM, *Jamaica: Struggle in the Periphery*, p. 97.
28. *Ibid.*, p. 101.
29. One of the Jamaican Government's leading economic advisers, Norman Girvan, a UWI economist, had written a study of the copper industry in Chile: *Copper in Chile: A Study in Conflict Between Corporate and National Economy* (Mona Jamaica: Institute of Social and Economic Research, 1972).
30. Rex Nettleford, 'News Analysis', 16 May 1974, PNP file 'General Correspondence, 1973–4.'
31. *DG*, 24 May 1974.
32. Hector Wynter, 'Manley Seen as Modern Robin Hood', *JWG*, 12 June 1974, p. 9. It is possible that Manley's success, which boded ill for JLP prospects, goaded Wynter into becoming an extreme opponent of the government.
33. *Vision Letter*, quoted in Associate Press story of 5 June 1974, reprinted in *JWG*. 12 June 1974, p. 30.
34. Carl Rowan, 'Jamaica Displays Guts and Decency', reprinted in *JWG*, 18 June 1974, p. 2.
35. *DG*, 28 December 1974.
36. See Kaufman, *Jamaica Under Manley*, p. 83., where government revenues from bauxite/alumina from 1954 to 1980 are given. Manley (*Jamaica: Struggle in the*

Periphery. p. 103) reports the increase in revenues as US$24.5 million (for 12.5 million tons) in 1973, rising to US$210 million (for 12 million tons) in 1980.

37. MM, *Jamaica: Struggle in the Periphery*, p. 103.
38. Stephens, *Democratic Socialism in Jamaica*, p. 127.
39. *Ibid* ., Table A. 20. p. 397. 'U.S. loans and grants to Jamaica, 1956–82'.
40. Kaufman, *Jamaica Under Manley*, p. 85.
41. MM, *The Politics of Change*, p. 262.
42. *JWG*, 5 December 1973.
43. Jean Fairweather, 'Michael Manley, Who is Leading Jamaica into Democratic Socialism, Is Many Things to Many People', *Commonwealth* (April-May 1975). p. 4.
44. MM, *Jamaica: Struggle in the Periphery*, p. 242. n. 4.
45. *JWG*, 14 September 1974, p. 2. Dunkley, however, could not resist observing that unspecified politicians of both major parties were enriching themselves at the people's expense.
46. Father Ingnatius Saldanha, 'Is Socialism Christianity in Action?', *JWG*, 25 March 1975. pp. 11, 15.
47. Alvin Nairne, 'The Politics of Treachery,' *Socialism* [Mona, Jamaica], (August 1974), p. 3, and review, *JWG*, 3 September 1974, p. 14.
48. *JWG*, 27 August 1974, pp. 27, 30, 1 October 1974, pp. 15, 29, and 8 October 1974, p. 13, respectively. 'Underdevelopment' is a complex matter. The instance of Jamaica's leader publishing a serious book while in office, the *Gleaner* serializing it and offering cash prizes for reviews, and the overall character of the reviews bespeaks a higher level of political development, in this particular, than that of the US.
49. Stephens, *Democratic Socialism in Jamaica*, pp. 64, 105.
50. PNP, *Principles and Objectives* (Kingston: PNP, 1979).
51. MM, *Jamaica: Struggle in the Periphery*, p. 122. Manley's version of the PNP's ideological debate is laid out here in detail on pp. 119–29.
52. Stephens, *Democratic Socialism in Jamaica*, p. 3. The Stephenses regard the PNP's lack of ideological clarity in the '70s as one of its key problems, if not *the* key problem, the source of many of its failures.
53. *Ibid*., p. 109.
54. MM, Speech to PNP Annual Conference, 12 September 1974, excerpts reprinted in MM, *The Search for Solutions* pp. 161, 162.
55. *Ibid*. pp. 160, 161.
56. *JWG*, 19 November 1974, p. 29.
57. 'Democratic Socialism: The Jamaican Model' (November 1974), excerpts in MM, *Search for Solutions*. pp. 155–60.
58. MM, *Jamaica: Struggle in the Periphery*. p. 125.

CHAPTER 15 **Polarization and Turning Points (1975)**

1. Kaufman, *Jamaica Under Manley*, p. 124.
2. Lindsay, 'The Myth of Independence', pp. 52–3.
3. Max Lerner, 'A Talk in Jamaica', and 'Third World Case History', *New York Post*, 19 and 21 February 1975. These sympathetic articles were reprinted in the *Gleaner*; see *JWG* of 4 and 11 March, 1975.
4. *DG*, editorial, 22 April 1975.
5. *JWG*, 13 May 1975.
6. Quoted in Hans J. Massaquoi, 'Kingston Summit: Third World Leaders Demand a "New Deal" or Else,' *Ebony* (July 1975), p. 33.
7. *DG*, editorial, 7 July 1975.
8. Fidel Castro, 'Translation of Speech by Prime Minister Fidel Castro at the Decoration Ceremony, Confirming the National Order of José Martí on Prime

Minister Michael Manley', (mimeo, [9 July 1975]), pp. 2 and 3, PNP file 'Heads of Government and Cuba'.

9. MM, 'Speech . . . on the Occasion of the Award of the National Order of José Martí, Havana, Cuba, 9th July 1975.' (PNP file 'Heads of Government and Cuba'.
10. 'Castro award speech'; plus 'Joint Communiqué of Cuba and Jamaica'. (Kingston: Agency for Public Information, 1975).
11. MM, *Jamaica: Struggle in the Periphery*, p. 109.
12. Kaufman, *Jamaica Under Manley*, pp. 122 and 127, n. 26. Manley does not mention this episode in his *Jamaica: Struggle in the Periphery*. In an interview with the author 9 September 1985, Manley said the speech had been carefully planned to make the point about the infeasibility of the get-rich-quick mentality, and that the 'five flights' was almost an afterthought which, attracting all the attention, totally overwhelmed the main point of the speech.
13. However, Stephens, *Democratic Socialism in Cuba*, p. 121, argue that this was not so.
14. *JWG*, 6 and 20 May 1975.
15. *JWG*, 15 July 1975, p. 1.
16. *JWG*, 29 July 1975, p. 27, and 12 August 1975, p. 1.
17. For example, 'A Lesson from Cuba's Road to Communism', by 'A Contributor', *JWG*, 4 November 1975, pp. 10, 29.
18. All quotations in this paragraph from Manley, *Jamaica: Struggle in the Periphery*, pp. 112–17.
19. Letter of Endre Marton, *JWG*, 20 February 1974, p. 21.
20. *JWG*, 30 July 1974, p. 3.
21. *JWG*, 18 March 1975, p. 27.
22. Stephens, *Democratic Socialism in Jamaica*, pp. 115–16: '. . . Things began to change at the *Gleaner* in 1975 when Hector Wynter, who had been chairman of the JLP in the late '60s, was made Executive Editor of the paper, and by mid–1975 the *Gleaner* had been transformed from a simply conservative to a blatantly partisan paper.' (See the extended analysis on pp. 115–18, 161–3, 229–32.) See also Kaufman, *Jamaica Under Manley*, 119–20: 'A fairly consistent anti-Manley line in Jamaica's leading paper from the mid–'70s blossomed into an intense and often hysterical diatribe at the end of the decade. Part of the anti-PNP focus was in the editorial content, which in some cases would be considered seditious'.

    See also the articles by Aggrey Brown, Carl Stone, and Sheila Nicholson in Stone and Brown (eds.) *Perspectives on Jamaica in the Seventies*, chs. 11–14, and by Brown in Stone and Brown (eds.), *Essays on Power and Change in Jamaica*, ch. 15. My own reading of the *Gleaner* tends to confirm these observations, with the important exception that for all its bias, one could still find truly informative columns among the propaganda, notably those of Carl Stone.
23. *JWG*, 18 February 1975.
24. *JWG*, 18 November 1975, p. 3.
25. Waters, *Race, Class, and Political Symbols*, pp. 200–201; Stone Poll, *JWG*, 8 May 1978, p. 11. In that poll RJR radio was regarded as the most honest and fair, the *Gleaner* second, and JBC a close third.
26. Stephens, *Democratic Socialism in Jamaica*, p. 307. Corina Meeks, the Chairperson of API during Manley's second administration, agreed that the agency was ineffective in getting the government's message across (interview with author, 21 December 1987).
27. *JWG*, 22 June 1976, pp. 6 and 26. Whylie, in turn, denied the JBC board's version (*Ibid*. 29 June 1976, p. 14).

28. *DG* editorial, 5 November 1975.
29. *JWG*, 19 August 1975, p. 7.
30. *JWG*, 23 December 1975, p. 6.
31. *DG*, 23 October 1975, p. 16.
32. See Chapters 9, 11, and 12.
33. Text of speech reprinted in *JWG*, 25 November 1975, p. 11.
34. *Voice at the Workplace*, p. 193. Manley returned to this theme again later in the book (p. 210); 'It is as if successive political parties in power have tended to assume that the union movement can look after the workers so that the government can be free to concentrate on the condition of the farmer, economic development, education, and so on. As a result there have been incredible examples of neglect such as the generation which passed before the Masters and Servants Law was repealed and the similar failures to deal with the Essential Services Law and a number of other matters of urgent concern to workers.'
35. *Ibid.*, pp. 198–9.
36. *Ibid.*, pp. 206–7.
37. *Ibid.*, pp. 207–16.
38. Undated letter [March, 1975?] to Manley, MPF.
39. *JWG*, 10 June 1975, p. 7.
40. *DG*, 3 October 1975, p. 20.
41. *SG*, 5 October 1975, p. 8.
42. *JWG*, 11 November 1975, p. 4; 9 December 1975, p. 13; and 16 December 1975, p. 22.
43. MM, 'From the Shackles of Domination and Oppression', *Ecumenical Review*, January 1976, pp. 49–65. This address was reproduced in successive issues of the *Sunday Gleaner* on 7, 14, 11 and 28 December 1975, and later issued by the government as a pamphlet (Kingston: Agency for Public Information, 1978). In the speech Manley used such phrases as 'we Christians' and 'our Lord' to identify himself with the Christian community.
44. See, for example, the cartoon in the *JWG* of 3 October 1973 in which a woman explains to a judge: 'I'm rape me in a de library yu honor and me couldn't scream.' Virtually every issue of the *JWG* during the '70s contained such cartoons.
45. *JWG*, 16 December, 1975.
46. [Tony Verity], 'Interview [with] Michael Manley', *West Indies Chronicle* (May 1972), p. 187.
47. *JWG*, 10 September 1974, p. 14.
48. [Untitled transcript of taped address to PNP], 12 September 1974, PNP speeches files. Natasha Manley was born in 1974.
49. MM, 'Speech . . . at Annual Dinner of the Soroptimist Club of Jamaica', 29 November [1973], speeches files, PNP archives.
50. *JWG*, 7 January 1975.
51. *JWG*, 17 December 1974, pp. 20, 15.
52. *JWG*, 14 January 1975, p. 10.
53. *Ibid.*, p. 23.
54. *JWG*, 1 July 1975, pp. 3, 24.
55. *JWG*, 11 February 1975, p. 4, 22 April 1975, p. 29: and 9 September 1975, p. 2.
56. Stone, 'A Second Reign of Terror', *DG*, 17 March 1986. Stone went on to describe the situation in 1986: 'The PNP leadership is now being as naive as they were before 1976 where their people were blown off the streets by JLP guns. And if they continue in that manner they are going to be clobbered by fascist interests intent on keeping them out of power.'

57. *JWG*, 28 January 1975, p. 23.
58. *Ibid.*, pp. 23, 30. Seaga provided details of 12 alleged attacks on JLP supporters between mid-November 1974 and mid-January 1975.
59. *JWG*, 8 April 1975.
60. *Ibid.*, 20 May 1975, p. 3.
61. Editorial, *DG*, 20 August 1975. Rema, like Trivoli Gardens, was a JLP stronghold.
62. *JWG*, 1 April 1975, pp. 15, 19, 29.
63. *JWG*, 22 April 1975, p. 23.
64. *JWG*, 2 December 1975, p. 17, and 30 December 1975, p. 7.

CHAPTER 16 **Destabilization, Violence, and Reelection (1976)**

1. John Barton Hopkin, 'Jamaican Children's Songs', *Ethnomusicology* (January 1984).
2. MM, *Jamaica: Struggle in the Periphery*, Chs. 9, 15, 16, Appendix, and pp. 3, 220, and 223.
3. *Ibid.*, pp. 131, 133, 134, 137.
4. *Ibid.*, p. 140. The police Special Branch is responsible for internal security.
5. *Caribbean Monthly Bulletin*, May-June-July 1976, pp. 14–15.
6. *DG*, 15 June 1976.
7. *JWG*, 22 June 1976.
8. *Toronto Star*, 12 July 1976. In the case of Kissinger, outright subterfuge was quite possible, as Seymour Hersh establishes in his book *The Price of Power*.
9. *JWG*, 10 August 1976, p. 6.
10. *JWG*, 21 September 1976, p. 9.
11. *JWG*, 30 November 1976, p. 17. Sharp compared the 'Big Lie' about destabilization to the JLP's Big Lie of 1962, when it claimed that Russian ships were about to descend on the island.
12. Ernest Volkman and John Cummings, 'Murder as usual', *Penthouse* (December, 1977), pp. 112–14, 182–90. The authors reportedly based their article on information from several senior American intelligence sources. They reported that Edward Seaga 'was spotted slipping off to a motel to meet with known CIA agents in what was believed to be strategy sessions,' which Seaga vehemently denied. A four-page flier summarizing the article, published by 'Garvey Star Press', was circulated in Jamaica.
13. Press association of Jamaica, 'Press Freedom for Whom?' (Kingston: Press Association of Jamaica, 1978), [3]. Sealy had held the following posts at the *Gleaner*: coordinating editor (1944–51), editor (1951–73), editor-in-chief (1973–77). He was a winner of the Norman Washington Manley Foundation's Award for Excellence, and had been honoured by the Jamaican government and the University of the West Indies.
14. *Caribbean Contact*, June 1985, p. 4.
15. Kaufman, *Jamaica Under Manley*, pp. 117–19.
16. Stephens, *Democratic Socialism in Jamaica*, pp. 131–7. Volkman and Cummings ('Murder as Usual', p. 183) reported that the CIA did use Cuban exiles in Jamaica. Penny Lernoux, 1980 winner of the Sigma Delta Chi and Maria Moors Cabot awards for her reporting on Latin America, and author of *Cry of the People* and *In Banks We Trust*, asserted in 1986 that 'almost every country in Latin America has been the target of destabilization efforts supported by the United States because a government crossed swords with U.S. business interests. . . .' (Lernoux, 'Tio Sam's Big Financial Stick', *The Nation* (New York), 22 March 1986, p. 338).
17. Kaufman, *Jamaica Under Manley*, p. 115, citing *DG* of 7 and 12 January 1976.
18. Stephens, *Democratic Socialism in Jamaica*, p. 133.

19. *JWG*, 20 January 1976; editorial, *DG*, 15 January 1976. In an interview with the author in September 1975, Manley claimed that never once did the *Gleaner* seek his opinion, as PM on stories affecting the government, after Hector Wynter became editor in 1975.
20. Kaufman, *Jamaica Under Manley*, p. 115. The 'Destabilization Diary' in Manley's *Jamaica: Struggle in the Periphery*, pp. 225–37 provides many specific details of violence in 1976.
21. John Dorschner, interview with Manley, *Miami Herald*'s 'Tropic Magazine', used the version reprinted in the *JWG* 27 April 1976, p. 30.
22. Some confusion exists about the political affiliation of the Orange Street residents; see Kaufman, *Jamaica Under Manley*, p. 115, and Stephens, *Democratic Socialism in Jamaica*, p. 132.
23. *JWG*, 25 May 1976, p. 6. Among the Manley government's positive achievements Reid listed the national minimum wage, low-income housing, the impact programme, the desire for real national independence, a more open foreign policy, the attempt to bring health care close to where people lived, land reform and the National Housing Trust.
24. MM, *Jamaica: Struggle in the Periphery*, pp. 232–3.
25. Kaufman, *Jamaica Under Manley*, p. 117, citing *DG*, 19 June 1976 and *Latin American Political Report*, (25 June 1976); see also *JWG*, 29 June 1976, p. 6.
26. *JWG*, 29 June 1976, pp. 6, 17.
27. *Ibid*.
28. MM, 'Statement by the Prime Minister', 19 June 1976, PNP files. For the actual State of Emergency document, The Emergency Powers Act [19 June 1976], see *The Jamaica Gazette Supplement*, Vol. 99, no. 69.
29. MM, *Jamaica: Struggle in the Periphery*, pp. 141–2; *Daily News*, 25 June 1976; *JWG*, 29 June 1976. Charles susbsequently wrote an account of his detention, *Detained*, which was highly critical of the entire process. Manley told me in an interview that he argued for two hours against the detaining of Charles and was very upset by it, but that the security chiefs insisted on it under threat of resignation. He also told me that he called Seaga, one of the very few times in which the political foes had conversed directly until recently and told him that he was perfectly free to go ahead with his election campaign. (Interview of 12 September, 1985). Volkman and Cummings ('Murder as Usual', p. 188) reported that one key JLP member detained 'was making Molotov cocktails in a mineral-[water] bottling plant he owned.'
30. MM, 'Statement by the Hon. Prime Minister to the House of Representatives, Tuesday, June 22, 1976', PNP files.
31. Except as noted, this paragraph is based on MM, 'Statement by the Prime Minister . . . to the House of Representatives, Tuesday, June 29, 1976', PNP files.
32. According to Kaufman, *Jamaica Under Manley*, p. 117, the briefcase was in the JLP candidate's hand when he was arrested at a JLP convention in Montego Bay.
33. Kaufman (*Jamaica Under Manley*, p. 117) erroneously states the tape of 14 May was made on the night of the fire, which occurred on 19 May.
34. MM, *Jamaica: Struggle in the Periphery*, p. 143.
35. *Daily News*, 3, 11 and 12 July, 1976; *Star*, 24 June 1976.
36. *Daily News*, 17 July 1976.
37. Cargill, *Jamaica Farewell*, p. 40.
38. Parnell and Bailey quoted in *JWG* 2 November 1976, p. 9.
39. *JWG*, 30 November 1975, p. 15.
40. Stephens, *Democratic Socialism in Jamaica*, p. 141.
41. *JWG*, 6 July 1976, p. 27 and 28 September 1976, p. 10.

42. *JWG*, 28 September 1976, p. 1. According to Stone's poll, 17% of those who voted for the JLP in 1972 switched to the PNP in 1976. They gave their reasons for doing so as:
Impressed with PNP leadership: 30%; Attracted by socialism: 20%; Impressed by PNP policies/projects: 27%; Personal benefits received: 14%; Association of JLP with violence: 9%.
Of PNP voters in 1976, 19% switched to the JLP in 1976, giving their reasons as: Impressed with JLP leadership: 8%; Hostile to socialism and communism: 29%; PNP economic management: 61%.
43. *JWG*, 9 November 1976.
44. *JWG*, 16 November 1976, pp. 6, 17.
45. 'Memorandum to the Prime Minister on Unemployment Policy and Proposals', by Gladstone Bonnick, Gary Laughton, M. G. Smith and Danny Williams, 18 November 1976, in PNP File 213: 'Hurricane Disaster'.
46. MM, The Prime Minister . . . at U.W.I.' 22 November 1976), PNP speeches files.
47. *JWG*, 30 November 1976, p. 15.
48. Volkman and Cummings, 'Murder as Usual', p. 188.
49. *JWG*, 28 September 1976, p. 28.
50. Malcolm Sharp, *JWG*, 10 August 1976, p. 28.
51. *JWG*, 24 August 1976, p. 10. Morgan criticized the PNP's problems with ideology, its attempts to do too many things at once, and its weakness in planning.
52. The speech is preserved in MM, *Not for Sale*, supplement by C. L. R. James (San Francisco: Editorial Consultants [1977]). The printed word scarcely does justice to the speech, which also exists as a two-record set. 'We Know Where We Are Going' (Kingston: PNP, 1976).
53. C. L. R. James commented that 'in fifty years of political activity and interest in all sorts of politics, I have never heard or read a speech more defiant of oppression and in every political way more suitable to its purpose' (Manley, 'Not For Sale', p. [7]).
54. *JWG*, 21 September 1976, pp. 18 and 30. Further background on Seaga may be found in Waters, *Race, Class, and Political Symbols*, pp. 63–5, and the *JWG*, 12 December 1973, p. 28, and 22 October 1974, p. 25.
55. The 1976 election manifestos of both parties can be found in Stone and Brown, *Perspectives on Jamaica in the Seventies*, pp. 275–8.
56. *JWG*, 30 November 1976, p. 29. 'Petras' described himself as an independent socialist (*ibid.*, 18 January 1977, p. 13).
57. *JWG*, 30 November 1976, p. 15.
58. Kaufman, *Jamaica Under Manley*, p. 127.
59. Letter to the editor, *JWG*, 26 October 1976, p. 13.
60. MM, 'The Prime Minister . . . at UWI', p. 46.
61. BM, 'Change Must Come', (Kingston: Women's Auxiliary of the PNP, 1976).
62. MM, 'Political Broadcast', 13 December 1976, PNP Speeches Files.
63. *JWG*, 30 November 1976, p. 17.
64. Waters, *Race, Class, and Political Symbols*, pp. 88, 138, 197, citing article by Stone, *DG*, 8 June 1980.
65. *JWG*, 21 December 1976, p. 1.
66. *JWG*, 1 March 1977, p. 14. Still later it added that 'the Government. . . . can be in no doubt of the electoral confidence it enjoys in the country' (*JWG*, 19 April 1977).
Subsequently as part of a plan to delegitimize the PNP government, the opposition charged that the PNP had stolen the 1976 election. The *Gleaner* itself noted that the opposition regularly charged fraud in the second election

of a two-term administration, that such charges had been levelled against the PNP in 1959 and the JLP in 1967 (*JWG*, 19 April 1977). For further information, see Appendix B.

CHAPTER 17 **The Descent Begins (1977)**

1. MM, interview with author, 12 September 1985.
2. *Ibid.*
3. Stephens, *Democratic Socialism in Jamaica,* pp. 158–9, quoting *DG*, 26 July 1978. Stephens and Stephens observe that the judicial inquiry 'indicated that Spaulding's account of events was more accurate than the JLP's charge of massive victimization.' They also see the Rema incident as 'the first act in the campaign of delegitimization conducted by the JLP during the PNP's second term.'
4. *JWG*, 8 March 1977, p. 3.
5. *JWG*, 1 March 1977, p. 6. The violence appears to have baffled many Jamaicans, as in this reflection (Lloyd A. Vermont, 'Manley's Eloquence, Charisma Decisive', *JWG* 25 January 1977, 15): 'Michael Manley and Edward Seaga don't shoot or attempt to murder each other and I would like to bet my bottom dollar that a part of each of them 'dies' when they hear that an opponent is shot and/or killed. . . . It is not believed that either of them think that they are worthy of the blood and life of another Jamaican. So why, why, why, should a JLP man kill a PNP man and a PNP man kill a JLP one? Why should one poor man kill another poor man because of politics?'
6. MM letter, 2 March 1977, MPF.
7. MM letter, of 20 January 1977, PNP file 'Executive 1977.'
8. Telegram, 10 February 1977; memorandum, 3 February 1977; and letter, 9 February 1977 in PNP file 'Executive 1977'. There is a large number of documents in this file illustrating such conflicts. One letter noted similar problems in the JLP. In 1984 internal fighting in the JLP led to an outbreak of murderous violence in Rema.
9. MM, 'Analysis of Contemporary Political Situation', 29 July 1977, PNP file 'Executive 1977'.
10. Letter to MM, 13 September 1977, PNP file 'Executive 1977', original emphasis. The writer announced that because of his faith in Manley's leadership he was rejoining the PNP.
11. MM to Ralph Brown, PNP General Secretary, 26 September 1977, PNP file 'Executive 1977'.
12. 'Report from the Accreditation Committee to the Executive', 16 September 1977, PNP file 'Executive 1977'.
13. Letter, accreditation committee to Manley, 15 September 1977, PNP file 'Executive 1977'.
14. Stephens, *Democratic Socialism in Jamaica,* Table A.10, p. 390.
15. Extensive accounts of the dealings with the IMF are in Kaufman, *Jamaica Under Manley* and Stephens, *Democratic Socialism in Jamaica*. Manley's own account is in *Jamaica: Struggle in the Periphery*, Ch. 11.
16. MM, *Jamaica: Struggle in the Periphery*, p. 151.
17. Quoted in Kaufman, *Jamaica Under Manley*, p. 243. On pages 243–5 is a brief history and critique of the IMF.
18. Kaufman, *Jamaica Under Manley*, p. 132, quoting Norman Girvan, Richard Bernal and Wesley Hughes, 'The IMF and the Third World: The Case of Jamaica, 1974–1980', *Development Dialogue*, 2 (1980), p. 122.
19. MM, 'National Self-Reliance, Phase I: Transcript of Statement . . . on Government's Economic Reconstruction Programme, Presented to Parliament and the Nation, January 19, 1977', (Kingston: API, 1977), p. 2.

20. When Manley's son Joseph, attending the University of Havana, asked for a bicycle, Manley replied that he really had to think about it 'because we are very broke since I gave up all that salary.' MM to JM 27 April 1977, MPF.
21. Kaufman, *Jamaica Under Manley*, pp. 130–37, and Stephens, *Democratic Socialism in Jamaica*, pp. 148–53.
22. MM, *Jamaica: Struggle in the Periphery*, p. 155.
23. Quoted in Stephens, *Democratic Socialism in Jamaica*, p. 167.
24. Stephens, *Democratic Socialism in Jamaica*, pp. 163–72.
25. MM, interview with author, September 1985.
26. *JWG*, 26 April 1977, p. 8.
27. Bea Lim, 'Outlaw Sex Discrimination', *JWG*, 19 April 1977, p. 16.
28. *JWG*, 19 April 1977, p. 21.
29. *JWG*, 12 April 1977, p. 6.
30. MM, 'Budget Speech' (May 1977).
31. 'Prime Minister w/ Govt Workers,' 23 October 1977 in PNP file of same name.
    According to Manley's 1977 budget speech, the objectives of worker participation were to raise the status of labour to equality with capital, to create more harmonious relations between them, and to promote more efficient production. In 1975 an advisory committee chaired by Rex Nettleford and a unit in the Ministry of Labour led by Stone were established to begin institutionalizing worker's participation.
32. MM, 'Address . . . to the Annual Award banquet of the Jamaica Manufacturers Association', 2 December 1977, PNP speeches files.
33. *JWG*, 15 March 1977, p. 20.
34. Letter from MM, 24 March 1977, MPF.
35. Rosalynn Carter, *First Lady From Plains* (Boston: Houghton Mifflin, 1984), p. 191.
36. *San Francisco Chronicle*, 1 June 1977.
37. Carter, *First Lady From Plains*, p. 192.
38. *The Washington Post*, 8 August 1977, quoted in Kaufman, *Jamaica Under Manley*, p. 140.
39. Hugh Small, 'The Tactical Position of the People's National Party with Respect to the United States of America', 28 July 1977, pp. 8, 10, in PNP file 'Executive 1977'.
40. Reverend C. S. Reid, 'Jamaica and the World Since 1962', *JWG*, 15 March 1977, p. 16.
41. *JWG*, 29 March 1977, p. 12. According to Manley (*Jamaica: Struggle in the Periphery*, p. 234), da Costa, along with John Hearne and Wilmot Perkins, was part of 'a formidable team in the unwavering and orchestrated media campaign directed against the PNP and its leadership' from 1976 to 1980.
42. *JWG*, 22 March 1977, p. 15, and 5 April 1977, p. 21.
43. Confidential source.
44. *Los pueblos abrazan la libertad y se unen* (La Habana: Editora Política, 1978), pp. 50–51.
45. MM to JM, 11 November 1977, MPF. Manley went on that he had just returned from a trip to New York, Sweden, Norway, and Yugoslavia, and found the last two particularly impressive for their internal achievements and outlook on the world.
46. Manley Letter, 11 November 1977, MPF. In response to Manley's comment about Castro's visit being the most important by a foreigner in Jamaica's history, his correspondent's spouse said, 'He's forgotten about Columbus.' Manley replied: 'Tell J— that I send to say "Touché" about Columbus. My remark to myself was: "That'll larn ya!" ' (Letter to Manley dated only 'December 1977'; Manley's reply, 20 December 1977).

47. JM to MM, 16 September 1977, MPF.
48. JM to MM, 21 October 1977, MPF.
49. JM to MM, dated 'Monday 12 1977', MPF. Joseph had read the Volkman and Cummings article, 'Murder as Usual', in *Penthouse*, which described assassination attempts on his father. He commented that 'despite certain sensationalist errors the article has a profound impact.'
50. MM, 'Broadcast . . . on December 15th, 1977', PNP speeches files.
51. MM letter, 8 December 1977, MPF. Unfortunately both the foreign policy triumphs were to prove short-lived. Mexico was forced by its economic crisis to pull out of the Javamex project. Some of Jamaica's colleagues in the IBA, in the grips of their own economic crises, undersold Jamaica.
52. Kaufman, *Jamaica Under Manley*, p. 141.

CHAPTER 18 **'Evil . . . Stalks the Land' (1978)**

1. *JWG*, 13 February, 23 January and 1 May 1978.
2. MM, 'Statement by the Prime Minister', 13 January 1978, PNP speeches files. The government proposed increasing the minimum wage from J$20 to J$24 per week, National Insurance Scheme pensions from J$6.90 to J$7.90 per week, and public assistance from J$2 to $J12 per month. The Jamaican dollar (J$), worth US$1.20 when established in 1969, shrunk to US$0.64 in May 1978 (*JWG*, 22 May 1978). According to his personal files, Manley paid a worker J$5 per day at his mountain retreat and farm, Nyumbani, on various dates in September and October 1978.
3. MM, *Jamaica: Struggle in the Periphery*, pp. 159–61. For further details and discussion see Kaufman, *Jamaica Under Manley*, pp. 146–9 and Stephens, *Democratic Socialism in Jamaica*, pp. 199–204.
4. *JWG*, 22 May 1978, pp. 1, 7, 5, 13, 14 and 29.
5. Stephens, 'Colonial Minds and Escapism', *ibid.*, p. 11.
6. In this context, it was only small comfort to Manley to hear from one of the island's capitalists that 'the Negril Beach Village became a reality because of the action you took in 1974. It has been an outstanding success, giving employment to many.' (Letter to MM, 24 July 1978, MPF).
7. *JWG*, 5 June 1978, pp. 1 and 29.
8. *Caribbean Monthly Bulletin* (July 1981), p. 16, and (February 1982), p. 17.
9. MM, personal communication with author, February 1988.
10. Kaufman, *Jamaica Under Manley*, p. 149, and Waters, *Race, Class, and Political Symbols*, p. 209. Manley does not discuss the event in his *Jamaica: Struggle in the Periphery*, nor is it mentioned in Stephens, *Democratic Socialism in Jamaica*.
11. Personal communication to author, February 1988.
12. *JWG*, 13 February 1978, p. 11.
13. David d'Costa, 'Green Bay': What Made Them Do It?', *JWG*, 5 June 1978, pp. 1 and 29.
14. *JWG*, 12 June 1978, p. 1.
15. MM, news release, 30 August 1978.
16. Personal communication with author, February 1988.
17. As a matter of historical perspective, it should be noted that in the '80s, summary executions by police (not the army) became common, involving possibly hundreds of victims: see Americas Watch, *Human Rights in Jamaica* (New York and Washington: Americas Watch, 1986).
18. *JWG*, 16, 23, and 30 January and 6 February 1978.
19. *JWG*, 13 and 27 February, 6 and 20 March, 24 April, and 1 and 15 May 1978.
20. *JWG*, 1 May 1978, p. 2.
21. *Ibid.*, p. 11.
22. Kaufman, *Jamaica Under Manley*, p. 149.

23. *JWG*, 29 March 1977, p. 12. For background on Hearne, see Frank M. Birbalsingh, ' "Escapism" in the Novels of John Hearne', *Caribbean Quarterly* (March 1970), pp. 28–38, where it is argued that after the promise of Hearne's first novel, *Voices Under the Window*, his subsequent four novels were works of 'exceedingly trivial' escapism. One of the latter, *Faces of Love*, featured a central character named Michael.
24. *DG*, 2 August 1984, p. 8.
25. MM, interview with author, 12 September 1985.
26. *JWG*, 9 January 1978.
27. *JWG*, 20 March 1978, p. 9. Clarke's statement was untrue because the state owned only 25 percent of radio RJR; the rest was owned by popular organizations such as unions and teachers' associations, which had a majority on the station's board of directors. According to Manley (*Jamaica: Struggle on the Periphery*, p. 231), it had been public knowledge since 1963 that the Inter-American Press Association to which Clarke spoke was connected to the CIA. Kaufman (*Jamaica Under Manley*, p. 190) also points to a tie between IAPA and the CIA.
28. *JWG*, 30 January 1978, p. 11.
29. *JWG*, 13 February 1976, p. 11.
30. *JWG*, 6 March 1978, p. 11. On the Hungarian credits and agreements, see *Caribbean Monthly Bulletin* (February 1978), p. 5.
31. *DG*, 26 May 1978.
32. One of the four submachine guns was later used in the killing of two policemen in Runaway Bay. Two of the three suspects were of the group of fourteen discharged soldiers; the third had deserted the JDF earlier.
33. MM, news releases, 30 and 31 August 1978. Manley's statement was supported by a separate statement by Major General Green.
34. Quoted in the *Daily Challenge* (Brooklyn, New York), 1 February 1985. In addition, according to the *City Sun* (Brooklyn, New York) of 7 February 1985, the *Gleaner* expressed regret for a paid advertisement 'which may have suggested that Mr. Manley was selling out the country,' and issued apologies to former Minister of Justice Carl Rattray, former Minister of Agriculture and Minister of National Security Keble Munn, and Manley adviser sociologist M. G. Smith.
35. *New York Times*, 21 January 1978.
36. *JWG*, 13 February 1978, p. 13. To balance its often rabid columnists and distorted news coverage, the *Gleaner* maintained an aura of 'impartiality' in its editorials.
37. John Hearne, 'New World Order—An Act of Faith', *JWG*, 1 May 1978, 15, 29. This article contains a lengthy analysis of NIEO negotiations. For a critique of the NIEO, see William Loehr and John P. Powelson, *Threat to Development: Pitfalls of the NIEO* (Boulder, Colorado: Westview, 1983).
38. *JWG*, 29 May 1978, p. 11.
39. MM, 'Excerpts from . . . Budget Speech—June 1978', PNP speeches files, p. 5.
40. Fidel Castro to MM, 2 February 1978, in PNP file 'Heads of Government, 1978:9.' This letter to 'Dear Friend Manley' was subsequently reprinted in the *Gleaner*.
41. *JWG*, 20 Februry 1978, p. 13.
42. *JWG*, 13 April 1978, p. 13.
43. The other six recipients were Kwame Nkrumah, Murtala Mohammed, Jawaharlal Nehru, Paul Robeson, Canon Lewis John Collins, and Olof Palme.
44. MM, 'The Time for Action Must be Now.' Speech of 11 October 1978 at the United Nations (pamphlet, Kingston: API [1978?].

45. MM, notes of speech to the Academic Council on the Cultural Training Centre, 1 February 1978, MPF. An accompanying letter dated 15 March 1978 identified these notes as based on the recollection of a member of Manley's staff present during the speech.
46. Letter to MM, 9 June 1978, MPF.
47. Manley's letter of 28 June 1978, MPF.
48. MM, interview with author, September 1985.
49. 'Rachel Manley's Poems—Another Convincing Voice', *The Weekend Star*, 21 April 1978.
50. Manley to Rachel [Manley] Ennevor, 6 April 1978, MPF.
51. Rachel Ennevor to MM, 25 April [1978], MPF.
52. MPF.
53. Manley, letter of 26 May 1978, MPF.
54. Receipts, MPF.
55. JM to MM, undated [April 1978?], MPF.
56. MM to JM, 29 May 1978, MPF.
57. Letter of 14 November 1978, MPF.
58. Letter to MM, 19 February 1978, MPF.
59. Letter to MM, 8 November 1978, MPF.
60. MM letter, 30 November 1978 (copy), MPF.

CHAPTER 19 **Deepening Crisis (1979)**

1. Minutes of PNP NEC meeting, 28 January 1979, PNP file 'Minutes NEC.'
2. MM, 'The Social Contract', (Kingston: API, 1979). It is apparent from reading Manley's speeches of 1977, 1978, 1979 that he kept saying and hoping that the economy would improve, but it did not. Over time this undermined his credibility and hence his ability to govern.
3. MM, *Jamaica: Struggle in the Periphery*, pp. 149–50.
4. Jamaica API, 'Why & How to Conserve Energy', (1979).
5. United Nations Energy Programme, 'Brief on Current Developments in Oil Prices and Supplies', a report prepared for the Government of Jamaica (31 May 1979).
6. MM, 'Address . . . to the 20th Annual Meeting of the Board of Directors of the Inter-American Development Bank'. (Montego Bay, 28 May 1979), PNP speeches files.
7. News release, 6 August 1979.
8. MM, 'We Must Tackle Structures,' (Lecture to the Third World Foundation, London, 29 October 1979): (Kingston: Agency for Public Information, 1979), pp. 6, 9, 12, and 23. Added to man-made problems, severe floods swept Jamaica in June and again in September.
9. Kaufman, *Jamaica Under Manley*, p. 153.
10. MM, 'Broadcast to the Nation . . . on Wednesday December 19, 1979', PNP speeches files.
11. Kaufman, *Jamaica Under Manley*, p. 150.
12. Minutes of PNP NEC meeting 28 January 1979, PNP file 'Minutes NEC'. Many of the internal problems of the PNP during this time are discussed in People's National Party, 'Report of the General Secretary, 1978/1979'.
13. See Appendix A.
14. See Appendix B.
15. MM, *Jamaica: Struggle in the Periphery*, pp. 128–9. *Principles and Objectives* was accepted too late for political education to be effective while the PNP was in power. For the PNP's analysis and assessment of political education in its early stages see People's National Party, 'Report of the General Secretary 1978/1979', pp. 18–22.

16. Letter of 23 January 1979, Manley's 'Constituency Matters' file, PNP archive.
17. Minutes, 14 June 1979, in Manley's 'Constituency Matters' file.
18. Manley's handwritten notes for speech at NW statue on 22 March 1979, PNP speeches files.
19. Carl Rattray, 'The Constitutional Role of the Opposition', (pamphlet. [Kingston]: Agency for Public Information, 1979).
20. For a detailed analysis, see Vaughn A. Lewis, 'The Small State Alone: Jamaican Foreign Policy, 1977–80', *Journal of Interamerican Studies and World Affairs* (May 1983), 139–70.
21. MM, 'Speech . . . at the Official Dinner Given by the Government of the USSR, Grand Kremlin Palace, Moscow', 10 April 1979, PNP speeches files.
22. MM, 'Statement . . . on . . . Return from . . . Official Visit to the Soviet Union', 16 April 1979, PNP speeches files. The first shipment of Jamaican alumina bound for the USSR, 26,000 tons was loaded in early March 1980 on a Norwegian vessel (*Caribbean Monthly Bulletin*, [March/April 1980], p. 21). Notably, the anti-communist Seaga government of the '80s continued the Soviet trade arrangements.
23. MM, two letters dated 20 April 1979, MPF.
24. MM, letter of 24 July 1979, MPF.
25. MM, personal communication to author, February 1988.
26. The references are to the role of the United States in overthrowing the democratically-elected governments of Guatemala and Chile in 1954 and 1978, respectively, and invading during The Dominican Republic's civil war of 1965.
27. MM, 'Our Movement is Irreversible . . . Because Our Cause is Just', speech at the Sixth Summit of the Non-Aligned Nations, Havana, Cuba, 4 September 1979 (pamphlet, Kingston: API, 1979).
28. Kaufman, *Jamaica Under Manley*, p. 152.
29. Estrada, who had fought in the armed insurrection of the Cuban Revolution, had been an army major and also a journalist. Before becoming ambassador in Jamaica, his first diplomatic post, he had been Deputy Chief of the Department of the Americas of the Central Committee of the Cuban Communist Party [PCC], a post which involved maintaining relations between the PCC and national liberation movements in Latin America and the Caribbean (*Caribbean Monthly Bulletin*, [June-July 1979], p. 7).
30. For more details, see *Caribbean Monthly Bulletin*, (August-September 1979), pp. 16–17. According to Andrew Kopkind, ('Trouble in Paradise', *Columbia Journalism Review* [March/April 1980], p. 47), the *Gleaner* never produced evidence to support its claim that Estrada was a master spy.
31. MM, 'Political Broadcast', 30 September 1979, PNP speeches files.
32. PNP, 'Report of the General Secretary, 1978/1979', p. 12.
33. Maurice Bishop, 'Beat Back Destabilizers', national broadcast, 18 September 1979 (pamphlet [St. George's?], Grenada: People's Revolutionary Government, 1979).
34. MM to Bishop, [14 December 1979], in PNP file, 'Heads of Government' [1979].
35. MM to Carter, 24 December 1979, in PNP file 'Heads of Government' [1979].
36. MM, 'Speech Notes for . . . Address to the 35th Anniversary Function of the Jamaica Federation of Women', 30 May 1979 in PNP file 'PM's File: Mrs. Manley's Daily'.
37. Kaufman, *Jamaica Under Manley*, p. 175.
38. For further details see Jamaica API, 'Maternity Leave with Pay: Know Your Rights', (pamphlet, 1980).
39. Letter dated 7 December 1979 in PNP file, 'Citizens Mail' [1980]; spelling and punctuation as in original. This file, approximately two inches thick, contains

dozens of such letters which give an excellent feeling for how 'ordinary' people experienced the crisis of the late '70s.

40. BM, 'Address . . . at the Launching of the Jamaica-Soviet Association', 6 May 1979, PNP speeches files.
41. PNP Women's Movement, 'Good Friends . . . Good Neighbours: PNP Women's Movement in Solidarity with our Cuban Brothers & Sisters (pamphlet, [1979?]).
42. PNP Women's Movement, 'Convenor's Annual Report, 1978/1979,' pp. 17–18.
43. Letter of 10 May 1979, MPF.
44. Manley's letter of 29 August 1979, MPF.
45. Manley's letter of 17 January 1979 and receipt dated 17 October 1979, in MPF.
46. Manley's letters of 25 September and 20 December 1979, MPF.

CHAPTER 20 **'Hovering at the Edge of . . . Hell' (1980)**

1. 'The Owl', a poem by Rachel Manley in a letter to MM, 29 March 1980. Rachel made clear in the letter that Manley was the owl. EM recorded in her diary on 7 May 1963 that 'the owl is my great symbol—it always comes to me in moments of crisis.'
2. MM, *Jamaica: Struggle in the Periphery*, p. 124.
3. This paragraph is based on documents in MPF.
4. PNP, Economic Affairs Commission, 'The Non-IMF Alternative', (mimeo, 8 March 1980), p. 12, and 'Jamaica and the IMF—Some Basic Facts' (typescript, 13 February 1980), p. 2. According to Kaufman (*Jamaica Under Manley*, p. 186), between October 1976 and November 1980, real wages fell by 85% for men and 109% for women.
5. PNP, 'Jamaica and the IMF—Some Basic Facts' contains details.
6. MM, 'Statement by the Prime Minister', 3 February 1980 (mimeo, API, 1980).
7. In fact, the PNP lost 38 seats, retaining only 9 to the JLP's 51.
8. PNP, Economic Affairs Commission, The Non-IMF Alternative', pp. 1–3, 22, 25–7.
9. MM, *Jamaica: Struggle in the Periphery*, 189–92.
10. MM, 'Operation Reconstruction: Text of a Broadcast to the Nation . . . on March 30, 1980', ([Kingston]: API, [1980]).
11. Memorandum, George L. Beckford to D. K. Duncan, 26 March 1980, PNP file 'Executive 1980'.
12. *PNP Progress*, 10 May 1980.
13. Undated memo, 'Distribution of Food', in PNP file 'Executive 1980'.
14. MM, 'Independence Day Message to Jamaican Nationals Abroad', [August 1980], PNP speeches files.
15. PNP file, '213—Hurricane Disaster'.
16. API, 'News Release: Reports of Sabotage to Rural Pipelines', 2 September 1980.
17. MM, 'Statement . . . to News Conference on Alternative Energy Sources', (26 September 1980), PNP speeches files.
18. Stephens, *Democratic Socialism in Jamaica*, p. 238. Kaufman (*Jamaica Under Manley*, p. 188), gives a total for 1980 of 933 violent deaths—556 killed by gunmen, 234 by the security forces—compared to 334 in 1979. Manley dedicated *Jamaica: Struggle in the Periphery* 'to the memory of seven hundred and fifty people who died so needlessly, many in the first flower of youth.'
19. The author's research uncovered no such allegations against Manley personally, other than those discussed above, i.e. Lightbourne's question in 1976 and Hearne's libel in 1978.
20. MM, *Struggle in the Periphery*, pp. 193–4.

21. Kaufman, *Jamaica Under Manley*, p. 188; see also Stephens, *Democratic Socialism in Jamaica*, pp. 236–8.
22. Ellen Ray and Bill Schaap, 'Massive Destabilization in Jamaica: 1976 with a New Twist', *Covert Action Information Buletin*, (August-September 1980), pp. 7–17.
23. *JWG*, 21 July 1980, p. 4.
24. MM, 'Address to the Nation', 22 October 1980, PNP files.
25. *JWG*, 11 August 1980. Responses to the question 'Who do you think is behind the increase in violence and killing in Jamaica?' were: PNP (34%), JLP & PNP (20%), JLP (10%), unspecified politicians (14%), criminals (9%), communists (2%), CIA (1%), no views (10%). The poll had a maximum error of +/- 5%.
26. MM, *Jamaica: Struggle in the Periphery*, p. 195; see also Kaufman, *Jamaica Under Manley*, p. 114.
27. *PNP Progress*, 10 May 1980, p. 6.
28. *PNP Progress*, 17 May 1980.
29. Kaufman, *Jamaica Under Manley*, pp. 188–9; Stephens, *Democratic Socialism in Jamaica*, p. 240; and Ray and Schaap, 'Massive Destabilization in Jamaica', pp. 7–8. According to Ray and Schaap (p. 9), although Seaga said he did not know H. Charles Johnson, both men had dealt with George Corcho, a gun-runner and major JLP supporter.

    As in the Green Bay case, court proceedings were protracted. A Resident Magistrate ruled in September 1980 that there was sufficient evidence to support a *prima facie* case against 5 persons charged with intention to incite a revolt against the government. An order for a court martial of 6 JDF soldiers was posted in October, after which the proceedings were adjourned following the 1980 election. In February '81 5 out of 9 JDF members were found not guilty by a court martial. Chief Justice Kenneth Smith ordered a retrial for 4 persons charged with treason felony in the Supreme Court after the court failed to arrive at a unanimous verdict. In '82, H. Charles Johnson entered the trade union field. The author has been unable to determine the final disposition of the case, (*Caribbean Monthly Bulletin*, September-October '80, p. 27; November-December '80, p. 27; March-April '81, p. 12; May '81, p. 12; May-June '82, p. 18).
30. Stephens, *Democratic Socialism in Jamaica*, pp. 238–41; Kaufman, *Jamaica Under Manley*, p. 189; Manley, *Jamaica: Struggle in the Periphery*, p. 209.
31. See the account of JDF ex-major Abe Bailey, 'Why I Left Jamaica and the JDF' *JWG*, 7 and 14 July, 1980, which is, however, long on innuendo and short on facts, especially the second part.
32. *Covert Action Information Bulletin*, August-September 1980, pp. 2–4.
33. Unsigned letter, 5 July 1980, MPF.
34. *JWG*, 14 July 1980.
35. *Ibid*.
36. *JWG*, 4 August 1980.
37. Confidential source.
38. Stephens, *Democratic Socialism in Jamaica*, p. 238.
39. For an account of the killing, see Manley, *Jamaica: Struggle in the Periphery*, pp. 197–8; more details are provided in 'Prime Minister's Press Conference', 17 October 1980, PNP files. Ferdinand Neita, a successful PNP candidate in St Catherine in 1976, was shot in the stomach and survived after a long operation.
40. 'Prime Minister's News Conference', 17 October 1980, PNP files.
41. MM, 'Statement by the Prime Minister and President of the PNP', 13 October 1980, PNP files.
42. MM, 'Statement by the Prime Minister', 14 October 1980, PNP files.

43. MM, 'Address to the Nation', 22 October 1980, PNP files.
44. MM, interview with author, 12 September 1985.
45. MM, *Jamaica: Struggle in the Periphery*, pp. 201–2.
46. Andrew Kopkind, 'Trouble in Paradise', *Columbia Journalism Review* (March/-April 1980), pp. 49–50.
47. Fred Landis, 'Psychological Warfare in the Media: The Case of Jamaica', (32 p. pamphlet, Kingston: Press Association of Jamaica, 1980); quote from p. 17. For a defense of the *Gleaner''s* role see Ramesh Deosaran. 'The Role of the Press in the Caribbean', *Caribbean Review* (Autumn 1984), pp. 16–19, 45–6.
48. MM, interview with author, 12 September 1985.
49. Ray and Schaap, 'Massive Destabilization in Jamaica', p. 14.
50. This and the following paragraphs are based on Stone, 'Manley and Seaga', *JWG*, 4 and 11 August 1980.
51. Needless to say, Manley found Stone's comparison one-sided. He particularly protested against his supposed lack of workable economic strategies in the light of successes in the bauxite and small farm sectors. He also disputed his supposed inability to take unpopular actions, citing foreign exchange restrictions on luxury purchases, and Seaga's lack of same. Manley also denied that he had become victim of his own propaganda about his popularity (Manley, communication with author, February 1988). It should be added that in the '80s, Stone became critical of Seaga's performance and more appreciative of Manley (see Chapter 22).
52. MM, 'Message from Comrade Leader . . . to Canvassers', (mimeo, July 1980), PNP speeches files.
53. C. S. Reid, 'Our Present Crisis', *JWG*, 7 July 1980, p. 30.
54. *JWG*, 28 July 1980, p. 15.
55. *JWG*, 7 July 1980, p. 1.
56. Kaufman, *Jamaica Under Manley*, p. 188.
57. *DG* of 2 July 1980, reprinted in *JWG*, 14 July 1980, pp. 11, 27.
58. *JWG*, 21 July 1980, p. 11.
59. Kaufman, *Jamaica Under Manley*, p. 187.
60. Stephens, *Democratic Socialism in Jamaica*, pp. 244–9.

CHAPTER 21 **The Enigmatic Wheel of History (1981–3)**

1. *Rising Sun* (June 1982). p. 15. The subject of this observation was Björn Borg's retirement from tennis.
2. MM interview with author, December 1987.
3. PNP, 'Report to the 43rd Annual General Conference . . . October 10 and 11, 1981, p. 6.
4. *Jamaica: Struggle in the Periphery*, 207–13.
5. *Ibid.*, pp. 36–44. In a speech in 1985, Manley lent credence to PNP outsiders' view that consumer shortages played a key role in the defeat. 'Many people', he said, thought that import controls and loss of 'luxuries' was the main factor in the loss. (Manley, 'Michael Manley speaks on Transformation and Jamaica's Economy.' p. 6.
6. *Rising Sun*, January/February, April, June, and December 1982.
7. Winston Van Horne, 'Why Manley Lost:' Problems of Michael Manley's Politics of Change,' *Journal of Caribbean Studies* (Fall/Winter 1981), pp. 23–36.
8. Quoted in Charles E. Cobb, 'Jamaica: Hard Times, High Hopes', *National Geographic* (January 1985), p. 130.
9. Stone, 'Jamaica's 1980 Elections: What Manley Did Do; What Seaga Need Do.' *Caribbean Review* (Spring 1981), p. 40.
10. Michael Massing, 'The Jamaica Experiment', *The Atlantic Monthly* (September 1983), p. 42.

11. EM Diary, 14 March and 11 May 1981.
12. *Wall Street Journal*, 27 April 1981.
13. IDB, *Economic and Social Progress in Latin America*, 1983 report, p. 260; 1984 report, p. 332; and *Progresso Sócio-Econômico na América Latina*, 1986 report, p. 310. Figures for various indicators vary from year to year in the IDB reports; generally the most recent figures have been chosen.
14. *Rising Sun* (January/February 1982).
15. Banco Interamericano de Desenvolvimento, *Progresso Sócio-Econômico na América Latina*, 1986 Report, p. 310.
16. Massing, 'The Jamaican Experiment', pp. 46–8, 51.
17. *Rising Sun*, March 1982, pp. 1–2. The September 1982 issue of this PNP party paper carried a picture of a union negotiating session in which Shearer is more centrally depicted that Manley.
18. *Rising Sun*, June and July 1982.
19. *Miami Herald*, 20 June 1982.
20. IDB, *Economic and Social Progress in Latin America*, 1987 Report, p. 334.
21. *Wall Street Journal*, 11 March 1983, p. 32.
22. Stone, *DG*, 15 October 1984.
23. MM in *Rising Sun*, June, July, and August 1983.
24. Women's Resource and Outreach Centre, 'Annual Report March 1983-April 6 1984' (typescript, Kingston, 1984).
25. *Miami Herald*. 7 December 1981, p. 16; *Rising Sun*, July and September 1982.
26. Massing, 'The Jamaican Experiment', p. 45.
27. Stone, *DG*, 28 November 1984.
28. EM, Diary, 20 February 1981, 10 April and 14 July 1982.
29. PNP, 'Report to the 43rd Annual General Conference' (Kingston, 1981), p. 18.
30. *Miami Herald*, 8 April 1982.
31. EM, Diary, 9 April 1982.
32. MM 'Belafonte—No Pawn of Reagan', *Rising Sun* (August 1982), p. 16. In an interview with the *Miami Herald* (20 June 1982), Manley reportedly said, 'When President Reagan says that Seaga saved Jamaica from Communism, you only have two choices. He's either a fool or a liar. . . .'
33. *Rising Sun* (political education supplement), August 1982.
34. EM, Diary, 18 September 1982.
35. Speech of 14 April 1983, author's notes.
36. *Miami Herald*, 16 February 1983, p. 12A.
37. EM, Diary, 12 January 1983. She added, 'Of course, they're right in a way, one has to survive.'
38. *Rising Sun*, October 1983, p. 3.
39. *Rising Sun*, June 1983. In April, 1983, the SI consisted of 70 affiliated parties, including, at the time, the governing parties of France, Sweden, Spain, Portugal, Austria, Finland, Greece, Australia, Costa Rica, Grenada, Barbados, and the Dominican Republic. The SI Committee on Economic Policy, chaired by Manley, published its findings and recommendations as *Global Challenge: From Crisis to Cooperation: Breaking the North-South Stalemate* (London: Pan Books, 1985).
40. MM, 'Why Nonaligned Isn't Neutral', *The Nation* (New York), 30 April 1983, pp. 536–7.
41. *Maurice Bishop Speaks: The Grenada Revolution, 1979–83* (New York: Pathfinder Press, 1983), pp. 80–1, 90. 92, 113, 155, 215. See also Ch. 19 above.

    Gordon K. Lewis's *Grenada: The Jewel Despoiled* (Baltimore: Johns Hopkins University Press, 1987) is an outstanding account of the collapse of the revolutionary government and US invasion. The *Caribbean Review* of autumn 1983 carried several articles on the Grenada crisis.

42. MM, letter of 25 October 1983, MPF.
43. MM to Nyerere, 7 November 1983, MPF.
44. MM, 'Grenada in the Context of History: Between Neocolonialism and Independence', *Caribbean Review* (Autumn 1983), pp. 7–9, 45–7.
45. PNP, 'Report to the 43rd Annual General Conference,' p. 9.
46. *Rising Sun* (September 1982), p. 8.
47. *Ibid.*, pp. 10–14.
48. Political Analyst, 'Announcing Socialism', *Rising Sun*, September 1982. Manley confirmed that he wrote the article in a letter to the author. he also wrote another piece in which he styled NM a 'cautious revolutionary'.
49. *Rising Sun* (Political Education Supplement), August 1982.
50. MM, 'Candidate Criteria Set', *Rising Sun* (Political Education Supplement). June 1983. This issue contains provisional criteria for candidates and party groups.
51. MM, 'Our 21st Anniversary', *Rising Sun* (Political Education Supplement), August 1983.
52. *Rising Sun*, October 1983, p. 1.
53. EM, Diary, 5 August 1982.
54. *Rising Sun*, April 1982, p. 2.
55. Stone, 'Jamaica's 1980 Elections: What Manley Did Do; What Seaga Need Do', *Caribbean Review* (Spring 1981), pp. 40–1. For Seaga's views on the '70s, Jamaican culture, his background and plans, see Stephen Davis, 'Jamaican Politics, Economics and Culture: An Interview with Edward Seaga,' *Caribbean Review* (autumn 1981), pp. 14–17.
56. The Stone Poll in the *DG* of 3 November 1984 includes trends from October 1980 to September 1984.
57. MM, 'Electoral Reform', *Rising Sun* (June 1982).
58. Waters, *Race, Class, and Political Symbols*. p. 275.
59. MM, 'Seaga "Stone" Polled Dead', *Rising Sun* (December 1982).
60. *Rising Sun* (July 1983).
61. Waters, *Race, Class, and Political Symbols*, contains a full account of the 'election' of 1983.
62. MM, 'Why the PNP stayed out of Seaga's Election', *South* (February 1984).
63. EM, Diary, 29 and 30 November 1983.
64. DG, 10 September 1985, p. 2.
65. DG, 4 April 1985.
66. EM, Diary, 3 December 1983.

CHAPTER 22 **New Consciousness and Responsibility (1984–8)**

1. *EM*, Diary, 27 July 1977.
2. Stone in *DG*, 8 October 1984.
3. IDB, *Social and Economic Progress in Latin America*, 1987 Report, p. 334.
4. *DG*, 18 July 1984.
5. *DG*, 19 September 1984.
6. *DG*, 4 February 1985.
7. *DG*, 20 and 22 February 1985.
8. *Carib News*, 23 April 1985, and *DG* 10 May 1985.
9. *DG*, 15 April 1985 and Stone, 'Believing in Miracles', *DG*, 22nd April 1985.
10. *Caribbean Times*, 31 May 1985.
11. *DG*, 27 July 1984.
12. *DG*, 9 October 1984.
13. *DG*, 1, 6, and 21 July 1984: *City Sun*, 25 July 1984.

14. *DG*, 27 August 1984.
15. *DG*, 4, 5, and 8 September 1984.
16. *DG*, 25 October 1984.
17. Stone column, *DG*, 10 September 1984. A Stone Poll published 29 October 1984 indicated that only 17% of Jamaicans had been adversely affected by problems in education, while 44% had experienced adversities in health care, mostly long delays.
18. Stone column, 26 November 1984.
19. Matthew Beaubrun, interview with author, 19 December 1987.
20. *DG*, 8 August and 11 November 1984 (Stone Poll); 4 February, 28 April, and 1 May 1985; and Doris Kitson, 'Fed Up with Suffering Under "Savior" Seaga'. *The Guardian* (New York), 30 January 1985.
21. *The Star*, 31 August 1984.
22. *DG*, 11 July 1984.
23. *The Star*, 13 July 1984.
24. *DG*, 31 August 1984.
25. Americas Watch, *Human Rights in Jamaica*, pp. 10, 60. The report observed (pp. 15–16) that despite persistent rumours of political motiviations behind police killings, 'we could not establish a partisan motive in any except perhaps one of the individual cases we studied.'
26. *DG*, 17 August 1984.
27. Quoted in Barbara Crossette. 'The Caribbean After Grenada', *The New York Times Magazine* (18 March 1984), p. 64.
28. Stone Poll (September 1984), *DG*, 11 November 1984.
29. *DG*, 17 and 18 October 1984.
30. *DG*, 10 September 1984.
31. *The Star*, 7 September 1984, Manley, interview with author, December 1987.
32. *DG*, 17 August 1984.
33. Stone, 'Full Marks to Manley', *DG*, 20 August 1984.
34. Stone column, *DG*, 21 January 1985. For coverage of the gas protests, see *DG*, 16 to 21 January 1985 and the *New York Times* 16 to 18 January 1985.
35. EM diary, 7 April 1985; receipt, University of West Indies Hospital; 17 May 1985; and Manley letter, 15 May 1985, in MPF; and *Daily Challenge*, 15 April 1985.
36. Stone, 'Savaging Local Government,' *DG*, 13 February 1985.
37. *DG*, 13, 15, 20, 24 (Stone column) and 25 April, 4 May, and 17 October 1985.
38. Franklin McKnight, 'When Manley Lost His Nerve,' [interview with D. K. Duncan], *DG*, 5 May 1985.
39. MM, 'The International dimension of Transformation', (August 1984) in: *Michael Manley Speaks on 'Transformation and Jamacia's Economy', 'The International Dimension of Tranformation', and 'Jamaica/US Relations'*, (Kingston: PNP. 1985). p. 16. Manley went on to say in connection with transformation that 'the most fundamental question in the society is . . . who has power, and what are the relations of power within the society.'
40. MM 'Transformation and the Present Political Economy of Jamaica', in: *Ibid*.
41. MM, speech in Black River, 8 September 1985, tape-recorded by author.
42. For details see 'Strategy for Development: Leading Jamaica into the 21st Century', *DG* Supplement, 25 October 1985.
43. Stone, 'The PNP's Alternative', *DG*, 9 October 1985.
44. DG, 1 November 1985.
45. *DG*, 12 November 1985.
46. *DG*, 25 November 1985.
47. MM. '1985 in Retrospect—The Year of the Shadows', *SG*, 5 January 1986.
48. *DG*, 11 March 1986.

49. Stone column, *DG*, 17 March 1986. Stone also faulted the 'total failure' of the security forces to deal with the violence and the PNP for failing to arrange its own security following earlier violence against its supporters.
50. 'J'can [Jamaican] Human Rights Infringed', *Caribbean Contact*, April 1986. With respect to the *Gleaner*, Daly said that 'for most of Jamaica's history the printed media has been dominated by a single newspaper company with a vested interest in a particular political viewpoint. . . .' He added that the Jamaican Council on Human Rights 'has frequently been the victim of the newspaper's discriminatory approach to publication.' Daly said that the situation in the electronic media, and particularly in the government-owned television station, was 'much worse'. Daly's comments contrasted with the less reliable report of the US State Department which said that 'during 1985 human rights were generally respected in Jamaica': see *DG*, 29 March 1986.
51. Stone, 'The IMF On Hold', *JWG*, 12 May 1986. According to JLP's Pearnel Charles the government adopted the PNP's advice on how to spend the oil price windfall.
52. *DG*, 16 and 17 May 1986.
53. MM, 'Looking at the Budget Deficit', *DG*, 29 May 1986.
54. *DG*, 1 July 1986. On the day nomination, for candidates for local elections were held, JLP MP Anthony Abrahams resigned from the party citing interference from party chairman Bruce Golding. Abrahams expected his character to be assassinated 'as abuse now seems to be the method of response to those who dare to differ' and that the JLP had departed from many of the principles of the party's founder, Bustamante (*DG*, 7 July 1986).

    One classic departure by Seaga from Bustamante's legacy was his policy toward labour. Stone noted that Seaga 'has treated the labour movement as one would have expected the capitalist Jamaica Democratic Party of 1944 to have treated Bustamante then.' Stone also said Seaga's attack on higher education would make Bustamante turn over in his grave (*DG*, 28 July 1986).
55. Stone, 'Keeping the Peace', *DG*, 26 May 1986.
56. *Daily Challenge*, 9 July 1986.
57. MM letter of 26 September 1986, MPF.
58. Stone 'An Interesting Contest,' *DG*, 23 July 1986. Stone also said JLP activists had threatened some of his pollsters and misled the voters into thinking that Stone's most recent poll was being done for the PNP.
59. *DG*, 30 July 1986.
60. Stone, 'Counting the Constituencies', *DG*, 8 September 1986.
61. Stone columns, *DG*, 30 June, 23 July, and 4 August 1986.
62. After the election Seaga attempted to discredit the results by setting up a commission to enquire into irregularities that were committed by both sides. For a discussion see Stone, 'Commission of Enquiry', *DG*, 18 August 1986. where Stone observes: 'Mr Seaga should be the last person talking about election thuggery as his West Kingston constituency is a shining example of precisely the worst forms of that thuggery. His constitunecy is matched only by the atrocities committed by the PNP in South and South West St. Andrew.' For an analysis of overvoting in the 1976 and 1980 elections see Mark Figuera, 'An Assessment of Overvoting in Jamaica', *Social and Economic Studies* 34 (September 1985), 71–106; See also Appendix B herein.
63. MM, 'Mwalimu's Democratic Legacy', *South* (February 1986), p. 32.
64. MM, 'A Dictator We Wouldn't Miss', *South* (November 1986), p. 39.
65. MM, 'Red Herring on a Trail of Evil', *South* (December 1986), p. 112.
66. MM, 'View from a Small Island,' *The Nation* (New York), 22 March 1986.
67. MM, 'Third World Development and the International Economic System', *Transafrica Forum* (Summer 1986). pp. 83–93.

68. MM, 'Dominance or Equity: The Multilateral Institutions' (London: Third World Foundation and Lahore: Manzur Qadir Memorial Society, 1986).
69. Rex Nettleford, interview with author, 23 December 1987.
70. MM, interview with author, December 1987.
71. Matthew Beaubrun, interview with author, 19 December 1987. For the debate over Manley's health, see *JWG* 14 March 1986, p. 8.
72. Stone, 'Mr. Seaga's Correct Priorities', *JWG*, 2 May 1988, p. 8.
73. *JWG*, 19 October 1987, p. 30.
74. *JWG*, 7 March 1988, p. 30.
75. *JWG*, 14 March 1988. p. 30.
76. *JWG*, 15 August 1988, p. 30.
77. *JWG*, 7 March 1988, p. 4.
78. University of the West Indies, Trade Union Education Institute, 'The Jamaican Workers views of the Industrial and Political Environment,' (November 1987), The survey was done by Stone. Some of its other findings were that 72% of unionized workers felt that the union movement had weakened since 1980; that 95% of unionized and 92% of non-unionized workers believed that Jamaican workers do not get a fair deal from employers; that only 15% of unionized workers think it is good for unions to be linked to political parties; that 50% of workers approve of workers' participation, but that 49% have no understanding of it; that workers feel overwhelmingly that the private sector treats workers better than the public sector; that identifying themselves as Jamaican matters more to the workers than being identified as either working class or black; that the most admired official national hero is Marcus Garvey, followed in order by Bustamante and Norman Manley.
79. *JWG*, 13 July 1987, p. 2.
80. *JWG*, 21 September 1987, p. 3.
81. *JWG*, 28 September 1987, p. 2.
82. *JWG*, 28 September 1987, pp. 3, 24; and 5 October 1987, pp. 6, 26.
83. *JWG*, 7 March 1988, pp. 12 and 20.
84. See Stone, 'CBS Propaganda', *JWG*, 4 March 1988, pp. 23 and 27, and Trevor Lawson, 'The Silly Season', *Friends for Jamaica Caribean Newsletter*, vol. 8. no. 4–5 [April-May 1988], p. 4.
85. *JWG*, 18 April 1988, pp. 4. 11–12, and 23. An irony of the reports from the US was the strong case made on the Public Broadcasting Service programme 'Frontline' on 17 May 1988 about the historic involvement of the CIA and its foreign and domestic allies in the drug trade in Southeast Asia and Central America.
86. Dana Priest, 'Manley Steps Toward the Middle on the Road Back to Power,' *Washington Post National Weekly Edition*, 6 April 1987, p. 17, and Joseph B. Treaster, 'From Politics' Fiery Man, Mea Culpa', *The New York Times*, 21 September 1987.
87. For examples of such criticism, see *Friends for Jamaica Caribbean Newsletter*, January and September/October 1987.
88. *JWG*, 14 September 1987, p. 14.
89. Taped interview with author, December 1987.
90. Dudley Stokes, interview of MM, *JWG*, 5 October 1987, pp. 10 and 12. The interview took up the entire two pages.
91. *JWG*, 2 November 1987, 1 and 3; 9 November 1987, p. 5; 1 January 1988, pp. 1 and 8; and 25 January 1988. At the National Prayer Breakfast Reverend Herbert Swaby praised Manley and Seaga for their efforts, but warned that 'Jamaica is bitterly divided. There is a rage which threatens Jamaica,' he said.
92. *JWG*, 11 April 1988, p. 10.
93. Stone, 'Managerialism Versus Populism', *JWG*, 25 April 1988, p. 8.

CHAPTER 23 **Michael Manley in Jamaican History**

1. Thomas Carlyle, *On Heroes, Hero-Worship, and the Heroic in History* (Boston: D. C. Heath, 1913), p. 1.
2. Leo Tolstoy, *War and Peace* (1869) (New York: New American Library, 1968), p. 1178.
3. Murray Edelman, *The Symbolic Uses of Politics* (Urbana: University of Illinois Press, 1967), pp. 73, 78.
4. David Morris, 'Call It the Luck of the Gipper', *Tallahassee Democrat*, 20 July 1988.
5. George Plekhanov, *The Role of the Individual in History* (New York: International Publishers, 1940), p. 41.
6. For detailed treatments, see Kaufman, *Jamaica Under Manley*, pp. 205–38, and Stephens, *Democratic Socialism in Jamaica*, pp. 270–345.
7. In a recent article, Stone argues that 'most of the economic hardships of the '70s and '80s are accountable to the country's impaired balance of payments situation and its failure to increase earnings to match the demand for imports rather than to bad PNP or JLP policies, although some bad policies made a bad situation worse.' Stone, 'Does Jamaica Have the Political Skills for Crisis Management, or Is This Just a Ceasefire?' *Caribbean Affairs* (July-September 1988), p. 144.
8. IDB, *Economic and Social Progress in Latin America*, 1980–1 Report, p. 400. Jamaica's per capita GDP declined 19.2% in the '70s while Barbados's rose 14.6%. Costa Rica's increased 32.7%, the Dominican Republic's rose 52.2%, and all of Latin America's (including Barbados, Guyana, Jamaica, and Trinidad and Tobago) gained 36.2%.
9. Arias is quoted in Morris J. Blachman and Ronald G. Hellman, 'Costa Rica,' in: Blachman, William M. LeoGrand and Kenneth Sharp, *Confronting Revolution: Security Through Diplomacy in Central America* (New York: Pantheon, 1986), p. 165. Arias was elected President of Costa Rica in 1986 and won the 1987 Nobel Peace Prize for his Central American peace initiative.
10. Tom Barry, Beth Wood, and Deb Preusch, *The Other Side of Paradise: Foreign Control in the Caribbean* (New York: Grove, 1984), pp. 264–7 and 289–304.
11. IDB, *Social and Economic Progress in Latin America*, 1987 Report, p. 426.
12. MM, *Jamaica: Struggle in the Peripherey*, p. 208.
13. DM, interview with author, 6 September 1985.
14. See Omar Davies, 'An Analysis of the Management of the Jamaican Economy: 1972–85.' *Social and Economic Studies* (March 1986). pp. 73–109.
15. See Honor Ford-Smith, 'Sistren: Exploring Women's Problems Through Drama', *Jamaica Journal* (February-April 1986), pp. 2–11, where it is reported that 10,000 of the 14,000 persons in the Special Employment Programme were women. When in 1977 Ford-Smith asked the working-class 'teacher/actresses' of Sistren what kind of play they wanted to do for Workers' Week they said: 'We want to do plays about how we suffer as women. We want to do plays about how men treat us bad.' See also: Sistren, with Honor Ford-Smith (ed.), *Lionheart Gal: Life Stories of Jamaican Women* (London: The Women's Press, 1986).
16. MM, *Up the Down Escalator*, Ch. 11.
17. *New African Development* (September 1986), pp. 9–11 notes: 'It is a misconception that Britain has led the Commonwealth, morally or otherwise, in the last 20 years. The development of the Commonwealth has been the work of a number of non-British leaders—notably Presidents Nyerere of Tanzania and Kaunda of Zambia, and Prime Ministers Pierre Trudeau of Canada, Malcolm Fraser of Australia and Michael Manley of Jamaica.'

18. Michael Harrington, 'Socialists Help Shape the Best in Our Politics', *In These Times* (24 February–8 March 1988), p. 16.
19. Carlos Fuentes, 'Against Intervention in Latin America', commencement address at Harvard University, 9 June 1983.
20. Stone, *The Political Opinions of the Jamaican People* (1976–81) (Kingston: Blackett Publishers, 1982), pp. 54, 64, and 78.
21. University of the West Indies, Trade Union Education Institute. 'The Jamaican Workers Views of the Industrial and Political Environment', (November 1987). 21% identified Reagan as their most admired foreign leader, compared to 5% for Castro, and 3% for Jesse Jackson.
22. MM, 'We Must Tackle Structures', (Kingston: API, 1979).
23. In addition to works cited in Chs. 21 and 22, Manley's major addresses and writings in the '80s include: *Up the Down Escaltor: Development and the International Economy: A Jamaican Case Study*, completed in 1980, with final touches in 1984, but not published until 1987; 'Rebuilding for the Future', (Kingston: PNP, 1981); a series of speeches at Columbia University in 1984 (forthcoming); 'Aspects of a Caribbean Development Strategy', in: *Proceedings of a Regional Seminar on Caribbean Sovereignty, Mobilization and Self-Reliance, and the Tasks of Political Education*, December 1984 (Kingston: PNP, [1984?]: 'Michael Manley Speaks on Transformation and Jamaica's Economy. The International Dimension of Transformation, and Jamaica/U.S. Relations', (Kingston: PNP [1985?]; speech at Riverside Church, New York, 16 October 1987; and 'The Integration Movement', *Caribbean Affairs* (January-March 1988), pp. 6–15. In addition his *History of West Indies Cricket* (1989) contains interesting comments on Jamaican social history.

Epilogue

1. *JWG*, 13 and 27 February, 3 March 1989.
2. Julio A. Muriente Pérez, 'Victória de Manley, Derrota de un Modelo', *Claridad*, 17–23 February 1989, p. 6.
3. 'PNP's Victory: Implications', *Caribbean Contact*, March 1989, p. 3.
4. *The New York Times*, editorial, 14 February 1989.
5. *JWG*, 27 February 1989, p. 28.
6. *Ibid.*, p. 5.
7. For a recent treatment of political opinion and attitudes toward Manley, see Carl Stone, 'Political Change in Jamaica: Life's Better But the Polls are for Manley not Seaga', *Caribbean Affairs* (April-June 1988), pp. 31–46.

Bibliography

Note: The following abbreviations are used below:

API Agency for Public Information
ISER Institute of Social and Economic Research
PNP People's National Party

No attempt is made here to list documents such as Manley's speeches which are available only in PNP files; many of these are cited in the notes.

ABENG GROUP, 'The Elections, Where From Here,' Kingston, 1972.

Ibid., 'Joshua, Aaron, Moses, Eli: Two Months of Selling Jamaica.' Kingston, [1972].

ABRAHAMS, Roger D., *The Man-of-Words in the West Indies: Performance and the Emergence of Creole Culture*, Baltimore: Johns Hopkins University Press, 1983.

AMERICAS WATCH, *Human Rights in Jamaica*, New York, 1986.

ANDERSON-MANLEY, Beverley. See MANLEY, Beverley.

ARAWAK, Christopher (pseud.), *Jamaica's Michael Manley: Messiah Muddler, or Marionette?* Miami: Sir Henry Morgan Press, 1980.

The Award of the National Order of José Marti to Prime Minister Michael Manley: Speeches by Prime Minister Fidel Castro and Prime Minister Michael Manley, Havana, Cuba, 9th July 1975; plus Joint Communiqué of Cuba and Jamaica, Kingston: Agency for Public Information, 1975.

BARRY, Tom and Deb Preusch, *The Central America Fact Book*, New York: Grove Press, 1986.

BARRY, Tom, Beth Wood and Deb Preusch, *The Other Side of Paradise: Foreign Control in the Caribbean*, New York: Grove Press, 1984.

BEAUBRUN, Matthew, Interview with Author. 19 December 1987.

ETBELL, Wendell, *Jamaican Leaders: Political Attitudes in a New Nation*. Berkeley: University of California Press, 1964.

BERGER, Peter L., 'Can the Caribbean Learn from East Asia? The Case of Jamaica,' *Caribbean Review*, Spring 1984.

BERNAL, Richard L., 'The IMF and Class Struggle in Jamaica, 1977–1980,' *Latin American Perspectives*, Summer 1984.

BIRBALSINGH, Frank M., ' "Escapism" in the Novels of John Hearne,' *Caribbean Quarterly*, March 1970.

BISHOP, Maurice, 'Beat Back Destabilizers.' [St. George's?], Grenada: People's Revolutionary Government, 1979.

BLACHMAN, Morris J., William M. LeoGrande, and Kenneth Sharpe, *Confronting Revolution: Security Through Diplomacy in Central America*, New York: Pantheon, 1986.

BOGUES, Tony, Interview with Author, 19 December 1987.

BOODHOO, Ken I. (ed.), *Eric Williams: The Man and the Leader*, Lanham, Maryland: University Press of America, 1986.

BOXER, David, 'Edna Manley: Sculptor,' *Jamaica Journal*, February-April 1985.

Idem., *Jamaican Art, 1922–1982*, Washington, D.C.: Smithsonian Institution, and Kingston: National Gallery of Jamaica, 1983.

BRAITHWAITE, Edward, *The Development of Creole Society in Jamaica, 1770–1820*, Oxford: Clarendon Press, 1971.

BROWN, Wayne, *Edna Manley, the Private Years: 1900–1938*, London: André Deutsch, 1975.

Bulletin of Eastern Caribbean Affairs, May/June 1982.

CALDERA, Rafael and Michael Manley, *Venezuela and Jamaica: Speeches*, Caracas: Central Information Office, 1973.

CAMPBELL, Charles, *Memoirs of Charles Campbell*, Glasgow, 1828.

CAMPBELL, George, *First Poems*, New York: Garland, 1981.

Idem., Interview with Author, 13 September 1985.

CARGILL, Morris, *Jamaica Farewell*, Seacaucus, New Jersey: Lyle Stuart, 1978.

Caribbean Contact, 1984–8.

Caribbean Monthly Bulletin, various years.

CARLYLE, Thomas, *On Heroes, Hero-Worship, and the Heroic in History*, Boston: Heath, 1913.

CARR, Raymond, *Puerto Rico: A Colonial Experiment*. New York: Vintage, 1986.

CARTER, Rosalynn, *First Lady From Plains*, Boston: Houghton-Mifflin, 1984.

CASTRO, Fidel, 'Translation of Speech by Prime Minister Fidel Castro at the Decoration Ceremony, confirming the National Order of José Marti on Prime Minister Michael Manley.' (mimeo), [19 July 1975].

CHARLES, Pearnel, *Detained*, Kingston: Kingston Publishers, 1978.

CHEVANNES, Barry, 'The Rastafari and the Urban Youth,' in Carl Stone and Aggrey Brown (eds.), *Perspectives on Jamaica in the Seventies*, Kingston: Jamaica Publishing House: 1981.

COBB, Charles E, 'Jamaica: Hard Times, High Hopes,' *National Geographic*, January 1985.

COORE, David, Interview with Author, 22 December 1987.

CRILE, George, 'Our Man in Jamaica.' *Harper's*, October 1974.

CROSSETTE, Barbara, 'The Caribbean After Grenada,' *New York Times Magazine*, March 18 1984.

CURTIN, Philip, *Two Jamaicas: The Role of Ideas in a Tropical Colony, 1830–1865*, Cambridge: Harvard University Press, 1955.

Daily Gleaner.

DAVIES, Omar, 'An Analysis of Jamaica's Fiscal Budget (1974–1983, with Special Reference to the Impact of the Bauxite Levy,' University of the West Indies, Department of Economics, Occasional Paper Series No. 2, Mona, Jamaica, 1984.

Idem., 'An Analysis of the Management of the Jamaican Economy: 1972–1985,' *Social and Economic Studies*, March 1986.

DAVIS, Stephen, 'Jamaican Politics, Economics and Culture: An Interview with Edward Seaga,' *Caribbean Review*, autumn 1981.

DEANE, Herbert, *The Political Ideas of Harold Laski*, New York: Columbia University Press, 1955.

DEOSARAN, Ramesh, 'A Psychological Portrait of Political Power.' In: Ken I. Boodhoo, *Eric Williams: The Man and the Leader*, Lanham, Maryland: University Press of America, 1986.

Idem., 'The Role of the Press in the Caribbean,' *Caribbean Review*, Fall 1984.

DORSCHNER, John, Interview of Michael Manley, *Miami Herald*, 'Tropic Magazine,' 11 April 1976.

DUNKLEY, Carlyle, Interview with Author, 9 September 1985.

EATON, George E, *Alexander Bustamante and Modern Jamaica*, Kingston: Kingston Publishers, 1975.

Idem., (ed.) *Proceedings of Industrial Relations Conference Theory & Practice of Arbitration, November 11–12, 1962*, Mona, Jamaica: ISER, 1963.

EDELMAN, Murray, *The Symbolic Uses of Politics*, Urbana: University of Illinois Press, 1967.

'EDNA MANLEY: The Seventies: A Survey of the Artist's Work of the Past Decade.' Preface by David Boxer, [Kingston], National Gallery of Jamaica, [1980].

'EDNA MANLEY: Selected Sculpture and Drawings, 1922–1976.' Introduction by M. G. Smith, Kingston, National Gallery of Jamaica, [1977].

EISMANN, Christian, 'Laski,' *The McGraw-Hill Encyclopeadia of World Biography*, New York: McGraw-Hill, 1973. Volume 6.

EISNER, Gisela, *Jamaica, 1830–1930: A Study in Economic Growth*, Reprint of 1961 edition; Westport, Connecticut: Greenwood, [1974].

ELLMAN, Richard and Robert O'Clair, eds. *The Norton Anthology of Modern Poetry*, New York: Norton, 1973.

EVANS, Gaynelle, 'Jamaica's Plan to Levy Tax on Students Prompts Outcry from U. of West Indies,' *The Chronicle of Higher Education*, 9 July 1976.

EWART, Barclay, interview with author, 19 December 1987.

EWART, Glynne, interview with author, 29 December 1987.

FAIRWEATHER, Jean, 'Michael Manley, Who Is Leading Jamaica into Democratic Socialism, Is Many Things to Many People,' *Commonwealth* (April-May 1975).

FIGUEROA, Mark, 'An Assessment of Overvoting in Jamaica,' *Social and Economic Studies*, September 1985.

FORD-SMITH, Honor, 'Sistren: Exploring Women's Problems Through Drama,' *Jamaica Journal*, February-April 1986.

Friends for Jamaica Newsletter/Friends for Jamaica Caribbean Newsletter, New York, 1984–8.

FUENTES, Carlos, 'Against Intervention in Latin America,' Commencement Address, Harvard University, 9 June 1983. Tape recording in Author's Possession.

GILLIAM, Dorothy Butler, *Paul Robeson: All-American*, Washington, DC: New Republic, 1976.

GIRVAN, Norman, *Copper in Chile: A Study in Conflict Between Corporate and National Economy*, Mona, Jamaica: ISER, 1972.

Idem., Richard Bernal, and Wesley Hughes, 'The IMF and the Third World: The Case of Jamaica, 1974–1980,' *Development Dialogue*, 1980.

GLASSNER, Martin Ira, 'The Foreign Relations of Jamaica and Trinidad and Tobago, 1960–1965,' *Caribbean Studies*, October 1970.

GONSALVES, Ralph, 'The Trade Union Movement in Jamaica: Its Growth and Some Resultant Problems,' in: Carl Stone and Aggrey Brown, eds., *Essays on Power and Change in Jamaica* [Kingston]: Jamaica Publishing House, 1977.

HALL, Douglas, *Free Jamaica: An Economic History, 1838–1865*, New Haven: Yale University Press, 1959.

HARRINGTON, Michael, 'Socialists Help Shape the Best in Our Politics,' *In These Times* (24 February/8 March 1988).

HERSH, Seymour M., *The Price of Power: Kissinger in the Nixon White House*, New York: Summit Books, 1983.

'A Historia do Brasil, Escrita nos EUA,' *Veja*, 24 November 1971.

HOPKIN, John Barton, 'Jamaican Children's Songs,' *Ethnomusicology* (January 1984).

INGRAM, Derek, 'The Commonwealth: A Lucky Break?' *New Africa*, September 1986.

Inter-American Development Bank, *Social and Economic Progress in Latin America*, various years.

Jamaica, AP, 'Maternity Leave with Pay: Know Your Rights,' [1980].

Idem., 'Why and How to Conserve Energy,' 1979.
Jamaica Office of the Prime Minister, 'Michael Norman Manley: Prime Minister of Jamaica, A Political Biography,' Kingston, 1975.
Jamaica College Magazine, 1911–1912, 1941.
Jamaican Weekly Gleaner, 1971–1988.
'Jamaican Joshua,' *Time*, 2 August 1972.
JAMES, C. L. R, *The Future in the Present*, Westport, Connecticut: Lawerence Hill, 1977.
Idem., Spheres of Existence: Selected Writings, London: Allison and Busby, 1980.
JOHNSON, Caswell L., 'Emergence of Political Unionism in Economies of British Colonial Origin: The Cases of Jamaica and Trinidad,' *American Journal of Economics and Sociology*, April 1980. 151–164.
Idem., 'Political Unionism and Autonomy in Economies of British Colonial Origin: The Cases of Jamaica and Trinidad,' *American Journal of Economics and Sociology*, July 1980.
Idem., 'Political Unionism and the Collective Objective in Economies of British Colonial Origin: The Cases of Jamaica and Trinidad,' *American Journal of Economics and Sociology*, October 1975.
JOHNSON, Tony, 'Getting Out of the Poverty Cycle: An Interview with Sir Arthur Lewis,' *Jamaican Weekly Gleaner*, 22 March 1977.
KAUFMAN, Michael, *Jamaica Under Manley: Dilemmas of Socialism and Democracy*, Westport, Connecticut: Lawerence Hill, 1985.
KISSINGER, Henry, *For the Record: Selected Statements, 1977–1980*, Boston: Little, Brown, 1981.
KITSON, Doris, 'Fed Up With Suffering Under "Savior" Seaga,' *The Guardian* [New York] (30 January 1985).
KNOWLES, William H., *Trade Union Development and Industrial Relations in the British West Indies*, Berkeley: University of California Press, 1959.
KOPKIND, Andrew, 'Trouble in Paradise,' *Columbia Journalism Review*, March/April 1980.
KUPER, Adam, *Changing Jamaica*, London: Routledge and Kegan Paul, 1976.
LACEY, Terry, *Violence and Politics in Jamaica, 1960–70: Internal Security in a Developing Country*, Totowa, New Jersey: F. Cass, 1977.
LANDIS, Fred, 'Psychological Warfare in the Media: The Case of Jamaica,' Kingston: Press Association of Jamaicia, 1980.
LERNER, Max, 'A Talk in Jamaica,' *New York Post*, 19 February 1975.
Idem., 'Third World Case History,' *New York Post*, 21 February 1975.
LERNOUX, Penny, 'Tio Sam's Big Financial Stick,' *The Nation*, [New York], 22 March 1986.
LEWIS, Gordon, 'The Decline and Fall of Manley's Socialism,' *Daily Gleaner*, 22 October 1959.
Idem., The Growth of the Modern West Indies, New York: Modern Reader, 1968.
Idem., Grenada: The Jewel Despoiled, Baltimore: Johns Hopkins University Press, 1987.
Idem., Puerto Rico: Freedom and Power in the Caribbean, New York: Monthly Review, 1963.
Idem. Review of Cheddi Jagan, *The West on Trial: My Fight for Guyana's Freedom*. In *Caribbean Studies* (January 1968).
LEWIS, Vaughn A., 'The Small State Alone: Jamaican Foreign Policy, 1977–1980.' *Journal of Interamerican Studies and World Affairs*. May 1983.
LINDSAY, Louis, 'The Myth of Independence: Middle Class Politics and Non-Mobilization in Jamaica,' Working Paper No. 6. [Kingston]: ISER, 1975, 1981 Reprint.

LOEHR, William and John P. Powelson, *Threat to Development: Pitfalls of the NIEO*, Boulder, Colorado: Westview, 1983.

Los Pueblos Abrazan la Libertad y se Unen, La Habana, Cuba: Editora Politica, 1978.

MADDEN, R. R., *Twelve Months' Residence in the West Indies During the Transition from Slavery to Apprenticeship*, 2 Volumes, London, 1935.

MANLEY, Beverley, 'Change Must Come,' [Kingston]: Women's Auxiliary of the People's National Party, 1976.

Idem., 'The Ideological and Programmatic Platforms of the People's National Party (1938 to 1944).' M.Sc. thesis, University of the West Indies (Mona), 1985.

Idem., Interview with author, 5 September 1985.

MANLEY, Douglas, Interviews with author, 4 and 6 September 1985.

MANLEY, Edna, Diary, 1939–1986, possession of Michael Manley.

Idem., News clippings, 1938–1983, Institute of Jamaica.

MANLEY, Joseph, Interview with author, 15 September 1985.

MANLEY, Michael (pseud. 'Political Analyst'), 'Announcing Socialism,' *Rising Sun*, September 1982.

Idem., 'The Angry Egalitarian: Roger Mais and Human Rights,' 1968.

Idem., 'Aspects of a Caribbean Development Strategy,' in: *Proceedings of a Regional Seminar on Caribbean Sovereignty: Mobilisation and Self Reliance, The Tasks of Political Education*, Kingston: PNP, 1984.

Idem., 'Belafonte—No Pawn of Reagan,' *Rising Sun*, August 1982.

Idem., *Budget Debate Speech, May 1st 1972*, Kingston: Government Printer, [1972].

Idem., 'A Dictator We Wouldn't Miss,' *South*, November 1986.

Idem., 'Dominance or Equity: The Multilateral Institutions,' London: Third World Foundation, 1986.

Idem., 'Economic Criteria by Which Arbitrators May Be Guided,' in G. E. Eaton, ed., *Proceedings of Industrial Relations Conference on Theory & Practice of Arbitration, November 11–12 1962*, Mona, Jamaica: ISER, 1963.

Idem., 'From the Shackles of Domination and Oppression,' *Ecumenical Review*, January 1976.

Idem., 'Grenada in the Context of History: Between Neocolonialism and Independence,' *Caribbean Review*, autumn 1983.

Idem., *A History of West Indies Cricket*, London: André Deutsch, 1989.

Idem., 'The Integration Movement, the CBI, and the Crisis of the Ministate,' *Caribbean Affairs*, January-March 1988.

Idem., Interviews with author, 12, 15, and 21 August 1984; 3, 6, 12, and 16 September 1985; 17, 18, 22, and 24 December 1987.

Idem., *Jamaica: Struggle in the Periphery*. London: Writers and Readers, 1982.

Idem., Letters, 1943–1951, possession of Beverley Anderson-Manley.

Idem., 'Michael Manley Speaks on Transformation and the Jamaican Economy, the International Dimension of Transformation, Jamaica/US Relations,' Kingston: PNP, [1985].

Idem., 'Mwalimu's Democratic Legacy,' *South*, February 1986.

Idem., 'News Release—Reports of Sabotage to Rural Pipelines,' (Kingston: API, 1980).

Idem., '1985 in Retrospect—Year of the Shadows' *Sunday Gleaner*, 5 January 1986.

Idem., 'Not for Sale,' Supplement by C. L. R. James, San Francisco: Editorial Consultants, [1977].

Idem., 'Operation Reconstruction,' Kingston: API, [1980].

Idem., 'Our Movement is Irreversible Because Our Cause Is Just,' Speech to Sixth Summit of Non-Aligned Nations, Havana, 4 September 1979, Kingston: API, 1979.

Idem., 'Our 21st Anniversary,' *Rising Sun*, August 1983.

Idem., 'Overcoming Insularity in Jamaica,' *Foreign Affairs*, October 1970.

Idem., Personal Files, 1970–72, 1974–80, 1983–6.

Idem., *The Politics of Change: A Jamaican Testament*, London: André Deutsch, 1974.

Idem., 'The Prime Minister of Jamaica . . . 1972 Address to the United Nations,' Kingston: Jamaica Information Service, 1972.

Idem., 'Rebuilding for the Future,' Kingston: PNP, 1981.

Idem., 'Red Herring on a Trail of Evil,' *South*, December 1986.

Idem., 'Seaga "Stone" Polled Dead,' *Rising Sun*, December 1982.

Idem., *The Search for Solutions: Selections from the Speeches and Writings of Michael Manley*, edited, with Notes and Introduction by John Hearne, Oshawa, Ontario: Maple House, 1976.

Idem., 'The Social Contract,' Kingston: API, 1979.

Idem., Speech to Forty-Seventh Annual Conference, People's National Party, 22 September 1985. Author's tape recording.

Idem., 'Third World Development and the International Economic System,' *Trans-Africa Forum*, Summer 1986.

Idem., 'The Time for Action Must Be Now,' Keynote Address to UN Special Committee Against Apartheid at the UN Assembly, 11 October 1978, Kingston: API, [1978].

Idem., *Up the Down Escalator: Development and the International Economy: A Jamaica Case Study*, Washington, DC: Howard University Press, 1987.

Idem., 'View from a Small Island,' *The Nation* [New York], 22 March 1986.

Idem., *A Voice at the Workplace: Reflections on Colonialism and the Jamaican Worker*, London: André Deutsch, 1975.

Idem., 'We Know Where We Are Going.' Two long-playing record set of 19 September 1976 speech; Kingston: PNP, 1976.

Idem., 'We Must Tackle Structures,' Lecture to the Third World Foundation, London, 29 October 1979, Kingston: API, 1979.

Idem., 'Why Nonaligned Isn't Neutral,' *The Nation* [New York], 30 April 1983.

Idem., 'Why the PNP Stayed Out of Seaga's Election,' *South*, February 1984.

MANLEY, Norman Washington, 'The Autobiography of Norman Washington Manley,' *Jamaica Journal*, March/June 1973.

Idem., Collection of Papers Relative to Manley's Political Life, Institute of Jamaica, ms 1632.

Idem., 'A Message to My Constituents the Voters of Eastern St. Andrew,' December 1949, Institute of Jamaica.

Idem., 'My Early Years: Fragment of an Autobiography, in: *Norman Washington Manley and the New Jamaica: Selected Speeches and Writings, 1939–1968*, ed. Rex Nettleford, New York: Africana, 1971.

Idem., Norman Manley Papers, National Archive of Jamaica.

Idem., 'West Kingston—The Real Danger, Jamaica Beware,' PNP Newsletter, July-September 1966, Institute of Jamaica.

MANLEY, Rachel, *Poems 2*, St. Michael, Barbados: R. Manley, 1978.

MANLEY, Thelma, Interview with author, 24 September 1985.

MASSAQUOI, Hans J., 'Jamaica: A New Beginning Under New Leadership,' *Ebony*, October 1973.

Idem., 'Kingston Summit: Third World Leaders Demand a "New Deal" or Else,' *Ebony*, July 1973.

MASSING, Michael, 'The Jamaica Experiment,' *The Atlantic Monthly*, September 1983.

MAXWELL, John, Interview with author, 7 September 1985.

Maurice Bishop Speaks: The Grenada Revolution, New York: Pathfinder Press, 1983.

MCNEILL, William H., *Mythistory and Other Essays*, Chicago: University of Chicago Press, 1986.

MEEKS, Corina, Interview with author, 21 December 1987.

MELHADO, O. K., Interview with author, 23 December 1987.

MESA-LAGO, Carmelo, *Cuba in the 1970's: Pragmatism and Institutionalization*, revised edition, Albuquerque: University of New Mexico Press, 1977.

MUNROE, Trevor, *The Politics of Constitutional Decolonization: Jamaica, 1944–1962*. Mona. Jamaica: ISER, 1972.

NAIRNE, Alvin, 'The Politics of Treachery,' *Socialism* [Mona, Jamaica], August 1974.

NETTLEFORD, Rex M., *Caribbean Cultural Identity: The Case of Jamaica: An Essay in Cultural Dynamics*. Forward by Claudia Mitchell-Kernan. Los Angeles: Center for Afro-American Studies and UCLA Latin American Center for Publications, 1979.

Idem., interview with author, 23 December 1987.

Idem., *Mirror, Morror: Identity, Race and Protest in Jamaica*, New York: Morrow, 1972.

NKRUMAH, Kwame, *Ghana: The Autobiography of Kwame Nkrumah*, New York: Thomas Nelson and Sons, 1957.

NYERERE, Julius K., *Man and Development: Binadamu na Maendolo*, London: Oxford University Press, 1974.

O'GORMAN, Pamela, Interview with author, 24 December 1987.

PADMORE, George, *Pan-Africanism or Communism*. Forward by Richard Wright. Introduction by Azinna Nwafor. Garden City, New York: Doubleday, 1971.

PEELER, John A., *Latin American Democracies: Colombia, Costa Rica, Venezuela*, Chapel Hill: University of North Carolina Press, 1985.

People's National Party, Files, PNP Archive: Citizens Mail (1980); Commonwealth Heads of Government (1979); Executive (1977); Heads of Government and Cuba; Heads of Government (1978: 9); Heads of Government ([1979]: 8); Hurricane Disaster (1980, no. 213); Michael Manley on Policy; Ministry of National Security (1973–4); Miscellaneous (March 1973); Miscellaneous (May 1973); Miscellaneous (1975); Personal, Douglas (1970); PM's File: Mrs. Manley's Daily; Police (1980); 2nd Caricom Heads of Government Conference (1975); Speeches (1968–1986); State of Emergency (no. 83); State of Emergency, *Daily News* Clippings (19 June–19 August 1976); State of emergency, *Star* Clippings (22 June–20 July 1976); Women's Affairs (no. 114).

Idem., *Man of Destiny: Norman Washington Manley*. Kingston, 1962.

Idem., *PNP Progress*. Special issues of 10 and 17 May 1980.

Idem., *Principles and Objectives*. Kingston, 1979.

Idem., *Proceedings of a Regional Seminar on Caribbean Sovereignty: Mobilisation for Development and Self Reliance: The Tasks of Political Education*, Kingston, 1984.

Idem., 'Report of the General Secretary, 1978–79.'

Idem., 'Report of the PNP Women's Movement, PNP Youth Organization, [and] National Workers Union to the 43rd Annual General Conference, 10th October and 11th October, 1981.'

Idem., 'Report to the 43rd Annual General Conference,' (1981).

Idem., 'Strategy for Development: Leading Jamaica into the 21st Century,' Supplement, *Daily Gleaner*, 25 October 1985.

Idem., Women's Movement, 'Convener's Annual Report, 1978, 1979.' (1979).

Idem., 'In Solidarity with Our Cuban Brothers and Sisters. Good Friends . . . Good Neighbours.' (1979).

PLEKHANOV, George, *The Role of the Individual in History*, 1886; New York: International Publishers, [1940].

Press Association of Jamaica. 'Press Freedom for Whom?' Kingston: Press Association of Jamaica, 1978.

PRIEST, Dana, 'Manley Steps Toward the Middle on the Road Back to Power.' *Washington Post National Weekly Edition*, 6 April 1987.

Public Opinion (Kingston), 1951–4.

RATTRAY, Carl, 'The Constitutional Role of the Opposition,' Kingston: API, 1979.
RAY, Ellen and Bill Schaap, 'Massive Destabilization in Jamaica: 1976 With a New Twist,' *Covert Action Information Bulletin*, August-September 1980.
REID, Stanley, 'An Introductory Approach to the Concentration of Power in the Jamaican Corporate Economy and Notes on Its Origin,' in: Carl Stone and Aggrey Brown (eds.), *Essays on Power and Change in Jamaica*, [Kingston]: Jamaica Publishing House, 1977.
REID, Victor Stafford, *The Horses of the Morning: About the Rt. Excellent N. W. Manley, Q. C., M. M. National Hero of Jamaica: An Understanding*, Kingston: Caribbean Authors Publishing, 1985.
RENNY, R., *A History of Jamaica*, London, 1807.
Rising Sun (PNP Newspaper), 1982–4.
ROBERTSON, Paul, Interview with Author, 29 December 1987.
RODNEY, Walter, *Groundings with My Brothers*, London: Bogle L'Ouverture, 1969.
Idem., *A History of the Guyanese Working People, 1881–1905*, Baltimore: John Hopkins University Press. 1981.
Idem., *A History of the Upper Guinea Coast, 1545–1800*, London: Oxford University Press, 1970.
Idem., *How Europe Underdeveloped Africa*, London: Bogle L'Ouverture, 1972.
ROHLEHR, Gordon, 'History as Absurdity,' in: Orde Coombs (ed.) *Is Massa Day Done? Black Moods in the Caribbean*, Garden City, New York: Doubleday, 1974.
RUBIN, Vera, 'His Vision of Caribbean Man,' in: Ken I. Boodhoo (ed.), *Eric Williams: The Man and the Leader*, Lanham, Maryland: University Press of American, 1986.
SENIOR, Clarence and Douglas Manley, 'The West Indian in Britain,' London: Fabian Colonial Bureau, 1956.
SENIOR, Olive, *The Message is Change: A Perspective on the 1972 General Elections*, Kingston: Kingston Publishers, 1972.
SHEARER, Hugh, Interview with Author, 29 December 1987.
SHERLOCK, Philip M., *Norman Manley*, London: Macmillan, 1980.
SINGHAM, A. W., *The Hero and the Crowd in a Colonial Polity*, New Haven: Yale University Press, 1968.
SISTREN, with Honor Ford-Smith, *Lionheart Gal: Life Stories of Jamaican Women*, London: The Women's Press, 1986.
SMITH, M. G., 'The Plural Framework of Jamaican Society,' in Lambros Comitas and David Lowenthal (eds.), *Slaves, Free Men, Citizens: West Indies Perspectives*, Garden City: New York: Anchor, 1973.
SMITH, Terry, 'Union and Political Party Go Hand in Hand,' *Caribbean Business*, September 1970.
Socialist International, Committee on Economic Policy (Michael Manley, Chairman), *Global Challenge: From Crisis to Cooperation: Breaking the North-South Stalemate*, London: Pan Books, 1985.
STEPHENS, Evelyn Huber and John D. Stephens, *Democratic Socialism in Jamaica: the Political Movement and Social Transformation in Dependent Capitalism*, Princeton, New Jersey: Princeton University Press, 1986.
STOKES, Dudley, 'Democracy Cannot Survive Politically-Induced Violence—Says Michael Manley,' *Jamaican Weekly Gleaner*, 5 October 1987.
STONE, Carl, *Class, Race and Political Behaviour in Urban Jamaica*, [Mona, Jamaica]: ISER, 1973.
Idem., *Class, State and Democracy in Jamaica*, New York: Praeger, 1986.
Idem., *Democracy and Clientelism in Jamaica*, New Brunswick, New Jersey: Transaction Books, 1980.
Idem., 'Does Jamaica have the Political Skills for Crisis Management? Or, Is This Just a Ceasefire?' *Caribbean Affairs*, July–September 1988.

Idem., 'Ideology, Public Opinion and the Media in Jamaica,' in: Carl Stone and Aggrey Brown (eds.), *Perspectives on Jamaica in the Seventies*, Kingston: Jamaica Publishing House, 1981.

Idem., Interview with Author, 23 December 1987.

Idem., 'Jamaica's 1980 Elections: What Manley Did Do; What Seaga Need Do,' *Caribbean Review*, Spring 1981.

Idem., 'The Long Road from 1938,' *Jamaican Weekly Gleaner*, 24 May 1978.

Idem., 'Manley and Seaga,' *Jamaican Weekly Gleaner*, 4 and 11 August 1980.

Idem., 'Political Change in Jamaica: Life's Better but the Polls Are for Manley not Seaga,' *Caribbean Affairs*, April-June 1988.

Idem., 'The Political Opinions of the Jamaican People (1976–1981),' Kingston: Blackett Publishers, 1982.

Idem., *Race, Class and Political Behaviour in Urban Jamaica*, Mona, Jamaica: ISER, 1973.

Idem., Review of Adam Kuper, *Changing Jamaica, Caribbean Studies*, July 1976.

STONE, Carl and Aggrey Brown (eds.), *Essays on Power and Change in Jamaica*, [Kingston]: Jamaica Publishing House, 1977.

Idem., *Perspectives on Jamaica in the Seventies*, Kingston: Jamaica Publishing House, 1981.

SUTTON, Paul, 'The Scholar: A Personal Appreciation,' in: Ken I. Boodhoo (ed.), *Eric Williams: The Man and the Leader*, Lanham, Maryland: University Press of America, 1986.

TOLSTOY, Leo, *War and Peace*, New York: New American Library, 1968.

TREATER, Joseph, 'From Politics' Fiery Man, Mea Culpa,' *The New York Times*, 21 September 1987.

United Nations, Energy Programme, 'Brief on Current Developments in Oil Prices and Supplies,' mimeo, prepared for the Government or Jamaica, 31 May 1979.

University of the West Indies (Mona), Trade Union Education Institute, 'The Jamaican Workers' Views of the Industrial and Political Environment,' (Typescript, November 1987).

VAN HORNE, Winston A., 'Why Manley Lost: Problems of Michael Manley's Politics of Change 1972–1980,' *Journal of Caribbean Studies*, autumn-winter 1981.

[VERITY, Tony], 'Interview [with] Mr. Michael Manley,' *West Indies Chronicle* [London], May 1972.

VOLKMAN, Ernest and John Cummings, 'Murder as Usual,' *Penthouse*, December 1977.

WALCOTT, Derek, 'The Muse of History,' in: Orde Coombs (ed.), *Is Massa Day Done? Black Moods in the Caribbean*, 'Garden City, New York: Doubleday, 1974.

WATERS, Anita, *Race, Class and Political Symbols: Rastafari and Reggae in Jamaican Politics*, New Brunswick, New Jersey: Transaction Books, 1965.

WILLEMS, Emilio, *Latin American Culture: An Anthropological Synthesis*, New York: Harper and Row, 1975.

WILLIAMS, Eric, *British Historians and the West Indies*, New York: Scribner's, 1966.

Idem., *Inward Hunger: The Education of a Prime Minister*. Introduction by Sir Denis Brogan, [Chicago]: University of Chicago Press, [1971].

Women's Resource and Outreach Centre, 'Annual Report, March 1983–April 6 1984,' Kingston, 1984.

INDEX

(Abbreviations used in the notes and in this index include: BAM (or BM) Beverley Anderson-Manley; DM Douglas Manley; EM Edna Manley; JM Joseph Manley; MM Michael Manley; MPF Manley's Personal Files; n note; NM (or NWM) Norman Washington Manley; NMP Norman Manley Papers, National Archive of Jamaica; Stephens: John D. and Evelyn Huber Stephens. A list of other abbreviations occurs on pp. ix–x)